Advances in Polymer Blends and Composites

Advances in Polymer Blends and Composites

Guest Editors

Eduard-Marius Lungulescu
Radu Setnescu
Cristina Stancu

Basel • Beijing • Wuhan • Barcelona • Belgrade • Novi Sad • Cluj • Manchester

Guest Editors

Eduard-Marius Lungulescu
National Institute for Research
and Development
in Electrical Engineering
ICPE-CA
Bucharest
Romania

Radu Setnescu
National Institute for Research
and Development
in Electrical Engineering
ICPE-CA
Bucharest
Romania

Cristina Stancu
National University of
Science and Technology
Politehnica Bucharest
Bucharest
Romania

Editorial Office
MDPI AG
Grosspeteranlage 5
4052 Basel, Switzerland

This is a reprint of the Special Issue, published open access by the journal *Materials* (ISSN 1996-1944), freely accessible at: www.mdpi.com/journal/materials/special_issues/Polymer_blends_composites.

For citation purposes, cite each article independently as indicated on the article page online and using the guide below:

Lastname, A.A.; Lastname, B.B. Article Title. *Journal Name* **Year**, *Volume Number*, Page Range.

ISBN 978-3-7258-2998-9 (Hbk)
ISBN 978-3-7258-2997-2 (PDF)
https://doi.org/10.3390/books978-3-7258-2997-2

© 2025 by the authors. Articles in this book are Open Access and distributed under the Creative Commons Attribution (CC BY) license. The book as a whole is distributed by MDPI under the terms and conditions of the Creative Commons Attribution-NonCommercial-NoDerivs (CC BY-NC-ND) license (https://creativecommons.org/licenses/by-nc-nd/4.0/).

Contents

Preface . . . . **vii**

Radu Setnescu, Eduard-Marius Lungulescu and Virgil Emanuel Marinescu
Polymer Composites with Self-Regulating Temperature Behavior: Properties and Characterization
Reprinted from: *Materials* 2022, 16, 157, https://doi.org/10.3390/ma16010157 . . . . **1**

Pavlo Bekhta, Antonio Pizzi, Iryna Kusniak, Nataliya Bekhta, Orest Chernetskyi and Arif Nuryawan
A Comparative Study of Several Properties of Plywood Bonded with Virgin and Recycled LDPE Films
Reprinted from: *Materials* 2022, 15, 4942, https://doi.org/10.3390/ma15144942 . . . . **21**

Diego S. Melo, Idalci C. Reis, Júlio C. Queiroz, Cicero R. Cena, Bacus O. Nahime and José A. Malmonge et al.
Evaluation of Piezoresistive and Electrical Properties of Conductive Nanocomposite Based on Castor-Oil Polyurethane Filled with MWCNT and Carbon Black
Reprinted from: *Materials* 2023, 16, 3223, https://doi.org/10.3390/ma16083223 . . . . **36**

Hanjie Hu, Bing Du, Conggang Ning, Xiaodong Zhang, Zhuo Wang and Yangyang Xiong et al.
Milling Parameter Optimization of Continuous-Glass-Fiber-Reinforced-Polypropylene Laminate
Reprinted from: *Materials* 2022, 15, 2703, https://doi.org/10.3390/ma15072703 . . . . **53**

Hyungjin Cho, Ahyeon Jin, Sun Ju Kim, Youngmin Kwon, Eunseo Lee and Jaeman J. Shin et al.
Conversion of Polyethylene to Low-Molecular-Weight Oil Products at Moderate Temperatures Using Nickel/Zeolite Nanocatalysts
Reprinted from: *Materials* 2024, 17, 1863, https://doi.org/10.3390/ma17081863 . . . . **61**

Eduard-Marius Lungulescu, Cristina Stancu, Radu Setnescu, Petru V. Notingher and Teodor-Adrian Badea
Electrical and Electro-Thermal Characteristics of (Carbon Black-Graphite)/LLDPE Composites with PTC Effect
Reprinted from: *Materials* 2024, 17, 1224, https://doi.org/10.3390/ma17051224 . . . . **72**

Bogna Sztorch, Roksana Konieczna, Daria Pakuła, Miłosz Frydrych, Bogdan Marciniec and Robert E. Przekop
Preparation and Characterization of Composites Based on ABS Modified with Polysiloxane Derivatives
Reprinted from: *Materials* 2024, 17, 561, https://doi.org/10.3390/ma17030561 . . . . **96**

Despoina Tselekidou, Kyparisis Papadopoulos, Vasileios Foris, Vasileios Kyriazopoulos, Konstantinos C. Andrikopoulos and Aikaterini K. Andreopoulou et al.
A Comparative Study between Blended Polymers and Copolymers as Emitting Layers for Single-Layer White Organic Light-Emitting Diodes
Reprinted from: *Materials* 2023, 17, 76, https://doi.org/10.3390/ma17010076 . . . . **121**

Sotirios Pemas, Eleftheria Xanthopoulou, Zoi Terzopoulou, Georgios Konstantopoulos, Dimitrios N. Bikiaris and Christine Kottaridi et al.
Exploration of Methodologies for Developing Antimicrobial Fused Filament Fabrication Parts
Reprinted from: *Materials* 2023, 16, 6937, https://doi.org/10.3390/ma16216937 . . . . **140**

Tom Eggers, Sonja Marit Blumberg, Frank von Lacroix, Werner Berlin and Klaus Dröder
Influence Analysis of Modified Polymers as a Marking Agent for Material Tracing during Cyclic Injection Molding
Reprinted from: *Materials* **2023**, *16*, 6304, https://doi.org/10.3390/ma16186304 . . . . **156**

Seong Baek Yang, Jungeon Lee, Sabina Yeasmin, Jae Min Park, Myung Dong Han and Dong-Jun Kwon et al.
Blown Composite Films of Low-Density/Linear-Low-Density Polyethylene and Silica Aerogel for Transparent Heat Retention Films and Influence of Silica Aerogel on Biaxial Properties
Reprinted from: *Materials* **2022**, *15*, 5314, https://doi.org/10.3390/ma15155314 . . . . **178**

Ana Maria Lupu (Luchian), Marius Mariş, Traian Zaharescu, Virgil Emanuel Marinescu and Horia Iovu
Stability Study of the Irradiated Poly(lactic acid)/Styrene Isoprene Styrene Reinforced with Silica Nanoparticles
Reprinted from: *Materials* **2022**, *15*, 5080, https://doi.org/10.3390/ma15145080 . . . . **191**

Preface

Polymer blends and composites, including nanocomposites, have influenced the field of materials science and have found use in a wide range of engineering applications. These materials are derived as a combination of the useful properties of two polymers or polymer-filler systems and mainly do not have the weaknesses associated with theindividually used polymers. For instance, the incorporation of fillers such as graphite, carbon nanotubes, graphene, or metal nanoparticles increases the strength, stiffness, and thermal conductivity or antimicrobial properties of the polymeric composites. In the same manner, the fiber-reinforced composites developed using carbon fiber, glass fiber, and natural fiber also provide a remarkable strength-to-weight ratio, making them perfectly suitable for high-performance applications. In addition, functional polymer blends extend the range of application through the fusion of characteristics which creates products possessing enhanced mechanical strength, flexibility, and high-impact strength.

Because of their versatility, these materials have been able to make their way into almost every industry. Conductive polymer composite solutions will open the door for the creation of bendable, robust, and portable electronics. The use of antimicrobial polymeric composite materials in areas such as medical equipment, high-risk professional environments (e.g., hospitals), or food packaging can directly contribute to improving public health by reducing contamination risks and supporting global efforts to prevent infections. Also, polymer composites are utilized in renewable energy in relation to solar energy panels, wind turbine blades, and energy storage systems.

By addressing issues of sustainability, performance, and usefulness, polymer blends and composites are contributing to the development of a more ecologically friendly future while also advancing technology.

Eduard-Marius Lungulescu, Radu Setnescu, and Cristina Stancu
Guest Editors

Article

Polymer Composites with Self-Regulating Temperature Behavior: Properties and Characterization

Radu Setnescu [1,2], Eduard-Marius Lungulescu [1,*] and Virgil Emanuel Marinescu [1]

[1] National Institute for Research and Development in Electrical Engineering ICPE-CA, 313 Splaiul Unirii, 030138 Bucharest, Romania
[2] Department of Advanced Technologies, Faculty of Sciences and Arts, Valahia University of Târgoviște, 13 Aleea Sinaia, 130004 Târgoviște, Romania
* Correspondence: marius.lungulescu@icpe-ca.ro

Citation: Setnescu, R.; Lungulescu, E.-M.; Marinescu, V.E. Polymer Composites with Self-Regulating Temperature Behavior: Properties and Characterization. *Materials* **2023**, *16*, 157. https://doi.org/10.3390/ma16010157

Academic Editor: Anna Donnadio

Received: 26 November 2022
Revised: 12 December 2022
Accepted: 20 December 2022
Published: 24 December 2022

Copyright: © 2022 by the authors. Licensee MDPI, Basel, Switzerland. This article is an open access article distributed under the terms and conditions of the Creative Commons Attribution (CC BY) license (https://creativecommons.org/licenses/by/4.0/).

Abstract: A novel conductive composite material with homogeneous binary polymer matrix of HDPE (HD) and LLDPE (LLD), mixed with conductive filler consisting of carbon black (CB) and graphite (Gr), was tested against a HDPE composite with a similar conductive filler. Even the concentration of the conductive filler was deliberately lower for (CB + Gr)/(LLD + HD), and the properties of this composite are comparable or better to those of (CB + Gr)/HD. The kinetic parameters of the ρ-T curves and from the DSC curves indicate that the resistivity peak is obtained when the polymer matrix is fully melted. When subjected to repeated thermal cycles, the composite (CB + Gr)/(LLD + HD) presented a better electrical behavior than composite CB + Gr)/HD, with an increase in resistivity ($ρ_{max}$) values with the number of cycles, as well as less intense NTC (*Negative Temperature Coefficient*) effects, both for the crosslinked and thermoplastic samples. Radiation crosslinking led to increased $ρ_{max}$ values, as well as to inhibition of NTC effects in both cases, thus having a clear beneficial effect. Limitation effects of surface temperature and current intensity through the sample were observed at different voltages, enabling the use of these materials as self-regulating heating elements at various temperatures below the melting temperature. The procedure based on physical mixing of the components appears more efficient in imparting lower resistivity in solid state and high PTC (*Positive Temperature Coefficient*) effects to the composites. This effect is probably due to the concentration of the conductive particles at the surface of the polymer domains, which would facilitate the formation of the conductive paths. Further work is still necessary to optimize both the procedure of composite preparation and the properties of such materials.

Keywords: polymeric composites; PTC; NTC; self-regulating temperatures; conductive polymers

1. Introduction

Electrically conductive polymer composites (some remarkable pioneering work in this field should be mentioned, se for example [1–3]) attracted much attention due to their remarkable properties, which combine economic processing, good mechanical and chemical properties, and wide range of electrical properties, which can be finely controlled by adjusting the composition—polymer matrix and conductive filler [4,5]. Due to their properties, these materials exhibit high functionality and smartness in various technical applications comprising conductive coatings, electromagnetic shielding, electronic packaging flexible displays, sensors, etc. [6–10]. Various conductive powders or fiber materials/nanomaterials [7,11,12] can be used as conductive fillers, the carbon ones being of great interest due to low cost, good electrical properties, acceptable compatibility with polymer matrix, low density, and corrosion resistance. Carbon black [8,13,14], carbon fibers [15,16], graphite [17]), graphene [9,18], reduced graphene oxide [19], CNT [11,16,18], and other carbon materials were studied in order to obtain adequate properties for a wide range of applications. Among them, the composites exhibiting so-called *Positive Temperature*

Coefficient (of resistivity) effect, abbreviated PTC [1] or PTCR [13], are of special interest for specific applications claiming self-limitation (of current) or switching (conductive/resistive) behaviors, such as self-regulating heating elements, current or voltage protections or temperature sensors [15,17]. Another effect occurring in conductive composites is known as NTC (*Negative Temperature Coefficient*), i.e., decrease in resistivity as the temperature increase. Note that these effects may occur in the same conductive composite, depending on several factors, such as temperature, polymer state, distribution of the filler, etc. Even there are some applications for the NTC materials, for example as temperature sensors, the PTC effect have a wider field of applications due to resistance increase with temperature rise, resulting in limitation of both the current and temperature of the device [6].

Due to its high crystallinity, HDPE is a good candidate for high PTC effects, but the mechanical rigidity of the resulting composites would be high at higher filler contents (15–25%), specific for self-regulating heating materials, are used. Blends of HDPE with other (miscible, low crystallinity) polymers, such as EVA (Ethylene-vinyl acetate), LDPE (Low-Density Polyethylene), would improve the mechanical properties of the composites [20]. In general, improvement of the properties of PTC materials requires compositional optimization of both polymer matrix and conductive fillers [17]. Polymer blends and synergic mixtures of different conductive powders are of actual interest in this direction [5,17,21,22].

The aim of this work was to study the PTC properties of a composite with a matrix consisting of a miscible blend of two polymers, LLDPE and HDPE, in comparison to HDPE only. This type of blend was less studied in composites in general [23], but it would present some interesting properties for PTC composites. Even a decrease in PTC intensity due to blending with lower crystallinity LLDPE is expected if we consider the hypothesis that some amorphous non-crystalline regions of PE remain unperturbed in the vicinity of CB particles [24], and LLDPE would be of interest due to its regular structure. In addition, the studied composites were prepared using a procedure based on powder mixing in a dry state, which appears to present some advantages. The influence of the processing technology and matrix composition is discussed in comparison with a composite having similar conductive filler but HDPE matrix.

2. Materials and Methods

2.1. Materials

Blends of polymer and conductive particles were prepared by dry mixing of different amounts of polymer powders, antioxidants, and conductive carbon powders.

HDPE, type ELTEX A3180PN1852 from Ineos, Vienna, Austria, and LLDPE, type RX 806 Natural from Resinex, Bucharest, Romania were used, as received, for polymer matrices. The weight ratio LLDPE: HDPE was 0.6:1.

The list of abbreviations used in this manuscript are presented in Table 1.

Table 1. List of abbreviations.

Abbreviation	Full Name/Description
PTC	Positive Temperature Coefficient
NTC	Negative Temperature Coefficient
HDPE (HD)	High Density Polyethylene
LLDPE (LLD)	Linear Low-Density Polyethylene
FEF	Fast Extruder Furnace
SEM	Scanning Electron Microscopy
CB	Carbon Black
Gr	Graphite
ρ_v	Volume resistivity
ρ_s	Surface resistivity
T_s	Sample surface temperature
ρ-T	Resistivity-Temperature
X_c	Crystallinity degree
OOT	Onset Oxidation Temperature
$\Delta H_m / \Delta H_c$	Melting/Crystallization enthalpies
T_m / T_c	Melting/Crystallization temperatures

As conductive fillers, carbon black (FEF type) and natural graphite (CR10) were used (see [22] for more details).

The composite with HDPE matrix was codified (CB + Gr)/HD, meaning that it contains an HDPE matrix and carbon black (CB) and graphite (Gr) as conductive fillers. Similarly, the composite named below (CB + Gr)/(LLD + HD) has a binary blend matrix consisting in LLDPE and HDPE and a mixture of carbon black and graphite as conductive filler.

The procedures to prepare the above-mentioned composites (CB + Gr)/HD and (CB + Gr)/(LLD + HD) are described in applications [21,25]. In essence, the composite (CB + Gr)/HD is prepared by dry mixing of the powder components followed by extruder mixing (in melts state) and molding. The composite (CB + Gr)/(LLD + HD) was prepared by intensive dry mixing of the components followed by molding. To prove the effectiveness of this second procedure, the total concentration of the conductive filler was lower in (CB + Gr)/(LLD + HD), namely, 18.6%, as compared to 24.2% in (CB + Gr)/HD, but the weight ratio CB/Gr was the same, namely 4.5:1. Similar counterparts of these composites were prepared by melt extrusion + molding for (CB + Gr)/(LLD + HD) and intensive dry mixing + molding for (CB + Gr)/HD compounds.

2.2. Instruments and Methods

The pellets resulted from extrusion mixing or the powder mixture was formed as plates of 120 × 100 × 0.8 mm using a conventional heated mold with controlled temperature and a laboratory hydraulic press. Basically, the following molding conditions were used: temperature 170 °C (heating rate by 5 °C/min), holding time at maximum temperature 2 min, chilling under pressure (2.5 °C/min).

For radiation crosslinking, the samples were wrapped in aluminum foil and exposed to γ-rays in Ob-Servo Sanguis laboratory irradiator (Institute of Isotopes, Budapest, Hungary) equipped with ^{60}Co isotope (dose rate ~0.7 kGy/h, integral dose 150 kGy) in presence of air at room temperature.

SEM micrographs were recorded on FESEM scanning electron microscope dual beam type, model Auriga (Carl Zeiss SMT, Oberkochen, Germany). Additionally, a secondary electron detector Everhart-Thornley type within the chamber and In-Lens in column detector for ultra-topography images were employed. The magnification range was 1000x–50,000x with approximately 5 mm working distance. Various magnification SEM images (basically 1 kx, 5 kx and 20 kx) were taken out in different representative regions of each sample in order to better understanding their morphologies.

The volume (ρ_V) and surface (ρ_S) resistivities were measured at room temperature on 100 × 100 × 0.7 mm plates, using a Keithley electrometer, following a standard procedure.

The variation of the resistivity with temperature was measured on small samples (chips of 35 × 25 × 0.7 mm, with 25 mm distance between the flat electrodes, Figure 1) using a digital multimeter for electrical resistance and a thermocouple in contact with the sample surface.

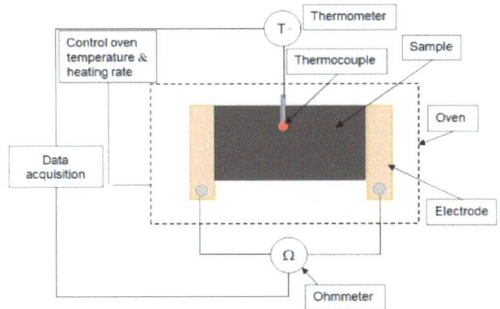

Figure 1. Schematic setup for ρ-T measurements.

The slope of ρ increase (the rate of resistivity increase on the heating curve) was calculated with the Formula (1):

$$\text{slope} = \frac{\rho_2 - \rho_1}{\rho_{max}} \cdot \frac{1}{T_2 - T_1} \quad (1)$$

Similar formula was used to describe the resistivity decrease (on NTC regions) or to calculate the slopes on the cooling curves.

A similar setup as above was used for measurements of the temperature on the sample surface (T_s, Figure 2). A thermo-insulating enclosure was used instead of the oven. An amperemeter was integrated within the circuit enabled the measurement of the absorbed electrical power.

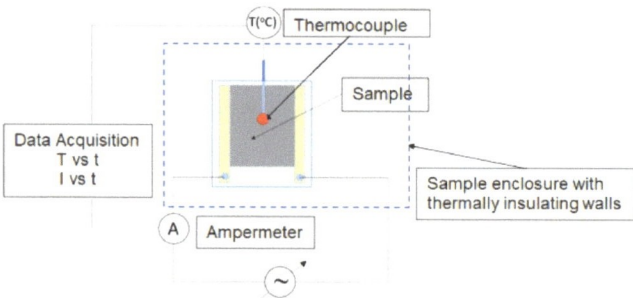

Figure 2. Schematic setup for T_s vs. time and I vs. time measurements.

A typical curve of ρ vs. T upon heating, with the parameters characterizing the kinetics of the process is shown in Figure 3. The cooling curve has similar shape, but is inversed, due to temperature decrease.

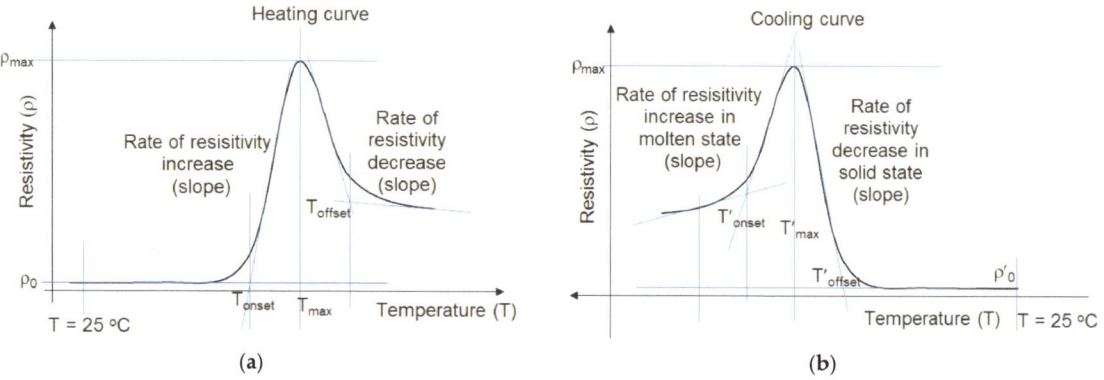

Figure 3. Typical ρ-T heating (**a**) and ρ-T cooling (**b**) curves and parameters used for kinetic characterization of resistivity vs. temperature.

DSC measurements were performed in non-isothermal mode, under either air (oxidation tests) or nitrogen (melting/crystallinity tests), using a DSC 131 evo instrument (Setaram, Lyon, France). The parameters describing the oxidation (OOT, oxidation rate, oxidation heat) were calculated from the thermograms as described in references [26,27]. The melting/crystallinity peaks, the melting/ crystallization temperatures, and the thermal effects associated to either melting or crystallization were calculated as described earlier in the reference [22]. The crystalline content (X_c) of the material was calculated by equations presented in reference [22], using a value of 279 J/g for melting enthalpy of totally crystalline polyethylene [28].

3. Results and Discussion

3.1. SEM Characterization

The Figure 4 presents the effect of long-term storage at room temperature (r.t.) on a (CB + Gr)/HD (24.2% conductive charge) sample prepared by melt extrusion followed by press molding (at 150 °C). It can be seen that the "luminosity", due to charge accumulation, on the sample surface during the SEM measurement decreased clearly from the freshly prepared sample to those stored for one year or more. It can be said that the stored material is better structured in the sense that for the long-term stored materials, and the conductive particles are more segregated from the polymer matrix, enabling the formation of more conductive channels and, as a result, higher conductivity. Resistivity measurements confirmed this interpretation: while the freshly prepared sample presented rather low conductivity [29], the stored samples presented considerably higher conductivity. This is seen, for example, in the composite (CB + Gr)/HD, which presented a decrease from 6.5×10^9 $\Omega \cdot$m for the freshly prepared sample [29] to 9.8×10^3 $\Omega \cdot$m after four years of storage. The surface temperature (T_s) of similar samples measured at different moments followed a similar trend (Figure 5): subjected to a same voltage, the stored sample presented significantly higher T_s values (on the plateau region) as compared to the freshly prepared one, indicating higher Joule effect due to increased current, which traverses the stored sample. The effects of irradiation and preparation procedure are also illustrated in Figure 4. The thermal effect of the studied composites is discussed in more detail in Section 3.3.

The Figure 6 shows the effect of processing on composite morphology: melt blending vs. room temperature physical blending of components in powder state [25]. For (CB + Gr)/HD, the morphologies of these samples appear significantly different (Figure 6a,b): the conductive particles in the case of the sample prepared by physical blending seem to be better segregated at the surface of the polymer phase. Hence, the formation of the conductive paths would be much easier. Indeed, the conductivity of the freshly prepared sample by dry physical mixing of powder components was considerably higher, which suggests a comparison of the thermal effects, namely, the curves (1) and (4) (Figure 5). Therefore, the same mixing procedure was applied for preparation of (CB + Gr)/(LLD + HD) composite (Figure 6c). It can be easily observed that the morphology of this sample is very similar to that of the sample (CB + Gr)/HD prepared by the same procedure (Figure 6b).

Another aspect of processing is comparatively illustrated in Figure 7 in the case of (CB + Gr)/HD composite prepared by melt blending: the sample in Figure 7a was subsequently pressed between two heated plates at 160 °C (press molding), using a spacer of 0.7 mm, while the sample in Figure 7b was prepared by injection molding [29]. While this sample presented conductivity (evolving from poor to high, as described above), the injected sample is practically non-conductive. When re-formed by pressing in the same conditions, the injected sample became conductive at the same level to the sample formed directly by press molding (from pellets of composite). This behavior would be understood if we compare the samples morphologies as they are seen from SEM analysis (Figure 7): the press mold sample appears as a continue material with moderate luminosity, while the injected mold sample appears as a stratified material, possibly due to the fine trepidations related to the injection molding. As a result, the sample present higher luminosity, suggesting that conductive channels may exist within each layer, but they are not extended between two neighboring layers due to the fine interlayer empty spaces. Again, for a same magnification, the aspect (morphology) of the samples (CB + Gr)/HD (Figure 7a) and (CB + Gr)/(LLD + HD) (Figure 7c) are similar due to similar processing procedures (dry mixing in powder state and hot molding).

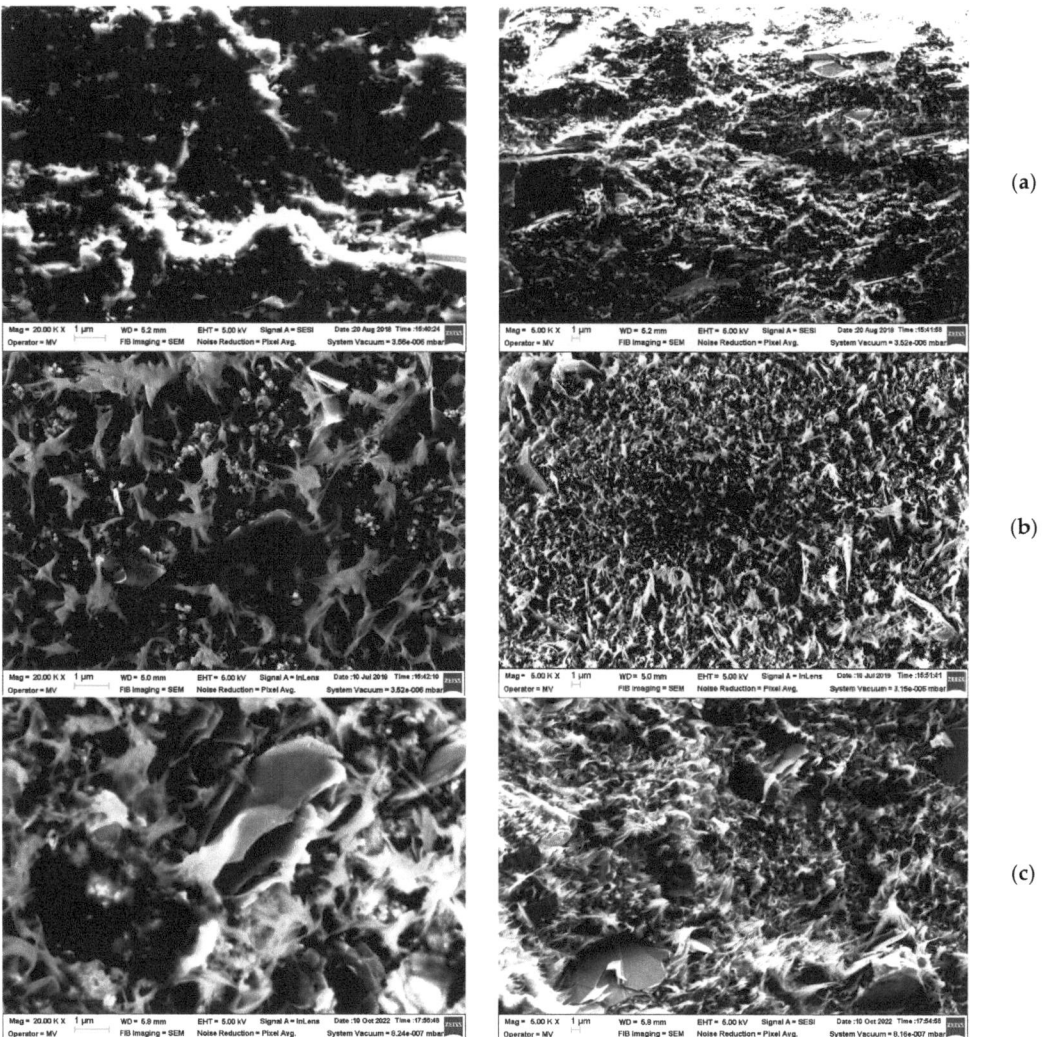

Figure 4. SEM images of (CB + Gr)/HD composite stored for different periods at r.t.: (**a**) freshly prepared sample; (**b**) stored for one year; (**c**) stored for four years. In the first column, the magnification is 20 kx while in the second one, it is 5 kx.

3.2. Resistivity vs. Temperature

3.2.1. Heating Curves

The first observation is that the heating curves ρ-T (resistivity-temperature curves) are sigmoidal, similar to non-isothermal oxidation curves in thermal analysis [26,27], with resistivity instead of the oxidation signal (heat flow, CL), hence similar kinetic parameters can be used to describe the ρ-T curves (Figure 3). The experimental observations are discussed below for each type of composite material, then a comparison of these materials is presented.

For both thermoplastic and crosslinked (CB + Gr)/HD composites, the ρ_{max} values decreased with the number of cycles (Figure 8, Table 2). The onset values for thermoplastic samples decreased with the number of cycles from 139 °C to 133 °C while for the crosslinked material, an inverse trend is observed. Hence, the onset values tend to reach a same value (of ~130 °C) for both thermoplastic and crosslinked materials submitted to repeated thermal cycles. T_{max} data in Table 2 suggest a similar behavior. The slope of resistivity increase appears lower at the first cycle, but it presents higher and comparable values for the further two cycles.

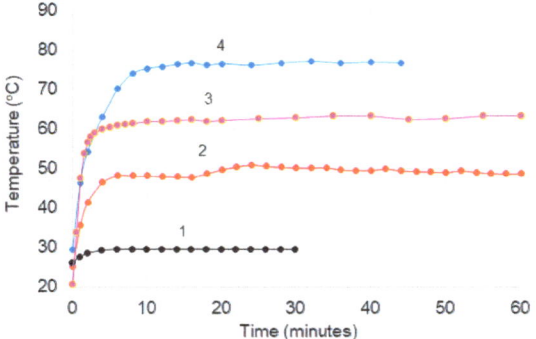

Figure 5. The surface temperature (T_s) vs. time for (CB + Gr)/HD composite with different histories/treatments: 1—freshly prepared by melt extrusion, unirradiated; 2—prepared by melt extrusion and irradiated after 2 months after preparation; 3—prepared by melt extrusion, stored as pellets for 3 years, then formed by press molding and irradiated; and 4—freshly prepared by physical mixing of the components and press molding, unirradiated.

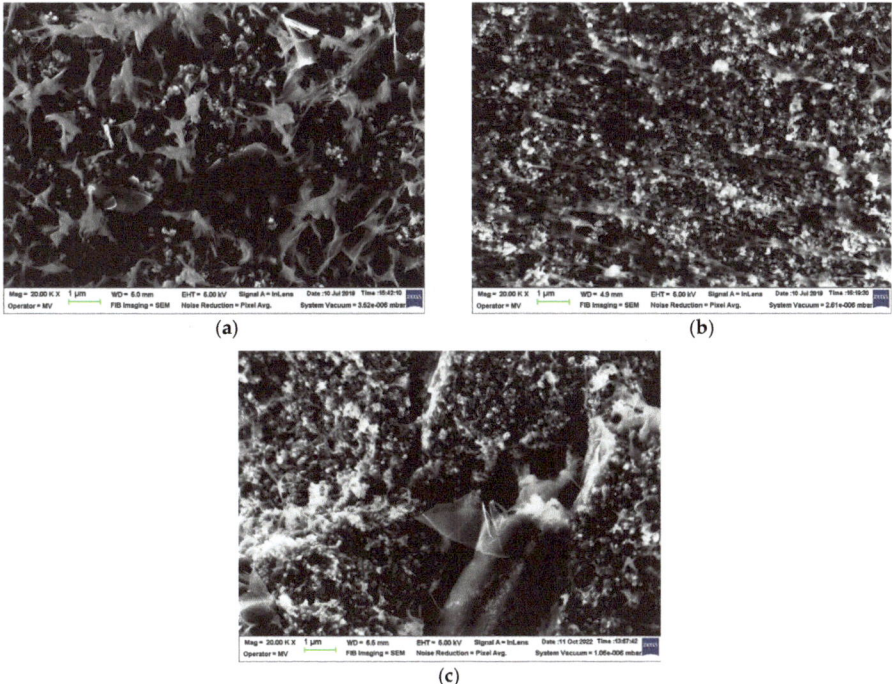

Figure 6. The effect of the preparation procedure on dispersion state of the conductive particles in a polymeric matrix: (**a**) (CB + Gr)/HD composite prepared by melt extrusion; (**b**) (CB + Gr)/HD composite prepared by mold pressing of a powder mixture; (**c**) (CB + Gr)/(LLD + HD) composite prepared by mold pressing of a powder mixture.

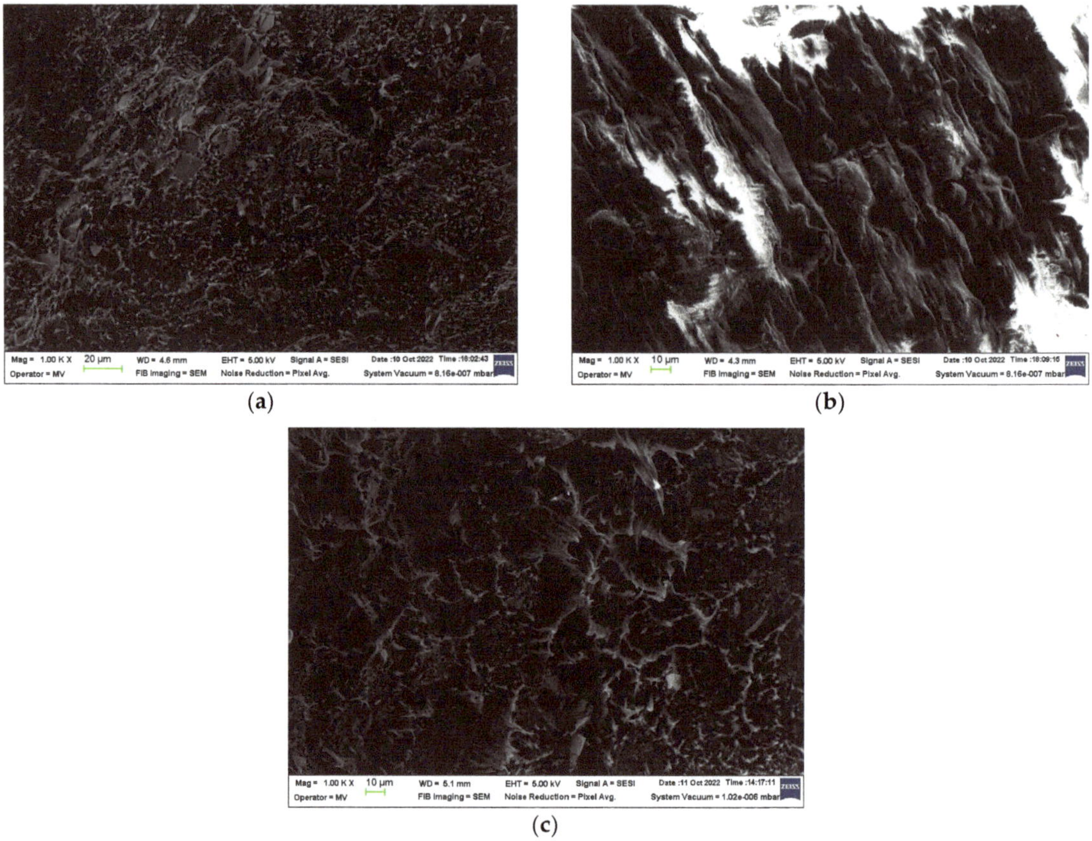

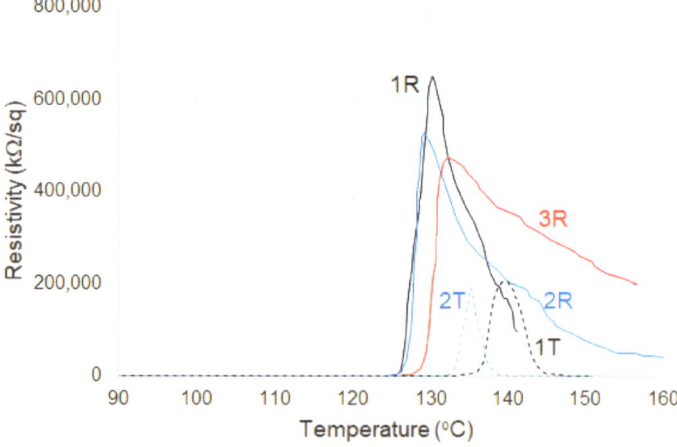

Figure 7. SEM micrographs of (CB + Gr)/HD composite prepared by melt blending followed by: (**a**) press molding; (**b**) injection molding; (**c**) micrography of sample CB + Gr/(LLD + HD).

Figure 8. Resistivity vs. temperature for (CB + Gr)/HD samples on repeated temperature cycles, heating portions: 1T, 2T—thermoplastic (D = 0 kGy) at first and second cycle, respectively; 1R, 2R, 3R—crosslinked (D = 150 kGy), cycles 1, 2 and 3, respectively.

Table 2. Kinetic data of resistivity increase upon the heating of (CB + Gr)/HD composites.

Dose (kGy)	Cycle Number	ρ_0 (kΩ/sq)	Onset Temperature (°C)	Slope of ρ Increase (K^{-1})	T_{max} (°C)	$\rho_{max} \cdot 10^{-5}$ (kΩ/sq)	PTC Intensity lg(ρ_{max}/ρ_0)	Slope of ρ Decrease (K^{-1})
0	1	3.50	137.0	0.54	139.6	2.32	4.83	−0.26
	2	1.49	133.2	0.49	135.2	1.92	5.11	−0.47
	3	1.23	133.1	0.47	135.1	1.90	5.19	−0.47
150	1	1.11	126.1	0.27	130.3	6.51	5.77	−0.14 / −0.07
	2	2.73	127.1	0.64	129.2	5.26	5.28	−0.09 / −0.03
	3	3.77	129.4	0.51	132.2	5.10	5.10	−0.04 / −0.03

The PTC effect tends to decrease slightly for the crosslinked material from the first to the third cycle due to increased values of ρ_0 and lower values of ρ_{max}. This behavior would be related to the thermo-oxidative degradation of the polymer matrix due to repeated exposure at elevated temperatures. However, for the thermoplastic material, the intensity of PTC effect increased with the number of cycles because the room temperature values of resistivity were lower after the former cycle. Possibly, the thermooxidative degradation is not the single factor affecting the electric properties of the materials subjected to multiple thermal cycles. The oxidation rate of HDPE was reported to be reduced in presence of CB and Gr mixture [22], through several possible mechanisms, namely, (i) direct annihilation of oxidation transient species produced by different active groups on CB surface [30,31], (ii) free radicals trapping by fullerene or fullerene-like structures on CB surface [30,32], and (iii) decrease in oxygen permeability within the amorphous phase induced by carbon particles [4].

In general, the ρ–T curves suggest (Figure 8) an increase in reproducibility with increasing the number of thermal cycles in agreement with previously reported data on different other PTC materials with HDPE matrix [2]. As compared to the other literature data on thermoplastic CB/HD composites, the stability and the reproducibility of the ρ–T curves (see for example [33]) appears higher with our materials (both HD and LLD + HD), possibly due to the benefic influence of the blend of conductive fillers used.

The ρ–T curves for thermoplastic and crosslinked (CB + Gr)/(LLD + HD) composites are shown in Figure 9, while the parameters describing the kinetics of resistivity increase with temperature are shown in Table 3. The curves at cycles 2 and 3 are closer to each other for the crosslinked material, suggesting more reproducibility, in agreement with previously reported data for radiation-crosslinked materials (see for example [2] for CB/HD and Gr/(LLD + HD) data reported by [17]).

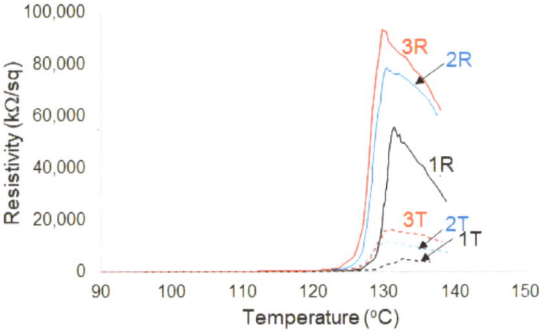

Figure 9. Resistivity vs. temperature for (CB + Gr)/(LLD + HD) samples on repeated temperature cycles, heating portions: 1T, 2T, 3T—thermoplastic (D = 0 kGy) cycles 1, 2, 3, respectively; 1R, 2R, 3R—crosslinked (D = 150 kGy), cycles 1, 2 and 3, respectively.

Table 3. Kinetic data of resistivity increase upon the heating of (CB + Gr)/(LLD + HD) composites.

Dose (kGy)	Cycle Number	ρ_0 (kΩ/sq)	Onset Temperature (°C)	Slope of ρ Increase (K^{-1})	T_{max} (°C)	$\rho_{max} \cdot 10^{-4}$ (kΩ/sq)	PTC Intensity $\lg(\rho_{max}/\rho_0)$	Slope of ρ Decrease (K^{-1})
0	1	4.12	122.2	0.25	132.8	0.481	3.07	−0.05
	2	6.85	126.3	0.26	130.5	1.143	3.22	−0.05
	3	8.53	128.1	0.35	130.7	1.610	3.28	−0.04
150	1	4.49	129.0	0.32	131.5	5.560	4.09	−0.04
	2	11.2	126.5	0.37	130.5	7.860	3.85	−0.03
	3	14.47	126.7	0.38	130.0	9.320	3.81	−0.04

T_{max} and T_{onset} are shifted toward lower temperatures, especially after the first cycle. This effect (attributable to either thermo-oxidative degradation or other structural changes as already mentioned above for (CB + Gr)/HD) would be seen as favorable for device security in limitation/switching applications if we take into account the increase in ρ_{max} with the number of cycles. In any case, these changes are lower as compared to the other literature data, suggesting a more stable network in our case.

For (CB + Gr)/(LLD + HD), the rate of resistivity increase, calculated as the slope of the leading edge of the resistivity peak, is significantly higher for the crosslinked samples as compared to the thermoplastic ones (Table 3).

It is obvious that the resistivity peaks are significantly higher for the crosslinked materials compared to the corresponding thermoplastic ones. The ρ_{max} values also increased with the number of cycles for both thermoplastic and crosslinked materials, but the ρ_{max} values are much higher for the crosslinked samples.

The intensities of the PTC effects are significantly higher (around one order of magnitude) for the crosslinked materials as compared to the thermoplastic ones. However, for the crosslinked materials, the PTC effect tends to slightly decrease as increasing the number of thermal cycles because of increased ρ_0 values, while an opposite trend is observed for both thermoplastic composites. These opposite behaviors are caused by increase in higher extent of room temperature resistivities with the number of cycles for crosslinked composite as compared to the thermoplastic one (Tables 2 and 3).

For the crosslinked (CB + Gr)/HD composite, the decrease in the resistivity after exceeding the T_{max} presented two slopes suggesting the occurrence of two processes: one is more rapid and is produced immediately after T_{max}, while the second is slower and covers a wider temperature range (Table 2). The slopes of both processes decreased with the number of cycles, and their values became comparable and considerably lower for cycles 2 and 3 as compared to the first cycle. This behavior reflects a decrease in NTC effect after the first thermal cycle. In the case of the thermoplastic material, the resistivity decreased sharply after T_{max}, (Figure 8) until a flat region with low resistivities (~100 kΩ) is reached, suggesting a strong NTC effect. Due to this behavior, the resistivity peak of the thermoplastic material appears more symmetric as compared to the crosslinked one (Figure 8).

The onset temperature appears a little higher for HDPE-matrix samples as compared to the blend ones, especially at the first cycle and for the unirradiated samples, while for the irradiated samples, the onset temperature values are practically similar for HD and (LLD + HD) composites, possibly due to increased similarity of both matrices induced by crosslinking.

The slopes values of ρ increase are significantly higher for HDPE composites than for (LLD + HD) ones, and the ρ_{max} values are considerably higher as well. The intensities of the PTC effects are, therefore, much higher with HDPE composites, especially with the radiation-crosslinked material.

Note that the slope values as calculated by formula (1) correspond to temperature coefficient of resistivity (see for example [34]). Unless the NTC effect, for PTC materials'

TCR has negative values. The slope values represent in our case the maximum of TCR values because the variation of ρ or (ρ/ρ$_{max}$) with temperature is typically not linear for PTC materials.

In addition, the resistivity peak of thermoplastic (CB + Gr)/(HD) appears shifted to lower temperatures to a greater extent than for (CB + Gr)/(LLD + HD) (Figures 8 and 9). For the (CB + Gr)/(LLD + HD) crosslinked samples, the peak becomes wider as the number of cycles increased (Figure 8), while for CB + Gr)/HD, this effect is considerably weaker (the ratio height/width remains practically constant).

The peak intensities (ρ$_{max}$) are lower for (CB + Gr)/(LLD + HD) composite than the (CB + Gr)/(HD) by more than one order of magnitude. For both thermoplastic and crosslinked (CB + Gr)/HD material, the ρ$_{max}$ value tends to decrease by thermal cycling, a behavior which is different to that observed for (CB + Gr)/(LLD + HD) composites and also differing to the above-mentioned literature data ([2] for CB/HD and [15] for CF/HD systems). After three thermal cycles, the ρ$_{max}$ value of crosslinked (CB + Gr)/(LLD + HD) composite became comparable to that of thermoplastic (CB + Gr)/HD.

The intensity of the NTC effects seem to be higher than in the above-mentioned literature cases, where a flat portion of high temperature heating curve is described ([2,15,17]), illustrating the possible role of the composition, conductive phase, type, and blending conditions on the PTC and NTC behavior of conductive composites.

It can be observed that the resistivity of the thermoplastic (CB + Gr)/HD material decreased strongly as the temperature increased (strong NTC effect), while the crosslinked sample presented only limited decrease in resistivity in molten state (Figure 8). For example, the resistivity at 150 °C (upon heating, 2nd cycle) was ~150 kΩ/sq for the thermoplastic material vs. 74,800 kΩ/sq for the crosslinked sample. This behavior illustrates that crosslinking suppressed significantly the NTC effect for (CB + Gr)/HD material. The wider peaks of resistivity observed for both thermoplastic and crosslinked CB + Gr)/(LLD + HD) and for crosslinked (CB + Gr)/HD, as compared to thermoplastic (CB + Gr)/HD, would be interpreted in a similar manner (see a comparison of resistivity values on heating in Figure 10).

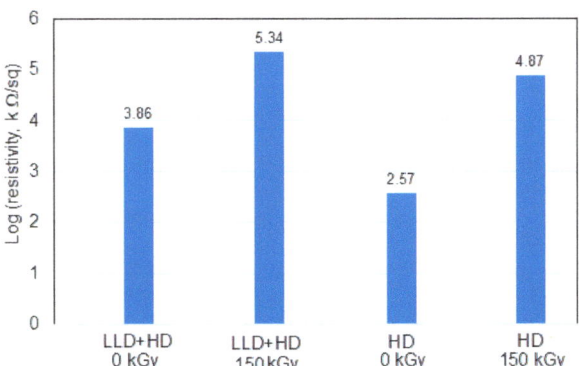

Figure 10. Resistivity at 10 °C from the T$_{max}$ on heating (2nd cycle) of the studied composites.

As compared to the similar composite samples with HDPE matrix, the blend LLD + HD induced lower ρ$_{max}$ values, wider resistivity peaks and lower PTC (Tables 2 and 3, Figures 8 and 9), but lower NTC also, even in thermoplastic state (Figure 10).

3.2.2. Cooling Curves

The cooling curves are presented in Figures 11 and 12 for (CB + Gr)/HD and (CB + Gr)/(LLD + HD) composites, respectively. The thermoplastic (CB + Gr)/HD presents again a distinct behavior as compared to other materials: the peak is sharp and symmetric while, for crosslinked (CB + Gr)/HD and (CB + Gr)/(LLD + HD), the peaks are clearly asymmetric with slow increase in resistivity in molten state and sharp decrease in solid state.

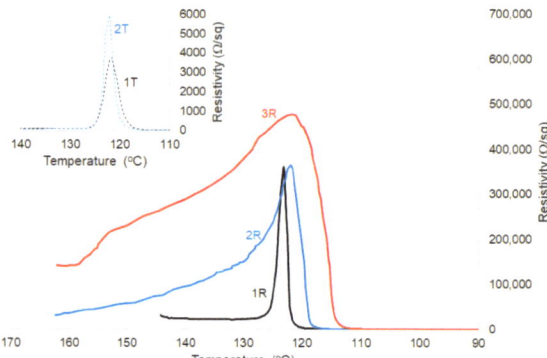

Figure 11. Resistivity vs. temperature for (CB + Gr)/HD samples on repeated temperature cycles, cooling portions: 1T, 2T– thermoplastic (D = 0 kGy) cycles 1 and 2 respectively; 1R, 2R, 3R—crosslinked (D = 150 kGy), cycles 1, 2 and 3, respectively. Note: because of weak intensity of the peaks of thermoplastic material. These curves are shown in the inset.

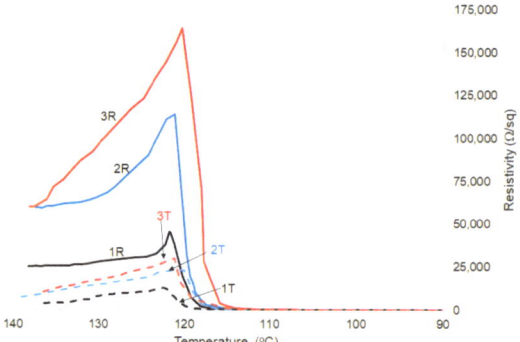

Figure 12. Resistivity vs. temperature for (CB + Gr)/(LLD + HD) samples on repeated temperature cycles, cooling portions: 1T, 2T, 3T—thermoplastic (D = 0 kGy) cycles 1, 2, 3, respectively; 1R, 2R, 3R—crosslinked (D = 150 kGy), cycles 1, 2 and 3, respectively.

It can be observed that the resistivity of the HD thermoplastic composites remained low for a relatively long period after heating cease, while other ρs increased with a smoother slope at the beginning followed by a more abrupt region as the temperature approached the T_{max} value. Thus, an onset temperature (T'_{onset}) of ρ increase can be defined as the intersection point of the flat (or slightly inclined) region at higher temperatures and the sharply increasing portion of the leading edge (when approaching the peak). The T'_{onset} values of the studied materials are shown in Tables 4 and 5. While for thermoplastic HD composite, T'_{onset} values are practically unchanged for the first and second cycle, and a significant increase with the number of cycles is observed for crosslinked (CB + Gr)/HD one. Increased values of T'_{onset} signify that a melt material become resistive, on cooling, earlier than a material with lower T'_{onset} values; this behavior would be related with lower NTC and higher PTC properties of such a material. Hence, the increase in T_{onset} values for crosslinked (CB + Gr)/HD suggests an improvement of the electrical properties of this material induced by thermal cycling (assuming that NTC effect is undesired for our case). It can be observed that (CB + Gr)/(LLD + HD) composites present this behavior even in thermoplastic state (Figure 12).

Table 4. Kinetic data of resistivity change on cooling of (CB + Gr)/HD composites.

Dose (kGy)	Cycle Number	T'_{onset} (°C)	Slope of ρ Increase in Molten State (K^{-1})	$\rho'_{max} \cdot 10^{-4}$ (kΩ/sq)	T'_{max} (°C)	Slope of ρ Decay in Solid State (K^{-1})	T_{offset} (°C)	$\rho_{r.t.}$ at the Cycle End (kΩ/sq)
0	1	124.6	−0.41	0.375	121.5	0.28	118.7	1.54
0	2	124.2	−1.07	0.594	122.1	0.50	120.2	1.23
0	3	124.1	−1.10	0.602	122.2	0.55	120.3	1.19
150	1	125.0	−0.53	36.06	123.1	1.09	122.0	2.79
150	2	127.2	−0.10 / −0.01	36.50	121.81	0.33	119.0	3.83
150	3	134.8	−0.03 / −0.01	47.65	121.64	0.20	114.4	8.7

Table 5. Kinetic data of resistivity change upon the cooling of (CB + Gr)/(LLD + HD) composites.

Dose (kGy)	Cycle Number	T'_{onset} (°C)	Slope of ρ Increase in Molten State (K^{-1})	$\rho'_{max} \cdot 10^{-4}$ (kΩ/sq)	T'_{max} (°C)	Slope of ρ Decay in Solid State (K^{-1})	T_{offset} (°C)	$\rho_{r.t.}$ at the End of Cycle (kΩ/sq)
0	1	134.1	−0.06	1.255	122.6	0.43	119.8	6.85
0	2	>139	−0.04	2.340	121.0	0.47	118.9	8.94
0	3	>137	−0.04	3.030	121.1	0.41	118.9	13.15
150	1	133.7	−0.04 / −0.18	4.55	121.7	0.51	119.2	11.22
150	2	129.2	−0.02 / −0.05	11.40	121.1	0.53	119.2	16.02
150	3	135.9	−0.03 / −0.08	16.43	120.3	0.30	117.1	18.81

For thermoplastic HD composites, as well as for crosslinked (CB + Gr)/HD at the first cycle, the rate of resistivity increase in molten state can be described by a single slope, while for others, two slopes can be defined for each process, suggesting two mechanisms of decay of conductive paths. This behavior could be related to the existence of two conductive powders with different aspect ratios, CB particles are spherical, while graphite ones are platelike [22], hence they would impart conductivity by different mechanisms [5]. Another factor would be the nature of the polymer matrix: the linear macromolecules allow easier movement of conductive particles and thus allow easier restoration/interruption of conductive paths while, in the case of crosslinked polymers, the mobility of the conductive particles is lower.

It is obvious also that ρ_{max} values are considerably higher for the crosslinked materials, a behavior which is similar to that observed upon cooling. Repeated cycles produced increased and wider resistivity peaks for crosslinked (CB + Gr)/HD and (CB + Gr)/(LLD + HD).

After the ρ_{max} value is reached, the resistivity dropped abruptly until values of hundred kΩ/sq, then the resistivity decreased slowly until few kΩ s were observed at r.t. The slopes of ρ decay (the rate of resistivity decrease) increased with the number of cycles for thermoplastic (CB + Gr)/HD, which remained practically unchanged for thermoplastic (CB + Gr)/(LLD + HD), but tended to decrease for crosslinked composites, especially for the (CB + Gr)/HD one (Tables 4 and 5). This behavior would be related, as well, to limited mobility of conductive particles in crosslinked polymers. Excepting thermoplastic (CB + Gr)/HD, all other materials presented slightly higher values of r.t. resistivity after each thermal cycle suggesting a certain "ageing" process.

3.3. DSC Measurements

DSC measurements aimed to check if the structural changes induced by repeated cycles in air in DSC furnace would be related to the above-discussed parameters of resistivity vs. temperature curves. The typical recorded heating and the cooling curves are presented in Figure 13. Note that, because the aim of these measurements was to correlate the parameters of the DSC curves to ρ-T curves to melting and crystallization data from DSC, and to detect eventual changes induced by repeated thermal cycles, the samples were measured in their initial state, as resulted from molding, without any treatment for erase their initial thermal history (as usual when the intrinsic melting and crystallization behaviors are assessed, see for example reference [35]).

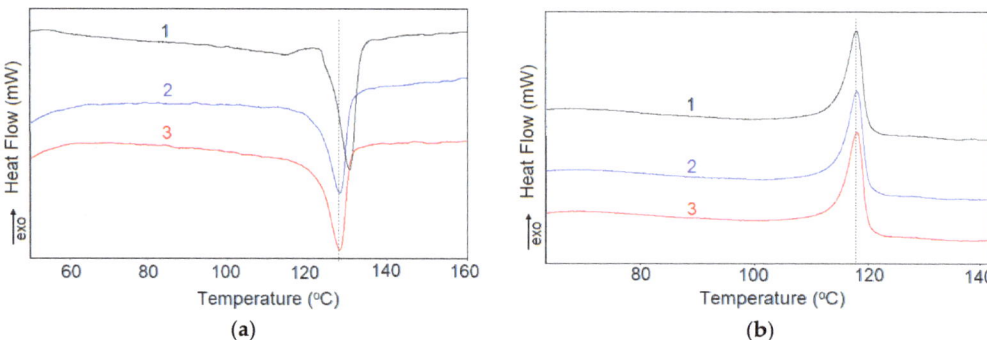

Figure 13. DSC curves from three repeated heating/cooling cycles between r.t. and 170 °C of thermoplastic (CB + HD)/(LLD + HD) composite: (**a**) heating curves (5 °C/min, air); (**b**) cooling curves (5 °C/min, air).

It was observed that for both thermoplastic and crosslinked (CB + Gr)/(LLD + HD) composites, the heating curves at the first cycle differs from others by presence of a peak at ~114.5 °C (it disappeared to further 2nd and 3rd cycles) and a T_{max} value of ~130.7 °C which subsequently decreased to ~128.1 °C. As the melting peak (for the 2nd and 3rd cycles) is unique, without shoulders or secondary peaks, it can be concluded that the LLD/HD blend is homogeneous. The small peak at ~114.5 °C, on the peak at first cycle, was related to a pseudo-crystalline phase possibly resulted on composite molding [22]. It seems to be related to the presence of HDPE. In the case of thermoplastic (CB + Gr)/HD composite, the shoulder on the main peak persists to further cycles while, for the crosslinked material, this shoulder is visible in the cooling curve at first cycle only, and not in the further ones. A diminution in crystallinity (calculated from DSC) of ~10% is also produced after the first cycle for crosslinked (CB + Gr)/HD as compared to less than 2% for (CB + Gr)/(LLD + HD) composites (either crosslinked or thermoplastic). For thermoplastic (CB + Gr)/HD composite, the drop in crystallinity after the first cycle is ~15%, but the decrease continued to further cycles, suggesting that thermoplastic HDPE network would be less stable than (CB + Gr)/(LLD + HD) one.

In general, the parameters of the heating curves of (CB + Gr)/(LLD + HD) shown in Table 6 were practically the same for the 2nd and the 3rd cycles, and differed slightly from those of the first cycle. In the case of (CB + Gr)/HD composites, the previous statements are especially valid for the crosslinked material, while the thermoplastic one appears less stable at repeated cycles test (Table 7). However, as the melting temperature does not practically change with repeated DSC cycles, the observed changes cannot be attributed to oxidative (chemical) degradation, but rather to molecular rearrangements which affect the crystallinity content.

Table 6. Kinetic parameters of the melting curve of (CB + Gr)/(LLD + HD).

Dose (kGy)	Cycle	T_m (°C)	ΔH_m (J/g)	T_{onset} (°C)	T_{offset} (°C)
	1	130.66	122.5	124.6	133.24
0	2	128.14	117.5	121.4	130.82
	3	128.18	117.4	121.3	130.70
	1	130.72	122.2	123.7	133.30
150	2	128.43	119.9	121.6	130.35
	3	128.44	119.5	121.6	130.33

Table 7. Kinetic parameters of the crystallization curve of (CB + Gr)/(LLD + HD).

Dose (kGy)	Cycle	T_c (°C)	ΔH_c (J/g)	T_{onset} (°C)	T_{offset} (°C)
	1	117.96	−106.6	119.87	114.28
0	2	117.93	−106.6	119.85	114.17
	3	117.95	−106.7	119.96	114.22
	1	118.47	−106.4	120.21	114.56
150	2	118.43	−106.2	120.25	114.45
	3	118.47	−106.2	120.34	114.57

The behavior of the (CB + Gr)/(LLD + HD) samples at the first heating cycle (Table 6) are similar to those of the resistivity variations with the temperature (see T_{onset}, T_{max}, T_{offset} data in Table 3) in the sense that the parameters of the first cycle are different from those of other two cycles which are practically equal. The resistivity peak (T_{max}) values are close to the T_{offset} from DSC ones (Figure 14), suggesting that the maximum of the resistivity is reached when the crystallinity completely disappear.

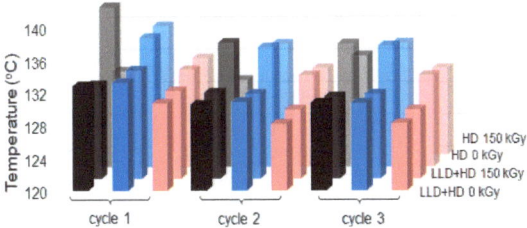

Figure 14. Correlation of T_{max} (■) from heating ρ-T curves with T_{offset} (■) and T_{max} (■) from heating DSC curves. The color intensities decrease, from front to back, in order to facilitate the comparison of the parameters.

In the case of thermoplastic HDPE composite, the temperature of reaching $ρ_{max}$ is closer to $T_{offset\ (DSC)}$, that is the maximum of resistivity corresponds to complete molten state of the matrix (Figure 14). This behavior resembles to (CB + Gr)/(LLD + HD) composites. For the crosslinked (CB + Gr)/HD composite, the temperatures of $ρ_{max}$ are better correlated to $T_{max\ (DSC)}$, meaning that there is still crystallinity within the system when the $ρ_{max}$ value is attained. In general, the resistivity data seem to be poorly correlated with DSC ones for (CB + Gr)/HD composites as compared to (CB + Gr)/(LLD + HD) ones.

The cooling curves are practically the same for all three cycles both in the case of thermoplastic and crosslinked (CB + Gr)/(LLD + HD) composites (Figure 13b, Table 7). In the case of crosslinked composite, the crystallization occurs slightly earlier, as the higher values of T_{max} and T_{onset} suggest, as compared to the thermoplastic material, due to lower mobility of crosslinked polymer chains. However, the duration of crystallization process remains practically the same, as suggest the values of difference between the average T_{onset} and T_{offset} values ($\Delta_c = \overline{T}_{onset} - \overline{T}_{offset}$), which are equal to 5.67 °C for thermoplastic and 5.72 °C for crosslinked composite. HDPE composites behave in general similarly (Tables 8 and 9), with the only difference of the above-mentioned shoulder. The Δ_c parameter has similar values to (CB + Gr)/(LLD + HD), namely, 5.67 °C for thermoplastic and 5.74 °C for crosslinked (CB + Gr)/HD composites, suggesting that crystallization processes are similar for both materials.

Table 8. Kinetic parameters of the melting curve of (CB + Gr)/HD.

Dose (kGy)	Cycle	T_m (°C)	ΔH_m (J/g)	T_{onset} (°C)	T_{offset} (°C)
0	1	131.92	183.4	124.56	135.84
	2	131.24	156.2	122.84	134.66
	3	131.24	149.4	122.68	134.85
150	1	131.92	175.5	124.56	135.84
	2	130.67	157.1	121.53	133.65
	3	130.64	157.2	121.57	133.75

Table 9. Kinetic parameters of the crystallization curve of (CB + Gr)/HD.

Dose (kGy)	Cycle	T_c (°C)	ΔH_c (J/g)	T_{onset} (°C)	T_{offset} (°C)
0	1	119.90	−128.7	122.73	112.31
	2	119.84	−122.4	122.52	112.59
	3	119.93	−122.2	122.69	112.28
150	1	119.47	−128.8	121.88	112.12
	2	119.42	−121.4	121.98	112.32
	3	119.41	−121.0	122.05	112.27

Concerning the correlation between the parameters of ρ-T and DSC cooling curves, for (CB + Gr)/(LLD + HD) composites, T'_{max} values (considered as relevant for PTC properties) are closer to T_{onset} ones from DSC (Figure 15). This result is consistent with that from heating curves (Figure 14) in the sense that the maximum of the resistivity corresponds to the start of crystallization process (the system does not contain crystallinity, but it is going to have it immediately). For (CB + Gr)/HD composites, T'_{max} values are also close to the T_{onset} (DSC), meaning that the melt HDPE matrix behave similarly to (LLD + HD) one. Note also that the values of peak temperatures (T_{max}) from DSC curves did not indicate the occurrence of some structural changes during the multiple thermal cycles, hence no chemical degradation can be supposed neither heating nor cooling.

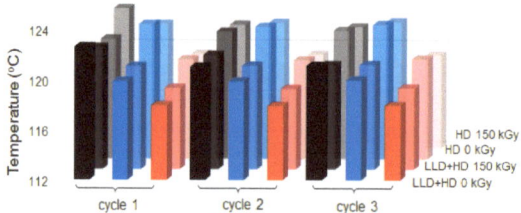

Figure 15. Correlation of T'_{max} (■) from cooling ρ-T curves with T_c (■) and T_{offset} (■) from cooling DSC curves. The color intensities decrease from front to back, in order to facilitate the comparison of the parameters.

The correlations of the behavior of the studied composites in DSC and ρ-T measurements would be related to major changes in conductive particles distribution within the liquid polymer matrix. However, it is not clear why the considerably rise of resistivity peak height in the case of crosslinked composites is observed. If radiation used for crosslinking bonds strongly, the CB particles on polymer chains, matrix dilatation in molten state should enable stronger dilatation effects. Thermally induced supplementary fixation of CB particles during the repeated cycles would result in higher resistivity values for (CB + Gr)/(LLD + HD), both in crosslinked and in thermoplastic state, as well as for crosslinked (CB + Gr)/HD composite. For Gr particles (where no significant interactions with the polymer chains are expectable), the rearrangement in the molten state is more probable mechanism. Hence, the higher values of the crystallinity and T_{max} at the first cycle can be assigned to molecular rearrangements during the sample molding and subsequent storage. Their evolution is in the same direction with the thermal parameters of the resistivity, but not with the resistivity values themselves.

3.4. Temperature Self-Regulation Behavior

The operation of the studied composites as self-regulating heating elements is illustrated in the Figures 5 and 16–18 and is based on the PTC effect shown in Figures 8 and 9. Practically, the jump of 4–5 orders of magnitude of the resistivity, from a few kΩ/sq to values of the order of 10^4–10^5 kΩ, enables a clear transition of the material from the state of semiconductor to that of electrical insulator, ensuring so the functionality of the element.

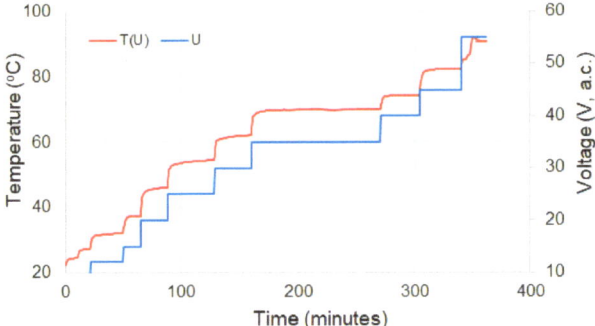

Figure 16. T_s vs. time and applied voltage for a (CB + Gr)/HD composite.

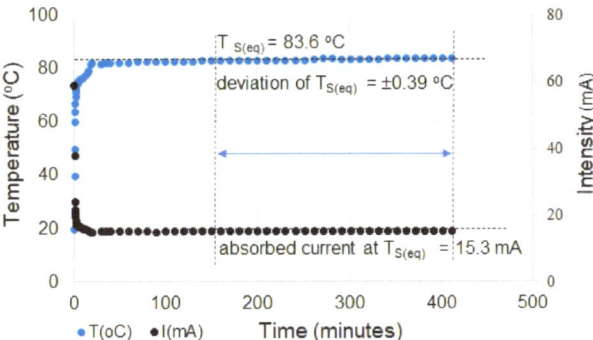

Figure 17. T_s vs. time and current intensity through the sample vs. time for a (CB + Gr)/HD composite (applied voltage, 55 V).

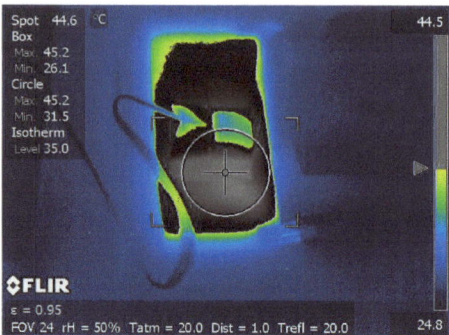

Figure 18. Thermal imaging image of a plate made of thermoplastic composite during heating (T_s not reached). The coil on the surface of the plate is a cable of a thermocouple attached to the sample.

The curves of surface temperature (T_s) vs. time (Figure 5) show, as already mentioned, the effect of some treatments on T_s. It can be noted that although the surface equilibrium (plateau) temperature ($T_{s(eq)}$) is considerably lower than T_{max} (from the ρ-T or DSC curves), and the materials show an obvious self-limiting effect of T_s.

Figure 16 shows that, apart from compositional effects and those regarding treatments applied during or after processing, the T_s value can also be controlled with the help of the voltage (electric field) applied between the electrodes of the element. This feature allows these materials to be used over a much wider temperature range, not just near T_{max}. In this range, where $T_s < T_{max}$, the material operates in a regime apparently similar to constant power devices. However, from Figure 17, it can be seen that the self-limiting properties are clearly manifested for very long, practically infinite periods, in which both T_s and the intensity of the current passing through the element remain constant.

Figure 18 shows the thermal image of the surface of a heating element in the form of a plate with dimensions of 120 × 100 × 0.8 mm made from (CB + Gr)/HD thermoplastic material (with the electrodes fixed on the 120 mm sides). The uniformity of the temperature distribution on the surface of the element is noticeable, even during heating, when the non-uniformity of the temperature field is expected to be higher, especially for samples with large distances between the electrodes.

4. Conclusions

A novel conductive composite material with homogeneous binary polymer matrix of HDPE and LLDPE and mixed conductive fillers (carbon black and graphite) was tested against a composite with similar conductive filler but with HDPE matrix. Even the concentration of the conductive filler was deliberately lower for (CB + Gr)/(LLD + HD), and the properties of this composite are comparable or better to those of (CB + Gr)/HD.

The kinetic parameters of the ρ-T curves (most relevant being T_{max} on heating and T'_{max} on cooling) correlate well with T_{offset} on heating or T_{onset} on cooling from the DSC curves, indicating that the resistivity peak is obtained when the polymer matrix is fully melted. However, for (CB + Gr)/HD, upon heating, the maximum of resistivity corresponds to T_{max} from DSC, i.e., when a certain degree of crystallinity still persists in the system.

When subjected to repeated thermal cycles, the composite (CB + Gr)/(LLD + HD) presented a better electrical behavior than CB + Gr)/HD, with an increase in $ρ_{max}$ values with the number of cycles, as well as less intense NTC effects, both for the crosslinked and thermoplastic samples.

Radiation crosslinking led to increased $ρ_{max}$ values, as well as inhibition of NTC effects in both cases, thus having a clear beneficial effect.

Limitation effects of surface temperature and current intensity through the sample were observed at different voltages, enabling the use of these materials as self-regulating heating elements at temperatures below the melting temperature.

The procedure based on physical mixing of components appears more efficient in imparting lower resistivity in solid state and high PTC effects to the composites, possibly due to the concentration of the conductive particles at the surface of the polymer domains. This heterogeneous distribution of the filler would facilitate the formation of the conductive paths, because a greater number of conductive particles are available on the surface of the polymer domains (the conductive filler would escape easier to be incorporated into the insulating polymer layer). Further work is still necessary to optimize both the procedure of composite synthesis and the properties of such materials.

Author Contributions: Conceptualization, R.S. and E.-M.L.; methodology, R.S. and E.-M.L.; investigation, R.S., E.-M.L. and V.E.M.; visualization, E.-M.L. and R.S.; writing—original draft preparation, R.S. and E.-M.L.; writing—review and editing, R.S., E.-M.L. and V.E.M.; project administration, E.-M.L.; funding acquisition, E.-M.L. All authors have read and agreed to the published version of the manuscript.

Funding: The financial support was provided by the Ministry of Research, Innovation and Digitization through contracts: 612PED/2022, and project number 25PFE/30.12.2021—Increasing R-D-I capacity for electrical engineering-specific materials and equipment regarding electromobility and "green" technologies within PNCDI III, Programme 1.

Institutional Review Board Statement: Not applicable.

Data Availability Statement: The data presented in this study are available on request from the corresponding author.

Acknowledgments: The authors thank to Alina Caramitu and Valerica Albu for the help in obtaining the polymer composites through extrusion and press molding, respectively. The help of Nicoleta Nicula (for procuring the polymeric materials), Adela Bara, and Cristina Banciu (for providing carbonic materials) is also acknowledged.

Conflicts of Interest: The authors declare no conflict of interest.

References

1. Meyer, J. Stability of polymer composites as positive-temperature-coefficient resistors. *Polym. Eng. Sci.* **1974**, *14*, 706–716. [CrossRef]
2. Narkis, M.; Ram, A.; Stein, Z. Effect of crosslinking on carbon black/polyethylene switching materials. *J. Appl. Polym. Sci.* **1980**, *25*, 1515–1518. [CrossRef]
3. Narkis, M.; Lidor, G.; Vaxman, A.; Zuri, L. Innovative ESD thermoplastic composites structured through melt flow processing. In Proceedings of the Electrical Overstress/Electrostatic Discharge Symposium Proceedings, 1999 (IEEE Cat. No.99TH8396), Orlando, FL, USA,, 28–30 September 1999; pp. 239–245.
4. Xiang, D.; Wang, L.; Zhang, Q.Q.; Chen, B.Q.; Li, Y.T.; Harkin-Jones, E. Comparative study on the deformation behavior, structural evolution, and properties of biaxially stretched high-density polyethylene/carbon nanofiller (carbon nanotubes, graphene nanoplatelets, and carbon black) composites. *Polym. Compos.* **2018**, *39*, E909–E923. [CrossRef]
5. Kiraly, A.; Ronkay, F. Temperature dependence of electrical properties in conductive polymer composites. *Polym. Test* **2015**, *43*, 154–162. [CrossRef]
6. Liu, Y.; Zhang, H.; Porwal, H.; Busfield, J.J.; Peijs, T.; Bilotti, E. Pyroresistivity in conductive polymer composites: A perspective on recent advances and new applications. *Polym. Int.* **2019**, *68*, 299–305. [CrossRef]
7. Bezzon, V.D.N.; Montanheiro, T.L.A.; de Menezes, B.R.C.; Ribas, R.G.; Righetti, V.A.N.; Rodrigues, K.F.; Thim, G.P. Carbon Nanostructure-based Sensors: A Brief Review on Recent Advances. *Adv. Mater. Sci. Eng.* **2019**, *2019*, 4293073. [CrossRef]
8. Li, J.; Chang, C.; Li, X.; Li, Y.; Guan, G. A New Thermal Controlling Material with Positive Temperature Coefficient for Body Warming: Preparation and Characterization. *Materials* **2019**, *12*, 1758. [CrossRef]
9. Shafiei, M.; Ghasemi, I.; Gomari, S.; Abedini, A.; Jamjah, R. Positive Temperature Coefficient and Electrical Conductivity Investigation of Hybrid Nanocomposites Based on High-Density Polyethylene/Graphene Nanoplatelets/Carbon Black. *Phys. Status Solidi (A)* **2021**, *218*, 2100361. [CrossRef]
10. Wang, X.; Liu, X.; Schubert, D.W. Highly Sensitive Ultrathin Flexible Thermoplastic Polyurethane/Carbon Black Fibrous Film Strain Sensor with Adjustable Scaffold Networks. *Nano-Micro Lett.* **2021**, *13*, 64. [CrossRef]
11. Luo, X.; Yang, G.; Schubert, D.W. Electrically conductive polymer composite containing hybrid graphene nanoplatelets and carbon nanotubes: Synergistic effect and tunable conductivity anisotropy. *Adv. Compos. Hybrid Mater.* **2022**, *5*, 250–262. [CrossRef]
12. Xue, F.; Li, K.; Cai, L.; Ding, E. Effects of POE and Carbon Black on the PTC Performance and Flexibility of High-Density Polyethylene Composites. *Adv. Polym. Technol.* **2021**, *2021*, 1124981. [CrossRef]

13. Alallak, H.M.; Brinkman, A.W.; Woods, J. I-V Characteristics of Carbon Black-Loaded Crystalline Polyethylene. *J. Mater. Sci.* **1993**, *28*, 117–120. [CrossRef]
14. Nagel, J.; Hanemann, T.; Rapp, B.E.; Finnah, G. Enhanced PTC Effect in Polyamide/Carbon Black Composites. *Materials* **2022**, *15*, 5400. [CrossRef]
15. Zhang, X.; Zheng, X.F.; Ren, D.Q.; Liu, Z.Y.; Yang, W.; Yang, M.B. Unusual positive temperature coefficient effect of polyolefin/carbon fiber conductive composites. *Mater. Lett.* **2016**, *164*, 587–590. [CrossRef]
16. Qiao, L.; Yan, X.; Tan, H.; Dong, S.; Ju, G.; Shen, H.; Ren, Z. Mechanical Properties, Melting and Crystallization Behaviors, and Morphology of Carbon Nanotubes/Continuous Carbon Fiber Reinforced Polyethylene Terephthalate Composites. *Polymers* **2022**, *14*, 2892. [CrossRef]
17. Zhang, P.; Wang, B.B. Positive temperature coefficient effect and mechanism of compatible LLDPE/HDPE composites doping conductive graphite powders. *J. Appl. Polym. Sci.* **2018**, *135*, 46453. [CrossRef]
18. Tung, T.T.; Pham-Huu, C.; Janowska, I.; Kim, T.; Castro, M.; Feller, J.-F. Hybrid Films of Graphene and Carbon Nanotubes for High Performance Chemical and Temperature Sensing Applications. *Small* **2015**, *11*, 3485–3493. [CrossRef]
19. Liu, G.; Tan, Q.; Kou, H.; Zhang, L.; Wang, J.; Lv, W.; Dong, H.; Xiong, J. A Flexible Temperature Sensor Based on Reduced Graphene Oxide for Robot Skin Used in Internet of Things. *Sensors* **2018**, *18*, 1400. [CrossRef]
20. Shebani, A.; Klash, A.; Elhabishi, R.; Abdsalam, S.; Elbreki, H.; Elhrari, W. The influence of LDPE content on the mechanical properties of HDPE/LDPE blends. *Res. Dev. Mater. Sci.* **2018**, *7*, 791–797. [CrossRef]
21. Setnescu, R.; Caramitu, A.; Lungulescu, M.; Mitrea, S.; Bara, A.; Stancu, N. Electroconductive composite presenting self-regulating temperature effect and process for making them. RO Patent Application OSIM A/01053/2018.
22. Setnescu, R.; Lungulescu, M.; Bara, A.; Caramitu, A.; Mitrea, S.; Marinescu, V.; Culicov, O. Thermo-Oxidative Behavior of Carbon Black Composites for Self-Regulating Heaters. *Adv. Eng. Forum* **2019**, *34*, 66–80. [CrossRef]
23. Ogah, A.; Afiukwa, J. The effect of linear low density polyethylene (LLDPE) on the mechanical properties of high density polyethylene (HDPE) film blends. *Int. J. Eng. Manag. Sci.* **2012**, *3*, 85–90.
24. Tang, H.; Piao, J.H.; Chen, X.F.; Luo, Y.X.; Li, S.H. The Positive Temperature-Coefficient Phenomenon of Vinyl Polymer Cb Composites. *J. Appl. Polym. Sci.* **1993**, *48*, 1795–1800. [CrossRef]
25. Setnescu, R.; Lungulescu, M.; Nicula, N.; Bara, A.; Caramitu, A. PTC material and process for manufacture a temperature self-regulating heating cable. RO Patent Application OSIM A/00503/2022.
26. Ilie, S.; Setnescu, R.; Lungulescu, E.M.; Marinescu, V.; Ilie, D.; Setnescu, T.; Mares, G. Investigations of a mechanically failed cable insulation used in indoor conditions. *Polym. Test* **2011**, *30*, 173–182. [CrossRef]
27. Lungulescu, E.-M.; Setnescu, R.; Ilie, S.; Taborelli, M. On the Use of Oxidation Induction Time as a Kinetic Parameter for Condition Monitoring and Lifetime Evaluation under Ionizing Radiation Environments. *Polymers* **2022**, *14*, 2357. [CrossRef]
28. Alamo, R.G.; Graessley, W.W.; Krishnamoorti, R.; Lohse, D.J.; Londono, J.D.; Mandelkern, L.; Stehling, F.C.; Wignall, G.D. Small angle neutron scattering investigations of melt miscibility and phase segregation in blends of linear and branched polyethylenes as a function of the branch content. *Macromolecules* **1997**, *30*, 561–566. [CrossRef]
29. Caramitu, A.-R.; Setnescu, R.; Lungulescu, M.; Mitrea, S.; Pintea, J. Dielectric Behaviour of some Composite Materials of HDPE/CB Type. *Electroteh. Electron. Autom.* **2018**, *66*, 73–79.
30. Cataldo, F. The role of fullerene-like structures in carbon black and their interaction with dienic rubber. *Fuller. Sci. Techn* **2000**, *8*, 105–112. [CrossRef]
31. Setnescu, R.; Jipa, S.; Setnescu, T.; Kappel, W.; Kobayashi, S.; Osawa, Z. IR and X-ray characterization of the ferromagnetic phase of pyrolysed polyacrylonitrile. *Carbon* **1999**, *37*, 1–6. [CrossRef]
32. Jipa, S.; Zaharescu, T.; Gigante, B.; Santos, C.; Setnescu, R.; Setnescu, T.; Dumitru, M.; Gorghiu, L.M.; Kappel, W.; Mihalcea, I. Chemiluminescence investigation of thermo-oxidative degradation of polyethylenes stabilized with fullerenes. *Polym. Degrad. Stab.* **2003**, *80*, 209–216. [CrossRef]
33. Ren, D.Q.; Zheng, S.D.; Huang, S.L.; Liu, Z.Y.; Yang, M.B. Effect of the carbon black structure on the stability and efficiency of the conductive network in polyethylene composites. *J. Appl. Polym. Sci.* **2013**, *129*, 3382–3389. [CrossRef]
34. Ngai, J.H.L.; Polena, J.; Afzal, D.; Gao, X.; Kapadia, M.; Li, Y. Green Solvent-Processed Hemi-Isoindigo Polymers for Stable Temperature Sensors. *Adv. Funct. Mater.* **2022**, *32*, 2110995. [CrossRef]
35. Zhang, J.J.; Rizvi, G.M.; Park, C.B. Effects of Wood Fiber Content on the Rheological Properties, Crystallization Behavior, and Cell Morphology of Extruded Wood Fiber/Hdpe Composites Foams. *Bioresources* **2011**, *6*, 4979–4989.

Disclaimer/Publisher's Note: The statements, opinions and data contained in all publications are solely those of the individual author(s) and contributor(s) and not of MDPI and/or the editor(s). MDPI and/or the editor(s) disclaim responsibility for any injury to people or property resulting from any ideas, methods, instructions or products referred to in the content.

A Comparative Study of Several Properties of Plywood Bonded with Virgin and Recycled LDPE Films

Pavlo Bekhta [1,*], Antonio Pizzi [2], Iryna Kusniak [1], Nataliya Bekhta [3], Orest Chernetskyi [4] and Arif Nuryawan [5]

1. Department of Wood-Based Composites, Cellulose, and Paper, Ukrainian National Forestry University, 79057 Lviv, Ukraine; kusnyak@nltu.edu.ua
2. LERMAB, Faculte des Sciences, University of Lorraine, Boulevard des Aiguillettes, 54000 Nancy, France; antonio.pizzi@univ-lorraine.fr
3. Department of Design, Ukrainian National Forestry University, 79057 Lviv, Ukraine; n.bekhta@nltu.edu.ua
4. "Shpon Shepetivka" LLC, 30400 Shepetivka, Ukraine; lokiorest@gmail.com
5. Department of Forest Products Technology, Faculty of Forestry, Universitas Sumatera Utara, Medan 20155, North Sumatra, Indonesia; arif5@usu.ac.id
* Correspondence: bekhta@nltu.edu.ua

Abstract: In this work, to better understand the bonding process of plastic plywood panels, the effects of recycled low-density polyethylene (rLDPE) film of three thicknesses (50, 100, and 150 µm) and veneers of four various wood species (beech, birch, hornbeam, and poplar) on the properties of panels were studied. The obtained properties were also compared with the properties of plywood panels bonded by virgin low-density polyethylene (LDPE) film. The results showed that properties of plywood samples bonded with rLDPE and virgin LDPE films differ insignificantly. Samples bonded with rLDPE film demonstrated satisfactory physical and mechanical properties. It was also established that the best mechanical properties of plywood are provided by beech veneer and the lowest by poplar veneer. However, poplar plywood had the best water absorption and swelling thickness, and the bonding strength at the level of birch and hornbeam plywood. The properties of rLDPE-bonded plywood improved with increasing the thickness of the film. The panels bonded with rLDPE film had a close-to-zero formaldehyde content (0.01–0.10 mg/m^2·h) and reached the super E0 emission class that allows for defining the laboratory-manufactured plastic-bonded plywood as an eco-friendly composite.

Keywords: plastic film-bonded plywood; recycled polyethylene film; formaldehyde release; physical-mechanical properties; bonding strength; wood species

1. Introduction

Despite the widespread use of traditional wood-based materials such as particle board, oriented strand board (OSB), and medium-density fiberboard (MDF), plywood is still a valuable material used in many industries. In 2020, the industrial production of wood-based panels in the world reached an output of 367 mln. m^3, including 118 mln. m^3 (32.2%) of plywood [1]. Synthetic adhesives based on formaldehyde are mainly used for the manufacture of wood-based panels, including plywood [2]. Among the synthetic adhesives, the urea-formaldehyde (UF) adhesives are the most widely used adhesives for the preparation of interior grade composites used for furniture and a wide variety of other applications, accounting for about 85% of the total volume worldwide [3,4].

Along with their many advantages such as chemical versatility, a high reactivity, excellent adhesion properties, solubility in water, relatively low curing temperatures, a short pressing time, ease of transportation, and a relatively low cost [2,5], these adhesives are characterized by certain problems associated with the release of hazardous volatile organic compounds (VOCs), including free formaldehyde, from finished wood composites. Formaldehyde is carcinogenic to humans and harmful to the environment, and its release indoors is associated with adverse human health problems [6,7]. The growing environmental problems and strict legal requirements for free formaldehyde emissions from wood-based panels have posed new challenges to researchers and industry in the development of environmentally friendly engineering wood products with close-to-zero formaldehyde emissions [8]. Therefore, the development of highly efficient ultra-low formaldehyde emissions is necessary for the sustainable production of wood-based panels. The current state of research and recent developments in the field of ultra-low formaldehyde emission wood adhesives and formaldehyde scavengers for manufacturing low-toxic, eco-friendly wood-based panels is summarized in the review [9].

On the other hand, it is well known that environmental pollution and the scarcity of natural resources are becoming pressing issues. Every year, a huge amount of plastic waste is generated worldwide ($\approx$6.3 billion tons) [10]. Only 9% of plastic waste is recycled, another 19% is incinerated, 50% ends up in landfill, and 22% evades waste management systems and goes into uncontrolled dumpsites, is burned in open pits, or ends up in terrestrial or aquatic environments [11]. Plastic waste generated annually per person varies from 221 kg in the United States and 114 kg in European OECD countries to 69 kg, on average, for Japan and Korea [11]. In Ukraine, on average, 250–270 kg of plastic waste is generated annually per person. Moreover, most of the waste goes to landfills, mainly industrial waste, of which only 2–3% is disposed of, others accumulate in landfills and local landfills [12]. Of the total amount of waste generated, almost 50,000 tons is polymer waste, of which low-density polyethylene (LDPE) and high-density polyethylene (HDPE) account for more than 30% of the total amount of polymer waste in Ukraine [13]. In Ukraine, only 10% of waste polymer materials are recycled, while the period of polymers biodegrading, such as plastic bags, is hundreds of years. The combustion of polymeric materials releases hazardous substances that pose a great danger to the environment. That is why the problem of recycling polymer waste is very relevant. Plastic is widely used in many applications, especially in the form of disposable products such as plastic bags, agricultural, and greenhouse films. These wastes mainly consist of polyethylene (PE), polypropylene (PP), polystyrene (PS), and polyvinyl chloride (PVC) [14]. The disposal of plastic is one of the major concerns for the environment due to its slow degradation. Nowadays, recycling waste by using it in production processes has also been in trend because it can prevent environmental pollution and reduce production costs.

Therefore, one of the promising directions for solving the problem of plywood toxicity is the use of thermoplastic polymers, especially recycled, instead of toxic thermosetting adhesives [4,15–20]. This not only improves the environmental performance of plywood and its production conditions, which mainly affects the quality and cost of plywood production, but also reduces the negative impact of such polymers on the environment due to the long process of their biodegradation. Recycled thermoplastic films can withstand more than two or three cycles of processing without compromising their physical and mechanical properties [21]. The application of thermoplastic film as an adhesive for the bonding of veneer, apart from the fact that the film is formaldehyde-free, has several other advantages compared with using liquid adhesives, which were described in our previous works [18–20]. Moreover, the recycled thermoplastic polymer materials are waste products and are cheaper than virgin polymers [22]. Therefore, the disposal of recycled thermoplastic synthetic waste allows economizing virgin polymer resources and simultaneously protecting the environment.

Nevertheless, recent studies on thermoplastic polymers used as adhesives have mainly focused on the application of virgin polymers for bonding wood veneers in plywood production [16,18–20,23]. However, there is already some experience in the use of secondary thermoplastic polymers for bonding veneers. For example, 1.5-mm industrial grade pine and birch veneers were bonded with waste polystyrene (PS) recovered from disposable plates and utensils at the bonding parameters as follows: thermoplastic load 750 g/m^2, press temperature

200 °C, maximum unit pressure 1.5 MPa, total pressing time 300 s [17]. The authors showed that birch wood having a higher hardness gave higher bonding strengths.

The recycled thermoplastic polymers produced from tetra package waste, domestic film waste, recycled synthetic textile fibers (polyurethane and polyamide-6), and recycled polypropylene are also used as adhesives for bonding the birch wood veneer [15]. It was found that the use of different recycled thermoplastic waste products for the bonding of birch wood veneer guaranteed the shear strength of the materials, which is higher than some virgin polymers and considerably exceeds the adhesive strength of industrial plywood based on phenol–formaldehyde adhesives. The optimal technological parameters for producing samples were noted: pressure 2 MPa, contact time 1–2 min, and a temperature for polyethylene of 130 °C, for polypropylene of 180 °C, and for polyamide-6 of 220 °C.

In another study [24], the recycled 500 g shopping plastic bags, mainly composed of polyethylene, polypropylene, polyvinyl chloride, and polystyrene were used as adhesive for bonding poplar veneer. Before using, the recycled plastic bags were cleaned, processed with a chemical reagent, dried, and shredded. A different amount of recycled plastic bags, 60, 80, 100, and 120 g/m^2, was spread between veneers. It was concluded that the optimal hot-pressing parameters are as follows: a plastic use of 100 g/m^2, a hot-pressing temperature of 150 °C, and a hot-pressing time of 6 min. The bonding strength of the plywood decreased markedly in response to the increase in the dosage of plastics.

Other authors [25] made the laminated veneer lumber (LVL), utilizing high-density polyethylene (HDPE) from supermarket plastic bags as an adhesive for bonding wood veneers from amescla wood (*Trattinnickia burseraefolia*) in a laboratory scale. Three HDPE amounts were evaluated: 150 g/m^2, 250 g/m^2, and 350 g/m^2. In general, composite boards showed good quality and mechanical properties were similar or higher than those found in LVL manufactured using thermosetting resin. The composite boards made with the 350 g/m^2 HDPE amount showed better mechanical and dimensional stability properties. Close contact between the HDPE plastics and the wood cell walls resulted in a stronger physical and mechanical bond [26].

One of co-authors mixed wood flour and low-density polyethylene (LDPE) for rerouting plastics waste to make wood composite plastics (WPC). Results of this study revealed that WPC could be produced with a predominant matrix of LDPE up to 95%. The role of the plastics matrix was beneficial in terms of a shortened degradation in nature, and even the mechanical properties were affected, particularly higher in MOE [27].

Another study [28] demonstrated the suitability of recycling waste milk pouches of 40 and 60 μm thickness made out of low-density polyethylene (LDPE) as the bonding agent in preparing plywood panels from veneers of *Melia dubia* wood. The panels were prepared with varying proportions of LDPE films amounting to a polymer content of 80 g/m^2, 105 g/m^2, 210 g/m^2, and 310 g/m^2. The polymer content of 210 g/m^2 was found to be the optimum level to ensure the satisfactory physical and mechanical properties of panels.

It was demonstrated that waste polyethylene in the form of granules between 0.7 mm and 1.3 mm in thickness could be also used in the manufacture of OSB panels, resulting in the enhancement of physical and mechanical properties [29].

In this study, the focus is on recycled LDPE as it is the most widespread type of plastics in the packaging industry (different films) and agriculture (crop protection films, haylage films). Recycled plastics contain a multitude of added chemical additives/contaminants (e.g., pesticide residues, pigments, flame retardants, etc.) [30]. The identification of these chemical additives is quite a difficult concern. It was established that the chemical composition of recycled LDPE was not more complex than that of virgin LDPE [31]. The authors explain this by the fact that in the process of recycling, organic compounds may be partially or selectively removed in the cleaning and/or extrusion steps of the recyclate [31]. Moreover, the recycled LDPE has higher tensile strength and shrinkage, but a lower mass flow index compared to virgin LDPE [32].

However, there is a lack of literature data comparing the adhesive ability of virgin and recycled LDPE polymers. Therefore, the purpose of this study was to obtain a better understanding of the bonding process of plastic plywood with recycled LDPE film when using various wood species and to compare the obtained properties with the properties of plywood panels bonded by virgin LDPE film.

2. Materials and Methods

2.1. Materials

In the experiments, the rotary-cut veneers of poplar (*Populus alba* L.), birch (*Betula verrucosa* Ehrh.), beech (*Fagus sylvatica* L.), and hornbeam (*Carpinus betulus* L.) with thicknesses of 0.75 mm, 1.55 mm, 0.45 mm, and 1.50 mm, respectively, and with a moisture content of 6 ± 2% were used. The recycled low-density polyethylene (rLDPE) film (LLC "Planet Plastic", Irpin, Ukraine) with the same dimensions as the veneers and thicknesses of 50 μm, 100 μm, and 150 μm, density of 0.92 g/cm^3, and melting point of 108 °C was used for the bonding of plywood samples. The amount of plastic rLDPE film at thicknesses of 50 μm, 100 μm, and 150 μm equals 46, 92, and 138 g/m^2, respectively. Virgin LDPE film with a melting point of 105–110 °C under the same conditions was used for the comparison.

2.2. Manufacturing and Testing of Plywood Samples

Three-layer plywood samples were made (Figure 1). Sheet of film was incorporated between the two adjacent veneer sheets. The prepared veneer assemblies were subjected to hot pressing in the lab press at a pressure of 1.4 MPa and temperature of 160 °C for 4.5 min. After hot pressing, the plywood samples were removed from the press and were subjected to the cold pressing at room temperature. The cold pressing was performed to release internal stresses and reduce the warping of samples. Then, the plywood panels were air conditioned at 20 ± 2 °C and 65 ± 5% (RH). Three plywood samples were prepared at each condition.

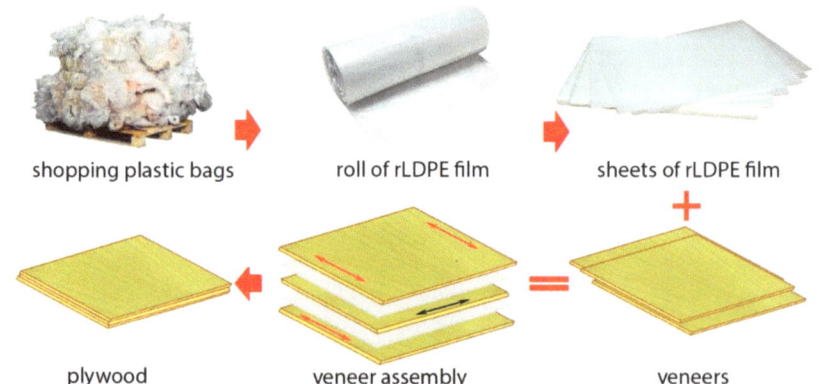

Figure 1. Schematic of plywood samples production.

Density, bending strength (MOR), modulus of elasticity in bending (MOE), shear strength, water absorption (WA), and thickness swelling (TS) of rLDPE film-bonded plywood samples were determined according to the standards [33–37]. The shear strength was measured after pre-treatment for bonding class 1—dry conditions—and plywood test pieces were immersed in water at 20 ± 3 °C for 24 h [35,36]. To determine the WA and TS, the samples were immersed in distilled water for 2 and 24 h according to the EN 317 standard [37]. For each variant, at least ten samples were used for the shear strength test and six samples were used to determine MOR, MOE, WA, and TS.

For each test series, one panel was randomly selected for analysis of formaldehyde content (FC) based on EN ISO 12460-3 standard (gas analysis method) [38]. In addition, urea-formaldehyde (UF) adhesive was used to manufacture plywood samples for the comparison of formaldehyde release from UF and plastic-bonded samples. UF resin with a 67% solid content, Ford cup (4 mm, 20 °C) viscosity of 117 s, spot life of 49 s, and a pH value of 8.2 was used in the experiments. For the preparation of UF adhesive, 20% solution of ammonium chloride as hardener and kaolin as filler were used. The plywood samples using UF adhesive were produced according to the pressing parameters usually used in practice: adhesive spread 110 g/m^2, pressing temperature, pressure, and time of 160 °C, 1.8 MPa, and 6 min, respectively.

Furthermore, the measurement of the core temperature inside the veneer package under given wood species and thickness of rLDPE film was undertaken. Temperature changes were measured using thermocouples connected to a PT-0102K digital multichannel device [39]. Statistical analysis of the obtained results was conducted using SPSS software program version 22 (IBM Corp., Armonk, NY, USA).

3. Results

The properties of rLDPE-bonded plywood samples were compared with the properties of virgin LDPE-bonded samples obtained by us in the previous work [20]. The effect of the veneer wood species on several physical and mechanical properties of plywood samples was found statistically significant. In addition, different thicknesses of the film caused the differences in the properties of samples. The average values of the thickness and density of plywood samples bonded by rLDPE and LDPE films are given in Table 1.

Table 1. Thickness and density of plywood samples.

Wood Species	Thickness of Veneer (mm)	Thickness of Film (μm)	Thickness of Plywood Samples (mm)		Density of Plywood Samples (kg/m^3)	
			rLDPE	LDPE	rLDPE	LDPE
Poplar	0.75	50	2.19 (0.02) *	2.19 (0.03)	435.36 (19.94)	463.86 (14.48)
		100	2.23 (0.06)	2.23 (0.04)	463.91 (22.45)	488.74 (15.80)
		150	2.22 (0.04)	2.23 (0.03)	484.48 (13.87)	505.11 (23.46)
Beech	0.45	50	1.39 (0.07)	1.32 (0.05)	618.25 (25.36)	661.16 (17.68)
		100	1.31 (0.03)	1.35 (0.02)	659.83 (14.94)	677.45 (19.69)
		150	1.42 (0.09)	1.39 (0.06)	673.05 (39.87)	720.20 (27.75)
Birch	1.55	50	4.57 (0.09)	4.41 (0.11)	649.93 (13.65)	668.21 (18.90)
		100	4.62 (0.07)	4.52 (0.11)	644.95 (36.16)	651.82 (43.34)
		150	4.69 (0.09)	4.58 (0.06)	632.36 (39.23)	659.53 (13.06)
Hornbeam	1.50	50	4.47 (0.09)	4.35 (0.13)	749.41 (17.66)	784.63 (15.08)
		100	4.49 (0.16)	4.18 (0.03)	765.83 (34.76)	790.09 (15.82)
		150	4.33 (0.13)	4.52 (0.03)	771.01 (15.96)	802.13 (10.94)

* Values in parenthesis are standard deviations.

3.1. Density of Plywood Samples

Since the veneer of different wood species was used for the manufacture of plywood samples, it was natural that the density of the samples would depend on the wood species. Plywood samples had a lower density using low-density wood species under the same pressing conditions. The lowest average density of 461 kg/m^3 was recorded for the poplar and the highest of 762 kg/m^3 for the hornbeam plywood samples bonded with rLDPE film (Figure 2a). The densities of birch and beech plywood samples were 662 and 650 kg/m^3, respectively, and they differ insignificantly ($p > 0.05$) based on the Duncan's test. A similar trend in the density values was observed for samples bonded with virgin LDPE film (Figure 2a). The values of the density of the plywood samples were related with the initial density of the veneers. The densities of veneer used in this study were: for poplar wood 390 kg/m^3, for beech wood 605 kg/m^3, for birch wood 655 kg/m^3, and for hornbeam wood 730 kg/m^3. However, it should be noted that not only the density of the veneer, but also its thickness affects the density of the finished plywood samples. The thickness of the veneer used was different for various wood species (Table 1).

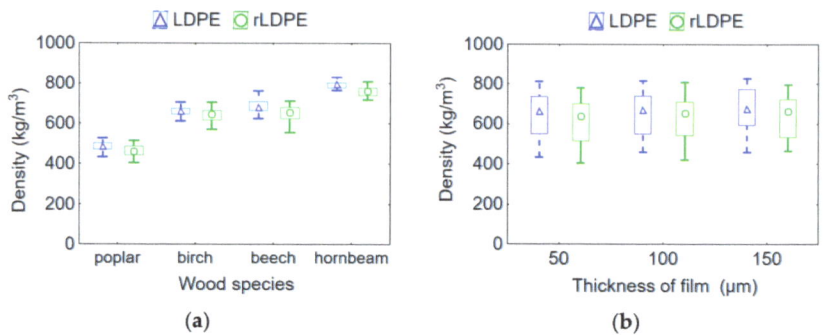

Figure 2. Mean plots of density of plywood samples depending on wood species (**a**) and thickness of rLDPE/LDPE film (**b**).

The film thickness has a much smaller effect ($F = 13.690$) on the density of plywood samples compared to wood species ($F = 803.682$) according to the F-values of the ANOVA analysis, but this effect was also significant ($p \leq 0.05$). This can be explained by the much smaller share of film compared to the share of wood in the volume of the plywood sample. As the thickness of the film used increases from 50 to 150 μm, the density of rLDPE and LDPE-bonded plywood samples increases by 4.2% and 4.1%, respectively (Figure 2b). The plywood samples bonded with a film thickness of 50, 100, and 150 μm differ insignificantly with each other in the density values ($p > 0.05$).

A non-strong significant difference in the values of the density of plywood samples bonded with LDPE and rLDPE films was found ($p = 0.043$). Since the films had the same thickness and density, this weak significance can be considered as a consequence of the heterogeneity of veneer among the groups. Wood has a very complex anatomical structure and different properties in various fiber directions. It is very difficult to select veneer sheets identical in structure and properties even within the same wood species. The average density values of the plywood samples that were bonded with LDPE film were higher (656.1 kg/m^3) than those for samples that was bonded with rLDPE film (629.0 kg/m^3). Thermosetting adhesives (urea-formaldehyde (UF), melamine-urea-formaldehyde (MUF), and phenol-formaldehyde (PF) also significantly affect the density, MOR, and MOE of plywood panels manufactured with eucalyptus, beech, and hybrid poplar veneers, as evidenced by the results obtained by other authors [40]. On the contrary, Shukla and Kamdem [41] studied laminated veneer lumber (LVL) manufactured from yellow poplar veneers using UF, MUF, melamine formaldehyde (MF), and cross-linked polyvinyl acetate (PVAc), and the results showed no differences.

3.2. Bending Strength and Modulus of Elasticity of Plywood Samples

ANOVA analysis showed that both wood species and film thickness and their interaction significantly affect MOR and MOE. The lowest MOR of rLDPE-bonded poplar plywood samples averaged at 61.5 MPa, the highest in hornbeam plywood was 101.6 MPa, and 81.3 and 85.4 MPa in beech and birch plywood, respectively (Figure 3a). There is no significant difference in MOR values between beech and birch plywood samples. There is also an insignificant difference in MOR values among rLDPE-bonded and LDPE-bonded plywood samples. The MOR of LDPE-bonded poplar plywood averaged at 62.5 MPa, 122.8 MPa in hornbeam plywood, and 84.2 and 82.2 MPa in beech and birch plywood, respectively. The average MOR values for the hornbeam plywood were 1.65 times higher than those values for poplar plywood. The density values of hornbeam plywood were also 1.65 times higher than those values of poplar plywood. This indicates a virtually linear dependence of MOR on the density of plywood samples. It is known that density strongly correlates with the strength of wood [42]—the greater the density the greater the strength. This is well confirmed by our research [43,44], and agreed well with the results of other researchers [40,41], who found that the MOR and MOE of plywood panels increase with increasing density.

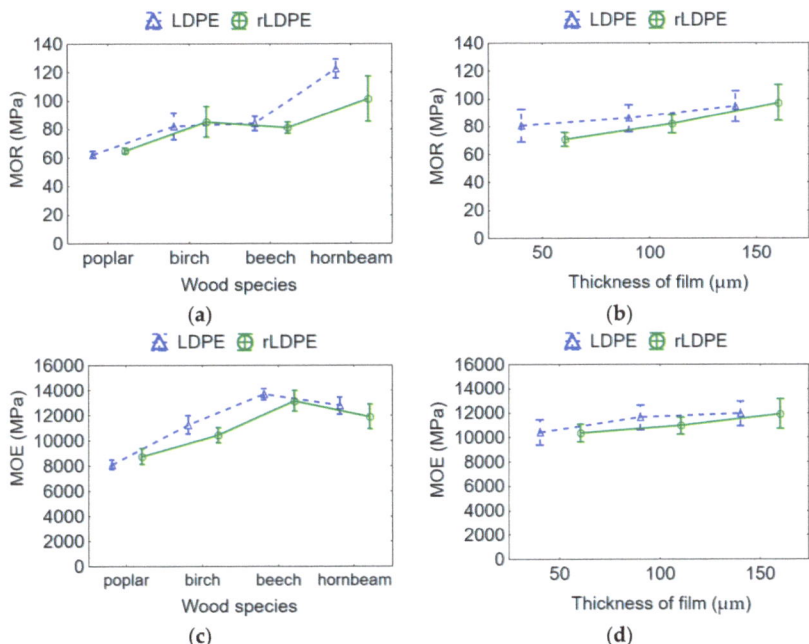

Figure 3. Mean plots of bending strength (a,b) and modulus of elasticity (c,d) of plywood samples depending on type of adhesive, wood species, and thickness of rLDPE/LDPE film.

However, when analyzing the impact of wood species on the MOR and MOE, it should be noted that veneers of different wood species and thicknesses were used in this study. As a result, the plywood samples had different thicknesses (Table 1). It is known that the thickness of the veneer affects the MOR of plywood samples [44]. Several researchers [45–47] found that the mechanical strength of plywood panels decreases with increasing veneer thickness due to an average depth of lathe checks. It was established that for a 3 mm thick veneer, the lathe check depth is about 70% of their thickness [48]. The thickness of the plywood panels also affects the MOR and MOE. These dependences are expressed linearly [43,44]—MOR and MOE decrease with increasing plywood sample thickness. This is in good agreement with the data of other authors [46,47], who also found a decrease in bending strength by increasing the thickness of the panel.

As the thickness of the rLDPE film used increases from 50 to 150 μm, the MOR of plywood samples increases by 37.1%, from 71.0 to 97.3 MPa (Figure 3b). For the LDPE-bonded plywood, the MOR of samples increases by 14.6%, from 80.9 to 94.7 MPa. Despite this, the insignificant difference in the values of MOR of the plywood samples bonded with rLDPE and LDPE films of different thicknesses was established. This is in good agreement with previous work [49], which showed a reduction in the properties of plywood panels with low amounts of polymer when using plastic film as an adhesive for bonding. Reducing the polymer content leads to the worse and less complete filling of wood cavities and as a result, fewer adhesive locks are formed, which provide bonding strength. This follows from the concept of mechanical adhesion, since mechanical locking is considered the most likely mechanism for bonding thermoplastic polymers to wood [15,16,18–20]. According to Pizzi [50], secondary forces are the dominant mechanism of wood bonding.

The lowest value of the MOE of rLDPE-bonded plywood samples is recorded for the poplar samples of 8772.2 MPa, then for the birch samples of 10,451.6 MPa, for the hornbeam samples of 11,922.7 MPa, and the highest for the beech samples of 13,165.5 MPa (Figure 3c). Thus, the poplar plywood samples had the greatest flexibility and elasticity, and the beech samples, the greatest rigidity. When increasing the film thickness from 50 to 150 µm, and, hence, with an increasing polymer content, the MOE of rLDPE-bonded plywood increases by 15.2% from 10,373.9 MPa for a 50 µm thickness to 11,949.7 MPa for a 150 µm thickness (Figure 3d); plywood becomes stiffer and less elastic. A similar tendency is observed for LDPE-bonded plywood (Figure 3d); under similar conditions, the MOE value increased by 12.8%. The difference between the MOE values for plywood bonded with LDPE and rLDPE films is insignificant ($p > 0.05$) based on the Duncan's test.

3.3. Shear Strength of Plywood Samples

It was established that the shear strength of plywood samples depends significantly on both wood species and film thickness and their interaction. The effect of film thickness was stronger ($F = 195.981$) than the effect of wood species ($F = 106.074$). All plywood samples met the requirements of the EN 314-2 standard [36] with the average values of shear strength (Figure 4a) that were above the limit value (1.0 MPa) indicated in the standard. Beech plywood samples showed the highest shear strength value of 1.77 MPa, and poplar samples showed the lowest shear strength value of 1.08 MPa. Poplar, birch, and hornbeam plywood samples differed insignificantly ($p > 0.05$) in terms of shear strength based on the Duncan's test. All plywood samples had satisfactory bonding strength for indoor applications. The difference between the values of shear strength for the plywood samples from investigated wood species bonded with LDPE and rLDPE films was insignificant ($p > 0.05$) based on the Duncan's test.

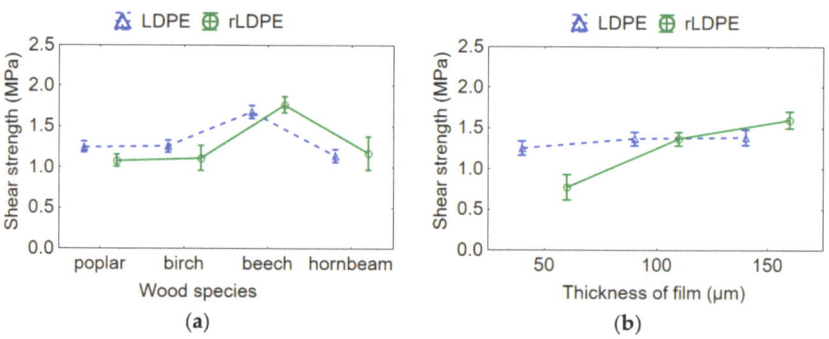

Figure 4. Mean plots of shear strength of plywood samples depending on type of adhesive, wood species (**a**) and thickness of rLDPE film (**b**).

One of the reasons for the higher bonding strength of beech plywood compared to other wood species used is the use of beech veneer of the smallest thickness (Table 1). Less veneer thickness, fewer lathe cracks in the veneer, and therefore the higher bonding strength. Several authors [51,52] also suggested that the shear strength reflects mainly the quality of veneer. Usually, the thicker the veneer, the deeper and more spaced the lathe checks [46]. The deep lathe checks and rough surfaces in thick veneers significantly reduce the shear strength of the plywood samples. Plywood samples from poplar veneer of lower density and thickness, as well as plywood samples from birch and hornbeam veneers of higher density and thickness, differed insignificantly in terms of bond strength based on the Duncan's test (Figure 4a). As was described in previous work "the amount of polymer that penetrates per unit volume of poplar veneer with a thickness of 0.75 mm is much greater than the amount of polymer that penetrates per unit volume of birch veneer with a thickness of 1.55 mm or hornbeam veneer with a thickness of 1.50 mm" [20]. In addition, less polymer penetrates the cavities of birch and hornbeam veneer, because their porosity is lower than that of poplar veneer. A linear relationship exists between the bonding strength and the porosity [53].

With an increase in the film thickness from 50 to 150 μm, and, hence, in the polymer content, the shear strength of plywood samples bonded with rLDPE film increases 2.1 times (Figure 4b). Since the bonding strength is ensured by the formation of mechanical locks [15,16], as the film thickness increases, more polymer penetrates the wood cavities and more such mechanical locks are formed. This positively affects the bonding strength. However, it was found that in the case of using a film thickness of 50 μm, the thermoplastic polymer is clearly insufficient to ensure a satisfactory bonding strength. The shear strength of plywood samples bonded with an rLDPE film thickness of 50 μm averaged at 0.77 MPa and was lower than the value of 1.0 MPa specified in standard EN 314-2 [36]. This low bonding strength is due to the low adhesive spread rate for this film thickness, which was equivalent to 46.0 g/m². This is almost three times less than the amount of adhesive used in practice for liquid thermosetting adhesives. On the contrary, a virgin LDPE film thickness of 50 μm provided a bonding strength 1.6 times higher (1.26 MPa) than an rLDPE film of similar thickness (Figure 4b). Whereas, a 150 μm rLDPE film provided a bonding strength (1.60 MPa) 1.2 times higher than an LDPE film of similar thickness. Nevertheless, the average values of shear strength for rLDPE- and LDPE-bonded plywood samples were 1.29 and 1.34 MPa, respectively, and differed insignificantly based on the Duncan's test.

The interaction of wood species and film thickness also significantly affects the bonding strength. As already mentioned, the bonding strength depends on the quality of the veneer, as well as the degree of penetration of the polymer into the veneer. Wood species with higher porosities provide a better penetration of polymer into the wood. According to several researchers [15,16], in the case of using thermoplastic polymers as an adhesive for the gluing of wood veneer, the concept of mechanical adhesion is suitable. This concept involves good penetration of polymer into the wood and the formation of adhesive locks. In this case, the bonding strength will depend on the number of such locks; the more of them and the deeper they are, the better the bonding strength. Therefore, the spread of polymer on the surfaces and in the structure of wooden elements is of great importance. For the melting of polymer and its spreading on a surface and in the structure of wooden elements, the pressing temperature should be sufficient. As can be seen from the heating curves of the core layer of the veneer package (Figure 5), the accepted pressing temperature was sufficient for this. The thickness of the film or the content of the polymer does not affect the heating rate and it can be neglected. Packages of poplar and beech veneer heat up faster than packages of birch and hornbeam veneer. This is due to both the density of wood and the thickness of the veneer used.

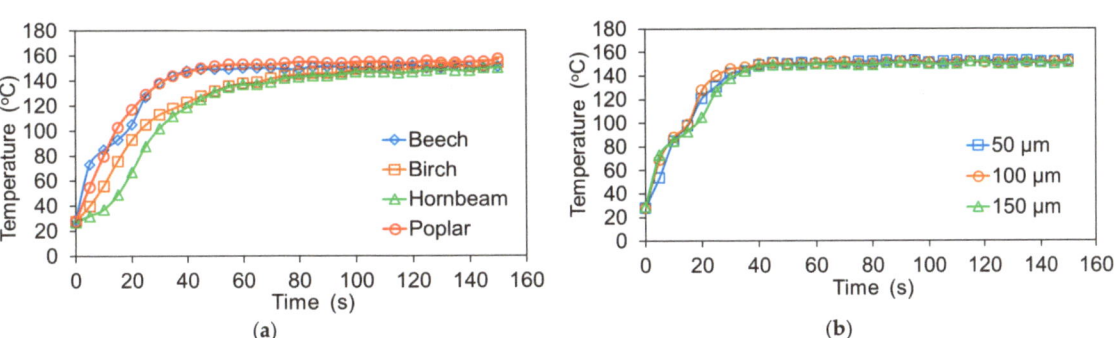

Figure 5. Core temperature curves of plywood samples made with (**a**) veneers of different wood species and rLDPE film of 150 μm thickness, and (**b**) beech veneers and different thicknesses of rLDPE film.

Moreover, unlike other wood composite materials, plywood has a continuous bondline. In the case of a significant porosity of wood or a low viscosity of the polymer and its low consumption, an excessive penetration of the polymer may occur, and it may not be enough to form a continuous bondline of uniform thickness, which leads to a decrease in bonding strength. That is why the film thickness of 50 μm was not sufficient to form a continuous adhesive layer. This caused starvation at the bondline and poor bonding (unsatisfactory bonding strength). Nevertheless, it should be noted that penetration is but one factor contributing to bond performance. The results of several authors [53] verified that compatibilizers and nanoparticles are suitable candidates to improve the bonding strength.

3.4. Water Absorption and Thickness Swelling of Plywood Samples

The ANOVA analysis was performed on the data to evaluate the effect of wood species and the thickness of the plastic film on the WA and TS of the plywood samples after 2 and 24 h immersion in water. It was found that both factors affect significantly the WA and TS of samples. The weaker WA (2 h) was observed in birch and hornbeam samples bonded with rLDPE film of 36.1% and 38.4%, respectively (the difference between them is insignificant), the stronger WA (2 h) in beech and poplar samples 46.0% and 48.8%, respectively (the difference between them is insignificant). After soaking in water for 24 h, the lowest WA of 48.7% was observed in birch plywood samples bonded with rLDPE film, and the largest WA of 95.5% in poplar samples (Figure 6a). Although the plywood samples bonded with virgin LDPE film showed a similar trend and less WA values than those bonded with rLDPE film (Figure 6a), but the difference between WA values for these two types of films was insignificant ($p > 0.05$).

The statistical analysis showed that wood species has a stronger effect on TS values of rLDPE-bonded plywood samples for both 2 h and 24 h than the thickness of plastic film. During the first 2 h of soaking in water, the least TS was observed in the poplar plywood samples of 5.0%, and the largest of 10.4% in the hornbeam samples (Figure 6c). Birch, beech, and hornbeam plywood samples did not differ in the values of TS for 2 h based on the Duncan's test. During the 24 h of soaking in water, the least TS was observed in the poplar plywood samples of 5.7%, and the highest of 11.4% in the hornbeam samples (Figure 6c). Birch and beech plywood samples did not differ in the values of TS for 24 h based on the Duncan's test. The plywood samples bonded with a virgin LDPE film demonstrated a similar trend in TS values as those bonded with rLDPE film (Figure 6c), and the difference between TS values for these two types of films was insignificant ($p > 0.05$).

As can be seen (Figure 6c), a very small differences between the values of TS for 2 h and 24 h exist for the rLDPE/LDPE-bonded plywood samples from veneer of all investigated wood species. Small differences between the values of WA for 2 h and 24 h were observed only for beech, birch, and hornbeam plywood samples (Figure 6a). This indicates that the samples absorb water and swell most intensively during the first 2 h of soaking in water. For poplar plywood samples bonded with rLDPE film, the WA increases almost twice with the duration of immersion in water from 2 h to 24 h (Figure 6a).

The obtained values of WA and TS are in good agreement with the generally accepted statement that these parameters are strongly related to density [54]. A higher density of plywood samples leads to fewer voids and, consequently, to lower WA, but higher TS. The wood species and the thicknesses of veneer and film affect the density of plywood samples, and therefore their WA and TS. This is well illustrated by the example of poplar plywood samples. A higher porosity (lower density) of poplar veneer compared to other investigated wood species leads to a higher WA of plywood samples bonded with rLDPE and LDPE films (Figure 6a). WA occurs by filling the voids and pores with moisture. On the other hand, for the plywood samples from poplar veneer, the lowest TS was observed (Figure 6c), which is due to the small thickness of the veneer used. As a result, more polymer penetrated per unit volume of this veneer than per unit volume of birch, beech, and hornbeam veneers.

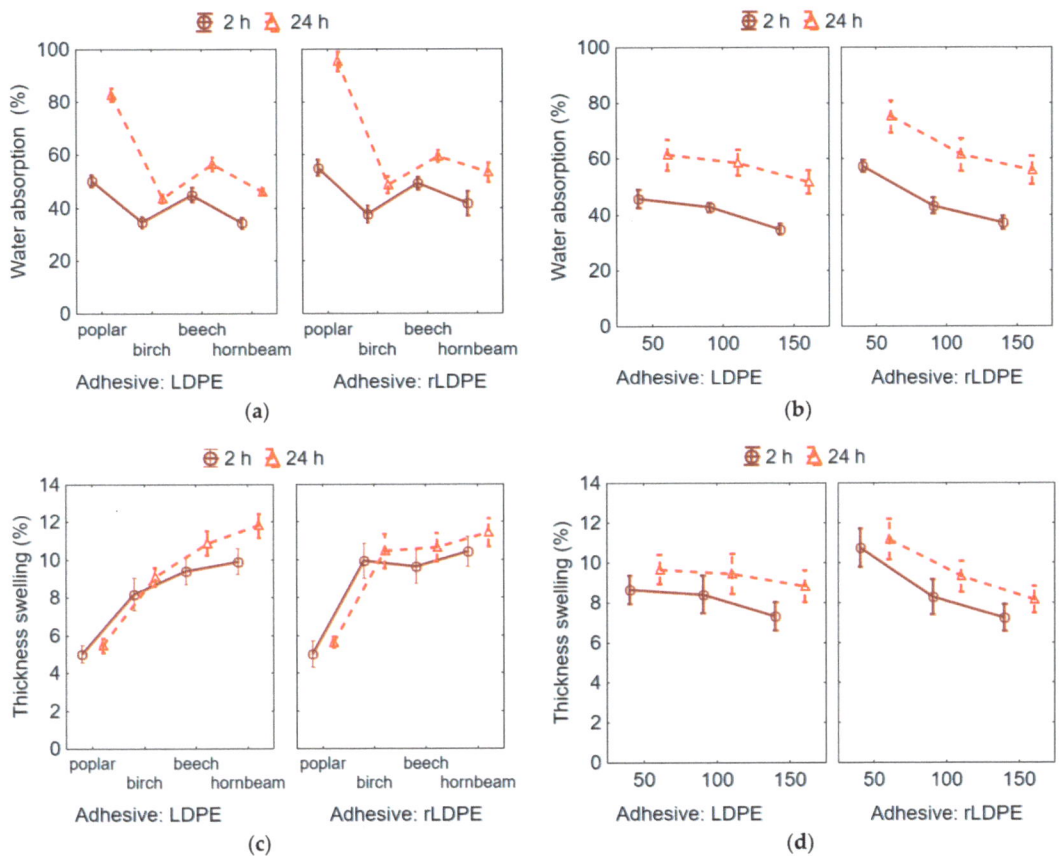

Figure 6. Mean plots of WA (**a**,**b**) and TS (**c**,**d**) of plywood samples depending on wood species (**a**,**c**) and thickness of film (**b**,**d**).

As the film thickness increases, and, hence, the amount of thermoplastic polymer, the WA and TS values of rLDPE-bonded plywood samples decrease for 2 h and 24 h. The lowest values of WA (Figure 6b) and TS (Figure 6d) were observed in the plywood samples bonded with 150 µm thick rLDPE film after 2 h of soaking in water, 34.4% and 7.3%, respectively, and 56.0% and 8.2%, respectively, after 24 h of soaking in water. The 50 µm thick rLDPE film demonstrated the worst WA and TS due to the low polymer content, which was insufficient to fill the wood cavities with the molten polymer.

A similar trend in the influence of wood species and thickness of film on WA and TS was observed for plywood samples bonded with virgin LDPE film. The average values of WA (24 h) for plywood samples bonded with virgin LDPE and rLDPE films were 57.3% and 64.2%, respectively. Based on the Duncan's test, this difference was significant ($p \leq 0.05$). On the contrary, the average values of TS (24 h) for samples bonded with virgin LDPE and rLDPE films were 9.3% and 9.6%, respectively, and this difference was insignificant ($p > 0.05$) based on the Duncan's test.

3.5. Pearson Correlation for Properties Versus Density

The obtained results showed that the investigated factors and their interaction affect significantly the properties of plywood samples bonded with rLDPE film. Information on the correlation of the investigated properties with each other and the density, as one of the parameters that can significantly affect the properties of plywood samples, would also be useful. Therefore, Pearson's correlation coefficient was applied to correlate properties among themselves and with density (Table 2). This coefficient indicates that density correlated significantly with all physical and mechanical properties. The highest correlation was found between density and WA (−0.834). This correlation was moderate and negative, which means that an increase in density causes a decrease in water absorption. The density also correlated well with TS (0.774), MOR (0.583), and MOE (0.617). These correlations were moderate and positive, which means that an increase in density causes an increase in bending strength, the modulus of elasticity, and thickness swelling. No correspondingly useful correlation was observed between density and shear strength (0.111). However, shear strength was positively correlated with MOR (0.498) and MOE (0.728), i.e., an increase in bonding strength means a corresponding increase in MOR and MOE. WA was negatively correlated with MOR (−0.657) and MOE (−0.621), which means that the greater the absorption of water by plywood samples, the lower the values of MOR and MOE.

Table 2. Pearson correlation (r) for properties versus density of plywood samples.

Property	Density	WA (24 h)	TS (24 h)	Shear Strength	MOR	MOE
Density	1					
WA (24 h)	−0.834 [a]	1				
TS (24 h)	0.774 [a]	−0.554 [b]	1			
Shear strength	0.111	−0.334	−0.287	1		
MOR	0.583 [b]	−0.657 [b]	0.113	0.498 [b]	1	
MOE	0.617 [b]	−0.621 [b]	0.325	0.728 [a]	0.704 [a]	1

[a] Correlation is significant at the 0.01 level; [b] Correlation is significant at the 0.05 level.

3.6. Formaldehyde Release

Table 3 shows that emissions of formaldehyde from plywood samples made from virgin and recycled LDPE films are very low (0.01–0.10 mg/m^2·h) compared to conventional UF-bonded (0.66–0.85 mg/m^2·h) plywood samples. The slight release of formaldehyde can be explained by its presence in natural wood [55]. This result suggests that plastic-bonded plywood reached super-E0 classification and can be considered a product with "zero formaldehyde emissions". This can be considered a significant contribution from the environmental point of view.

Table 3. Formaldehyde release of plywood samples.

	Formaldehyde Release (mg/m^2·h)			
	Wood Species			
Adhesive	Beech	Birch	Hornbeam	Poplar
UF	0.85 (0.03)	0.69 (0.02)	0.73 (0.03)	0.66 (0.03)
LDPE—50	0.03 (0.004)	0.06 (0.001)	0.10 (0.002)	-
LDPE—100	-	0.08 (0.001)	-	-
LDPE—150	-	0.04 (0.002)	0.06 (0.002)	0.07 (0.002)
rLDPE—50	-	0.05 (0.002)	-	0.09 (0.001)
rLDPE—100	-	-	0.10 (0.002)	0.03 (0.001)
rLDPE—150	0.06 (0.002)	0.01 (0.001)	0.10 (0.002)	-

Values in parentheses represent standard deviations.

4. Conclusions

Currently, eco-sustainability and the reuse of plastic wastes are highly topical issues. This study demonstrated that the rLDPE film is suitable for obtaining environmentally friendly plywood panels from veneers of different wood species with characteristics suitable for use successfully in indoor applications. Wood species and film thickness significantly affect the MOR, MOE, shear strength, and dimensional stability of rLDPE-bonded plywood samples. The use of high-density wood species (beech, birch, and hornbeam) compared to low-density wood species (poplar) resulted in higher MOR, MOE, and TS and lower WA of the samples. No clear effect of the density of wood species on the shear strength was found. With an increase in the thickness of plastic film, and, hence, in the polymer content, the values of MOR, MOE, and shear strength of rLDPE-bonded plywood samples increase by 37.1%, 15.2%, and 107.7%, respectively, whereas the values of WA and TS decrease by 25.6 % and 26.9%, respectively. Furthermore, the physical (except WA) and mechanical properties showed no statistical differences between the rLDPE-bonded and virgin LDPE-bonded plywood samples. The main advantage of these polymers is that the amount of formaldehyde emission from the plastic-bonded plywood is close-to-zero (0.01–0.10 mg/m^2·h). These results confirm the effectiveness and the environmental benefits in the use of the recycled plastic wastes.

Author Contributions: Conceptualization, P.B.; methodology, P.B., I.K. and O.C.; investigation, O.C., I.K., N.B. and A.N.; writing—original draft preparation, P.B. and A.P.; writing—review and editing, P.B., A.P., I.K., A.N. and N.B. All authors have read and agreed to the published version of the manuscript.

Funding: This research received no external funding.

Institutional Review Board Statement: Not applicable.

Informed Consent Statement: Not applicable.

Data Availability Statement: The data that support the findings of this study are available upon reasonable request from the authors.

Conflicts of Interest: The authors declare no conflict of interest.

References

1. FAO. Global Production and Trade in Forest Products in 2020. Available online: https://www.fao.org/forestry/statistics/80938/en/ (accessed on 20 May 2022).
2. Dunky, M. Adhesives in the wood industry. In *Handbook of Adhesive Technology*, 2nd ed.; Revised and Expanded; Pizzi, A., Mittal, K.L., Eds.; Marcel Dekker, Inc.: New York, NY, USA, 2003; 71p. [CrossRef]
3. Park, B.D.; Kim, J. Dynamic Mechanical Analysis of Urea-Formaldehyde Resin Adhesives with Different Formaldehyde-to-Urea Molar Ratios. *J. Appl. Phys.* 2008, *108*, 2045–2051. [CrossRef]
4. Pizzi, A.; Papadopoulos, A.N.; Policardi, F. Wood Composites and Their Polymer Binders. *Polymers* 2020, *12*, 1115. [CrossRef] [PubMed]
5. Mantanis, G.I.; Athanassiadou, E.T.; Barbu, M.C.; Wijnendaele, K. Adhesive systems used in the European particleboard, MDF and OSB industries. *Wood Mater. Sci. Eng.* 2018, *13*, 104–116. [CrossRef]
6. International Agency for Research on Cancer. *Monographs on the Evaluation of Carcinogenic Risk to Humans*; Formaldehyde, 2–Butoxyethanol and 1–tert–Butoxypropan–2–ol; World Health Organization—International Agency for Research on Cancer: Lyon, France, 2006; Volume 88.
7. Łebkowska, M.; Załęska–Radziwiłł, M.; Tabernacka, A. Adhesives based on formaldehyde—Environmental problems. *BioTechnologia* 2017, *98*, 53–65. [CrossRef]
8. Salthammer, T. Formaldehyde Sources, Formaldehyde Concentrations and Air Exchange Rates in European Housings. *Build. Environ.* 2019, *150*, 219–232. [CrossRef]
9. Kristak, L.; Antov, P.; Bekhta, P.; Lubis, M.A.R.; Iswanto, A.H.; Réh, R.; Sedliačik, J.; Savov, V.; Taghiayri, H.R.; Papadopoulos, A.N.; et al. Recent Progress in Ultra-Low Formaldehyde Emitting Adhesive Systems and Formaldehyde Scavengers in Wood-Based Panels: A Review. *Wood Mater. Sci. Eng.* 2022. [CrossRef]
10. Chaharmahali, M.; Mirbagheri, J.; Tajvidi, M.; Najafi, S.K.; Mirbagheri, Y. Mechanical and Physical Properties of Wood-Plastic Composite Panels. *J. Reinf. Plast. Comp.* 2010, *29*, 310–319. [CrossRef]
11. Plastic Pollution Is Growing Relentlessly as Waste Management and Recycling Fall Short. Available online: https://www.oecd.org/environment/plastic-pollution-is-growing-relentlessly-as-waste-management-and-recycling-fall-short.htm (accessed on 15 April 2022).
12. Waste Generation by Sources (1995–2020). State Statistics Service of Ukraine. Available online: https://www.ukrstat.gov.ua/ (accessed on 15 April 2022). (In Ukrainian)
13. Waste Generation and Management (1995–2020). State Statistics Service of Ukraine. Available online: https://www.ukrstat.gov.ua (accessed on 15 April 2022). (In Ukrainian)
14. Niska, K.; Sain, M. *Wood-Polymer Composites*; Woodhead Publishing Limited: Cambridge, UK, 2008.

15. Kajaks, J.; Reihmane, S.; Grinbergs, U.; Kalnins, K. Use of innovative environmentally friendly adhesives for wood veneer bonding. *Proc. Est. Acad. Sci.* **2012**, *61*, 207–211. [CrossRef]
16. Kajaks, J.A.; Bakradze, G.G.; Viksne, A.V.; Reihmane, S.A.; Kalnins, M.M.; Krutohvostov, R. The use of polyolefins-based hot melts for wood bonding. *Mech. Compos. Mater.* **2009**, *45*, 643–650. [CrossRef]
17. Borysiuk, P.; Mamiński, M.Ł.; Parzuchowski, P.; Zado, A. Application of polystyrene as binder for veneers bonding—The effect of pressing parameters. *Eur. J. Wood Prod.* **2010**, *68*, 487–489. [CrossRef]
18. Bekhta, P.; Sedliačik, J. Environmentally-Friendly High-Density Polyethylene-Bonded Plywood Panels. *Polymers* **2019**, *11*, 1166. [CrossRef]
19. Bekhta, P.; Müller, M.; Hunko, I. Properties of Thermoplastic-Bonded Plywood: Effects of the Wood Species and Types of the Thermoplastic Films. *Polymers* **2020**, *12*, 2582. [CrossRef]
20. Bekhta, P.; Chernetskyi, O.; Kusniak, I.; Bekhta, N.; Bryn, O. Selected Properties of Plywood Bonded with Low-Density Polyethylene Film from Different Wood Species. *Polymers* **2022**, *14*, 51. [CrossRef]
21. Nazarenko, V.V.; Bereznenko, N.M.; Novak, D.S.; Skrypnyk, S.P. Investigation of compositions based on secondary polyethylene with improved properties. *Electron. Sci. J. Technol. Des.* **2018**, *4*, 9. Available online: https://nbuv.gov.ua/UJRN/td_2018_4_20 (accessed on 15 June 2022).
22. Kajaks, J.; Kalnins, K.; Reihmane, S.; Bernava, A. Recycled Thermoplastic Polymer Hot Melts Utilization for Birch Wood Veneer Bonding. *Prog. Rubber Plast. Recycl. Technol.* **2014**, *30*, 87–102. [CrossRef]
23. Grinbergs, U.; Kajaks, J.; Reihmane, S. Usage of Ecologically Perspective Adhesives for Wood Bonding. *Sci. J. Riga Tech. Univ. Mater. Sci. Appl. Chem.* **2010**, *22*, 114–117.
24. Cui, T.; Song, K.; Zhang, S. Research on utilizing recycled plastic to make environment-friendly plywood. *For. Stud. China* **2010**, *12*, 218–222. [CrossRef]
25. Lustosa, E.C.D.B.; Del Menezzi, C.H.S.; de Melo, R.R. Production and properties of a new wood laminated veneer/high-density polyethylene composite board. *Mater. Res.* **2015**, *18*, 994–999. [CrossRef]
26. Singh, A.P.; Anderson, R.; Park, B.-D.; Nuryawan, A. A novel approach for FE-SEM imaging of wood-matrix polymer interface in a biocomposite. *Micron* **2013**, *54*, 87–90. [CrossRef]
27. Nuryawan, A.; Hutauruk, N.O.; Purba, E.Y.S.; Masruchin, N.; Batubara, R.; Risnasari, I.; Satrio, F.K.; Rahmawaty; Basyuni, M.; McKay, D. Properties of wood composite plastics made from predominant Low Density Polyethylene (LDPE) plastics and their degradability in nature. *PLoS ONE* **2020**, *15*, e0236406. [CrossRef]
28. Arya, S.; Chauhan, S. Preparation of plywood panels using waste milk pouches as an adhesive. *Maderas-Cienc. Tecnol.* **2021**, *24*, 1–10. [CrossRef]
29. Yorur, H. Utilization of waste polyethylene and its effects on physical and mechanical properties of oriented strand board. *BioResources* **2016**, *11*, 2483–2491. [CrossRef]
30. Horodytska, O.; Cabanes, A.; Fullana, A. Non-intentionally added substances (NIAS) in recycled plastics. *Chemosphere* **2020**, *251*, 126373. [CrossRef]
31. Jemec Kokalj, A.; Dolar, A.; Titova, J.; Visnapuu, M.; Škrlep, L.; Drobne, D.; Vija, H.; Kisand, V.; Heinlaan, M. Long Term Exposure to Virgin and Recycled LDPE Microplastics Induced Minor Effects in the Freshwater and Terrestrial Crustaceans *Daphnia magna* and *Porcellio scaber*. *Polymers* **2021**, *13*, 771. [CrossRef]
32. Czarnecka-Komorowska, D.; Wiszumirska, K.; Garbacz, T. Films LDPE/LLDPE made from post–consumer plastics: Processing, structure, mechanical properties. *Adv. Sci. Technol. Res. J.* **2018**, *12*, 134–142. [CrossRef]
33. *EN 323*; Wood-Based Panels—Determination of Density. European Committee for Standardization: Brussels, Belgium, 1993.
34. *EN 310*; Wood-Based Panels—Determination of Modulus of Elasticity in Bending and of Bending Strength. European Committee for Standardization: Brussels, Belgium, 1993.
35. *EN 314-1*; Plywood—Bonding Quality—Part 1: Test Methods. European Committee for Standardization: Brussels, Belgium, 2004.
36. *EN 314-2*; Plywood—Bonding Quality—Part 2: Requirements. European Committee for Standardization: Brussels, Belgium, 1993.
37. *EN 317*; Particleboards and Fibreboards: Determination of Swelling in Thickness after Immersion in Water. European Committee for Standardization: Brussels, Belgium, 1993.
38. *EN ISO 12460-3*; Wood-Based Panels—Determination of Formaldehyde Release—Part 3: Gas Analysis Method. European Committee for Standardization: Brussels, Belgium, 2015.
39. Bekhta, P.; Salca, E.-A. Influence of veneer densification on the shear strength and temperature behavior inside the plywood during hot press. *Constr. Build. Mater.* **2018**, *162*, 20–26. [CrossRef]
40. Bal, B.C.; Bektas, I. Some mechanical properties of plywood produced from eucalyptus, beech, and poplar veneer. *Maderas-Cienc. Tecnol.* **2014**, *16*, 99–108. [CrossRef]
41. Shukla, S.R.; Kamdem, D.P. Properties of laboratory made yellow poplar (*Liriodendron tulipifera*) laminated veneer lumber: Effect of the adhesives. *Eur. J. Wood Prod.* **2009**, *67*, 397–405. [CrossRef]
42. Kretschmann, D.E. Mechanical properties of wood. In *Wood Handbook—Wood as an Engineering Material*; General Technical Report FPL-GTR-190; U.S. Department of Agriculture, Forest Service, Forest Products Laboratory: Madison, WI, USA, 2010; Chapter 5; pp. 5.1–5.46
43. Bekhta, P.; Salca, E.-A.; Lunguleasa, A. Some properties of plywood panels manufactured from combinations of thermally densified and non-densified veneers of different thicknesses in one structure. *J. Build. Eng.* **2020**, *29*, 101116. [CrossRef]

44. Salca, E.-A.; Bekhta, P.; Seblii, Y. The Effect of Veneer Densification Temperature and Wood Species on the Plywood Properties Made from Alternate Layers of Densified and Non-Densified Veneers. *Forests* **2020**, *11*, 700. [CrossRef]
45. Kilic, Y.; Colak, M.; Baysal, E.; Burdurlu, E. An investigation of some physical and mechanical properties of laminated veneer lumber manufactured from black alder (*Alnus glutinosa*) glued with polyvinyl acetate and polyurethane adhesives. *For. Prod. J.* **2006**, *56*, 56–59.
46. Daoui, A.; Descamps, C.; Marchal, R.; Zerizer, A. Influence of veneer quality on beech LVL mechanical properties. *Maderas Cienc. Tecnol* **2011**, *13*, 69–83. [CrossRef]
47. De Melo, R.R.; Del Menezzi, C.H.S. Influence of veneer thickness on the properties of LVL from Parica (*Schizolobium amazonicum*) plantation trees. *Eur. J. Wood. Prod.* **2014**, *72*, 191–198. [CrossRef]
48. Pałubicki, B.; Marchal, R.; Butaud, J.-C.; Denaud, L.-E.; Bleron, L.; Collet, R.; Kowaluk, G. A method of lathe checks measurement; SMOF device and its software. *Eur. J. Wood. Prod.* **2010**, *68*, 151–159. [CrossRef]
49. Chen, Z.; Wang, C.; Cao, Y.; Zhang, S.; Song, W. Effect of Adhesive Content and Modification Method on Physical and Mechanical Properties of Eucalyptus Veneer–Poly-β-Hydroxybutyrate Film Composites. *For. Prod. J.* **2018**, *68*, 419–429. [CrossRef]
50. Pizzi, A. *Advanced Wood Adhesives Technology*; Marcel Dekker, Inc.: New York, NY, USA, 1994; 289p.
51. Darmawan, W.; Nandika, D.; Massijaya, Y.; Kabe, A.; Rahayu, I.; Denaud, L.; Ozarska, B. Lathe check characteristics of fast growing sengon veneers and their effect on LVL glue-bond and bending strength. *J. Mater. Proces. Technol.* **2015**, *215*, 181–188. [CrossRef]
52. Chow, S. *Lathe-Check Influence on Plywood Shear Strength*; Information Report VP-X-122; Canadian Forest Service: Vancouver, BC, Canada, 1974.
53. Zhou, Y.-G.; Zou, J.-R.; Wu, H.-H.; Xu, B.-P. Balance between bonding and deposition during fused deposition modeling of polycarbonate and acrylonitrile-butadiene-styrene composites. *Polym. Compos.* **2022**, *41*, 60–72. [CrossRef]
54. Del Menezzi, C.H.S.; Tomaselli, I. Contact thermal post-treatment of oriented strandboard to improve dimensional stability: A preliminary study. *Holz Roh-Werkst.* **2006**, *64*, 212–217. [CrossRef]
55. Roffael, E. Volatile organic compounds and formaldehyde in nature, wood and wood based panels. *Holz Roh Werkst.* **2006**, *64*, 144–149. [CrossRef]

Article

Evaluation of Piezoresistive and Electrical Properties of Conductive Nanocomposite Based on Castor-Oil Polyurethane Filled with MWCNT and Carbon Black

Diego S. Melo [1,2], Idalci C. Reis [3,*], Júlio C. Queiroz [3], Cicero R. Cena [4], Bacus O. Nahime [3], José A. Malmonge [1] and Michael J. Silva [2,*]

1. Department of Physics and Chemistry, Faculty of Engineering, São Paulo State University (UNESP), Ilha Solteira 15385-000, SP, Brazil
2. Department of Energy Engineering, Faculty of Engineering and Science, São Paulo State University (UNESP), Rosana 19274-000, SP, Brazil
3. Science and Technology Goiano, Federal Institute of Education, Rio Verde 75901-970, GO, Brazil
4. Institute of Physics, Federal University of Federal do Mato Grosso do Sul (UFMS), Campo Grande 79070-900, MS, Brazil
* Correspondence: idalci.reis@ifgoiano.edu.br (I.C.R.); michael.silva@unesp.br (M.J.S)

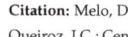

Citation: Melo, D.S.; Reis, I.C.; Queiroz, J.C.; Cena, C.R.; Nahime, B.O.; Malmonge, J.A.; Silva, M.J. Evaluation of Piezoresistive and Electrical Properties of Conductive Nanocomposite Based on Castor-Oil Polyurethane Filled with MWCNT and Carbon Black. *Materials* 2023, 16, 3223. https://doi.org/10.3390/ma16083223

Academic Editors: Eduard-Marius Lungulescu, Radu Setnescu and Cristina Stancu

Received: 23 February 2023
Revised: 17 March 2023
Accepted: 20 March 2023
Published: 19 April 2023

Copyright: © 2023 by the authors. Licensee MDPI, Basel, Switzerland. This article is an open access article distributed under the terms and conditions of the Creative Commons Attribution (CC BY) license (https:// creativecommons.org/licenses/by/ 4.0/).

Abstract: Flexible films of a conductive polymer nanocomposite-based castor oil polyurethane (PUR), filled with different concentrations of carbon black (CB) nanoparticles or multiwall carbon nanotubes (MWCNTs), were obtained by a casting method. The piezoresistive, electrical, and dielectric properties of the PUR/MWCNT and PUR/CB composites were compared. The dc electrical conductivity of both PUR/MWCNT and PUR/CB nanocomposites exhibited strong dependences on the concentration of conducting nanofillers. Their percolation thresholds were 1.56 and 1.5 mass%, respectively. Above the threshold percolation level, the electrical conductivity value increased from 1.65×10^{-12} for the matrix PUR to 2.3×10^{-3} and 1.24×10^{-5} S/m for PUR/MWCNT and PUR/CB samples, respectively. Due to the better CB dispersion in the PUR matrix, the PUR/CB nanocomposite exhibited a lower percolation threshold value, corroborated by scanning electron microscopy images. The real part of the alternating conductivity of the nanocomposites was in accordance with Jonscher's law, indicating that conduction occurred by hopping between states in the conducting nanofillers. The piezoresistive properties were investigated under tensile cycles. The nanocomposites exhibited piezoresistive responses and, thus, could be used as piezoresistive sensors.

Keywords: conductive nanocomposite; castor-oil polyurethane; multiwall carbon nanotube; carbon black; piezoresistive sensor

1. Introduction

Owing to the increasing environmental concerns and decreasing fossil fuel sources, the scientific community and industry search for sustainable sources to produce new materials. Materials obtained from renewable and sustainable sources, such as bio-based and natural polymers, have attracted considerable interest in recent decades. These include natural rubber [1], cellulose [2], polypropylene based on cane alcohol [3], and polyurethane based on castor oil (PUR) [4].

Among bio-based polymers, PUR attracts considerable attention because it uses vegetable oils as polyols obtained from natural sources [5]. Castor oil is attractive because it has a high quantity of hydroxyl groups (–OH) that react with isocyanate (–NCO), forming urethane bonds [6]. These hydroxyl groups are derived from ricinoleic acid, which constitutes approximately 80% of the oil. The hydrophobic nature of the triglycerides in the castor oil contributes to its excellent mechanical properties, including good elongation and tensile strength [6–8]. Owing to its good characteristics, a vegetable-oil-based PUR has been used as a polymeric matrix to obtain composite materials for different applications, such as adhesive resins [9,10], coatings [11,12], electromagnetic interference shielding materials [13], materials for sorption of oils and organic solvents [14], and conductive polymer composites (CPCs) [15,16].

A conventional polymer is combined with a conductive filler to form CPC, combining good mechanical properties with an easier processing of polymeric matrix and the excellent electrical properties of the conducting fillers [17]. Nonetheless, the electrical conductivity of CPCs is influenced by many factors, such as the preparation method (quantity and distribution of conductive fillers in the host matrix), as well as the volume fraction and electrical conductivity of the phase [18,19]. For instance, when the conductive phase concentration is extremely small, a considerable distance exists between each conducting filler, which results in a composite with a similar electrical conductivity as a polymer matrix [20,21]. However, if the concentration of conducting fillers is high, an insulator-to-conductor transition will occur, resulting in the formation of a continuous conducting path. It is at this point that percolation threshold can be observed, and the composite's electrical conductivity is approximately equal to the conductive fillers [21,22]. There are two main mechanisms that determine electrical conduction in CPCs: the percolation model for tunneling and hopping conduction [23]. The model of hopping conduction describes the charge carriers jumping between localized states in the conducting particles or aggregates in contact with each other forming a three-dimensional structure in the bulk of the composite [24]. In tunneling conduction, conductive particles and/or aggregates are separated by insulating layers or barrier potentials [25].

In this sense, for the production of a CPC-based PUR, several conductive fillers can be used, including carbon-based nanomaterials such as expandable graphite [26], multiwall carbon nanotubes (MWCNTs) [27,28], graphene oxide [29–31], and carbon black (CB) [16,32]. For example, Dai et al., 2020 obtained a MWCNT/PUR nanocomposite using a solvent-free method and showed that, owing to the interaction between the polymer matrix and MWCNT, the tensile strength and thermal stability were improved for MWCNT loadings up to 0.5 wt%. Kumar et al., 2022 evaluated the ameliorating properties of PUR-based nanocomposites via a synergistic addition of graphene and cellulose nanofibers. Consequently, these nanocomposites can be applied to protective coatings and automobile parts [33]. Min et al., 2023 conducted a study in which a highly electrically conductive composite film was fabricated by shearing the MWCNT/PDMS paste in two rolls. The authors report that the electrical conductivity and activation energy of the composite were 326.5 S/m and 8.0 meV, respectively, at 5.82 vol% of MWCNTs [34]. Another study examined how structural factors affected the electrochemical performance of carbon composites [35]. Carbonization of binary composites formed by graphene nanoplatelets and melamine (GNP/MM), multi-walled carbon nanotubes and melamine (CNT/MM), and trinary composites (GNP/CNT/MM) plays a crucial role in tailoring electrochemical properties of carbon hybrid materials, which are considered noble metal-free alternatives to traditional electrodes [36]. In a study conducted by Rozhin et al., 2023, single-walled carbon nanotubes (SWCNTs) were compared with double-walled carbon nanotubes (DWCNTs) as nanostructured additives for tripeptide hydrogels. It has been found that carbon nanomaterials (CNMs) can enhance the viscoelastic properties of tripeptide hydrogels, but their presence can also impair the self-assembling process [36].

Due to the excellent mechanical properties, lightness, and flexibility of polymer matrices and the good electrical properties of carbon-based nanofillers, composite materials obtained through this combination have the potential to be used as piezoresistive sensors. Piezoresistivity can be defined as an electromechanical phenomenon in which the electrical resistance of a material changes reversibly under a strain cycle [37]. In this sense, the piezoresistivity consists of a change in the conductive composite structure with strain, and it is a material property associated with a change in the structure and resistivity of material [37]. Changes in the degree of electrical continuity in the conductive composite are commonly associated with reversible microstructural changes. Many applications require the detection of strain or stress on structures, such as structural vibration control, traffic monitoring, weighing (including weighing in motion), and building facility management [37].

Several studies used PUR as a matrix and second-phase carbon-based nanofiller, while few studies investigated the electrical, dielectric, and piezoresistive properties of this type of nanocomposite. Using two different carbon-based nanofillers (CB and MWCNT), this project aims to obtain a castor oil-based polyurethane nanocomposite for use as a piezoresistive sensor. A simple and low-cost synthesis route was used to obtain the specimens through the casting method. A comparative analysis was conducted between nanocomposites with different concentrations of conductive nanofiller in order to determine which nanocomposites provide the best results in electrical, dielectric, and piezoresistive measurements. Additionally, the results showed that thin, flexible films with low percolation thresholds were obtained for the PUR/MWCNT and the PUR/CB nanocomposites, making them suitable for applications as sensors, anti-static (electrically conducting), shape memory alloys, and electromagnetic-wave shielding.

2. Materials and Methods

2.1. Materials

A bicomponent castor-oil-based polyurethane was purchased from Sinergia LTDA, city of Araraquara, SP, Brazil. A mixture of prepolymer (component A) and polyol (component B) was used to obtain a pure PUR film with a mass fraction B/A of 2/1.

Conducting CBs (Printex XE-2) were purchased from Degussa (Paulínia, Brazil). They had a surface area of 1000 m^2/g, average particle size of 35 nm, and bulk density of 0.1 g/cm^3. MWCNTs were supplied by CTNANO, Federal University of Minas Gerais (UFMG) in the city of Belo Horizonte, MG, Brazil. They had a surface area of 100 m^2/g, bulk density of 0.23 g/cm^3, and average diameter and length of 20 nm and 5–30 µm, respectively. Before use, both conductive fillers were milled.

2.2. Conducting Nanocomposite Preparation

A neat PUR was prepared by mixing components A and B in a mass proportion of 1 g/2 g (isocyanate component/polyol) in 1 mL of chloroform (Dynamics Analytical Reagents). PUR/CB and PUR/MWCNT nanocomposite samples were prepared at a constant PUR mass fraction (3 g) while varying the mass concentrations of MWCNT and CB nanoparticles. CB and MWCNT values used in the sample preparation can be found in Table 1.

Table 1. Masses of MWCNT and CB to produce PUR/CB and PUR/MWCNT nanocomposites (in 3.0 g of PUR).

Samples	MWCNT (g)	CB (g)
PUR/MWCNT1	0.030	0
PUR/MWCNT2	0.061	0
PUR/MWCNT3	0.093	0
PUR/MWCNT4	0.125	0
PUR/MWCNT5	0.158	0
PUR/CB1	0	0.030
PUR/CB2	0	0.061
PUR/CB3	0	0.093
PUR/CB4	0	0.125
PUR/CB5	0	0.158

Figure 1 illustrates the preparation diagram for nanocomposites specimens. In 1.0 mL of chloroform, polyol (2 g) was mixed with different concentrations (1–5 mass%) of the conductive filler (MWCNT or CB). After 8 h of stirring, the prepolymer was added to the polyol/conductive filler/chloroform dispersion and stirred for 10 min before it was sonicated for 10 min. Flexible samples, with thicknesses of 100–200 µm, were obtained by casting on glass slides and curing for 5 days at room temperature.

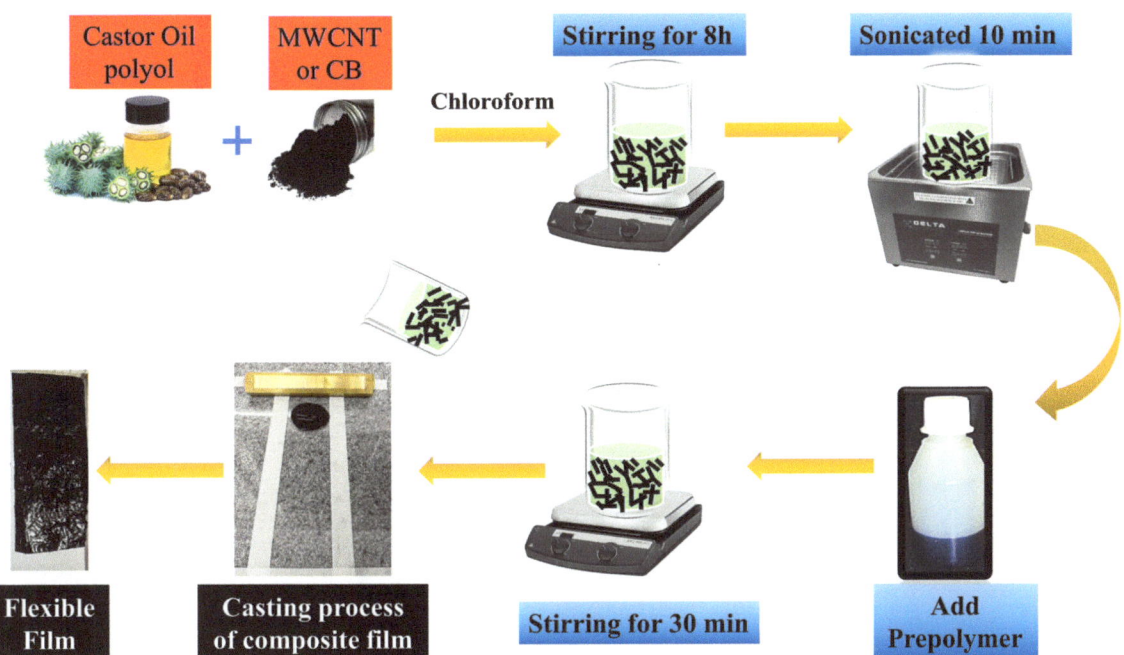

Figure 1. Schematic representation of the preparation of conducting nanocomposite samples.

2.3. Characterization

The morphology of PUR/CB and PUR/MWCNT nanocomposite films were analyzed by scanning electron microscopy (SEM) using an EVO LS15 Zeiss microscope. SEM analysis was performed after cryo-fracturing samples with liquid nitrogen and drying them under dynamic vacuum. Samples were coated with a thin layer of carbon (10 nm) using the sputtering method.

A voltage/current source from Keithley Instruments model 247 (high-voltage supply) was used to measure direct-current (DC). On both sides of the film, the electrodes were painted with conductive paint for electrical contact. The DC electrical conductivity was calculated by Equation (1):

$$\sigma_{dc} = \frac{I}{V}\frac{d}{A} \qquad (1)$$

where A is the electrode area, d is the sample thickness, and V and I are the applied voltage and measured current, respectively.

A Solartron SI 1260 impedance analyzer with a 1296 dielectric interface (0.05% basic accuracy) was used to measure the alternating-current (AC) electrical and dielectric properties of the PUR/MWCNT and PUR/CB nanocomposite samples. An impedance analysis of the nanocomposites was performed in a frequency range of 0.1–10^6 Hz, at a voltage of 1 V, at room temperature.

At room temperature, piezoresistive tests were conducted by measuring the electrical resistances of the PUR/MWCNT and PUR/CB nanocomposite samples, using Keithley 237 high-voltage (0.3% basic accuracy) source-connected copper electrodes, during a uniaxial mechanical deformation of the samples. This test was performed using a universal testing machine from Instron (model 3639, 0.15% and 0.2% basic accuracy of set speed and displacement) in accordance with the International Organization for Standardization (ISO) 37:2011 standard, with a 100-N load cell. The copper electrodes (10 mm × 5 mm × 0.2 mm) were wrapped around both upper and lower ends of the samples and fixed using clamps on the mechanical testing machine. Piezoresistive measurements were performed at an applied voltage of 10 V, deformation of 10%, and at a velocity of 100 mm/min. The samples were cut in accordance with ISO 1286:2006. Piezoresistive tests were carried out on PUR/MWCNT and PUR/CB nanocomposite samples containing 3, 4, and 5 mass% MWCNT and CB.

3. Results and Discussion

3.1. Morphological Analysis

Figure 2 shows SEM images of cryo-fractured surfaces of PUR/CB and PUR/MWCNT nanocomposite samples with different concentrations of conductive nanofillers. As shown in Figure 2A–D, the MWCNT fillers tended to cluster more than CB. The formation of MWCNT agglomerates has been attributed to van der Waals interactions between the fillers [15,38]. Nayak et al. observed a similar behavior when the CNT concentration was greater than 3% in the polyamide matrix [38]. On the other hand, the CB nanoparticles exhibited a more homogenous dispersion in the PUR matrix. Nevertheless, agglomeration of CB was also observed. Compared to MWCNTs, CB nanoparticles display better dispersion, and aggregates are easier to break due to their weak electrical interaction [39,40].

3.2. DC Electrical Conductivity

A CPC can be created by adding conductive particles to an insulating polymeric matrix. Various factors contribute to the electrical properties of composites and the electrical conduction process, including the preparation of CPC, electrical conductivity, and the volumetric fraction of the phase, particle size, and aggregate size, as well as their dispersion within polymeric matrixes [15,41]. Generally, nanoparticles homogeneously dispersed within a matrix produce a CPC with a lower percolation threshold [41].

At extremely low conductive particle concentrations, the distances between the conductive fillers or aggregates are considerable, and the CPC exhibits a similar electrical conductivity to the polymeric host. When the concentration of conductive filler reaches a critical level (or percolation threshold), an insulator–conductor transition occurs, resulting in the formation of a percolation network or continuous conductor path through which charge carriers move when an external electric field is applied.

Figure 3 illustrates this behavior for both PUR/MWCNT and PUR/CB nanocomposites containing different amounts of conductive nanofillers (MWCNT or CB). PUR/CB nanocomposite samples showed an insulator–metal transition at slightly lower filler concentrations than PUR/MWCNT nanocomposite samples, likely due to the fact that CB tends to cluster less than MWCNTs, as shown in the SEM images. There is a distinct difference between the dc electrical conductivity curve profiles of the two types of nanocomposites. PUR/MWCNT nanocomposite reaches maximum conductivity at around 2 mass% MWCNT in the PUR matrix, while PUR/CB reaches maximum conductivity at around 5 mass%. The reason for this behavior can be attributed to the morphology and dispersion of the nanoparticles within the PUR matrix. In the PUR/CB and PUR/MWCNT nanocomposite samples, the percolation thresholds calculated were 1.56 and 1.50 mass% of MWCNT or CB, respectively. For both types of nanocomposites, the maximum conductivity was approximately 10^{-2} S/m, which is approximately 9-fold greater than the conductivity of PUR.

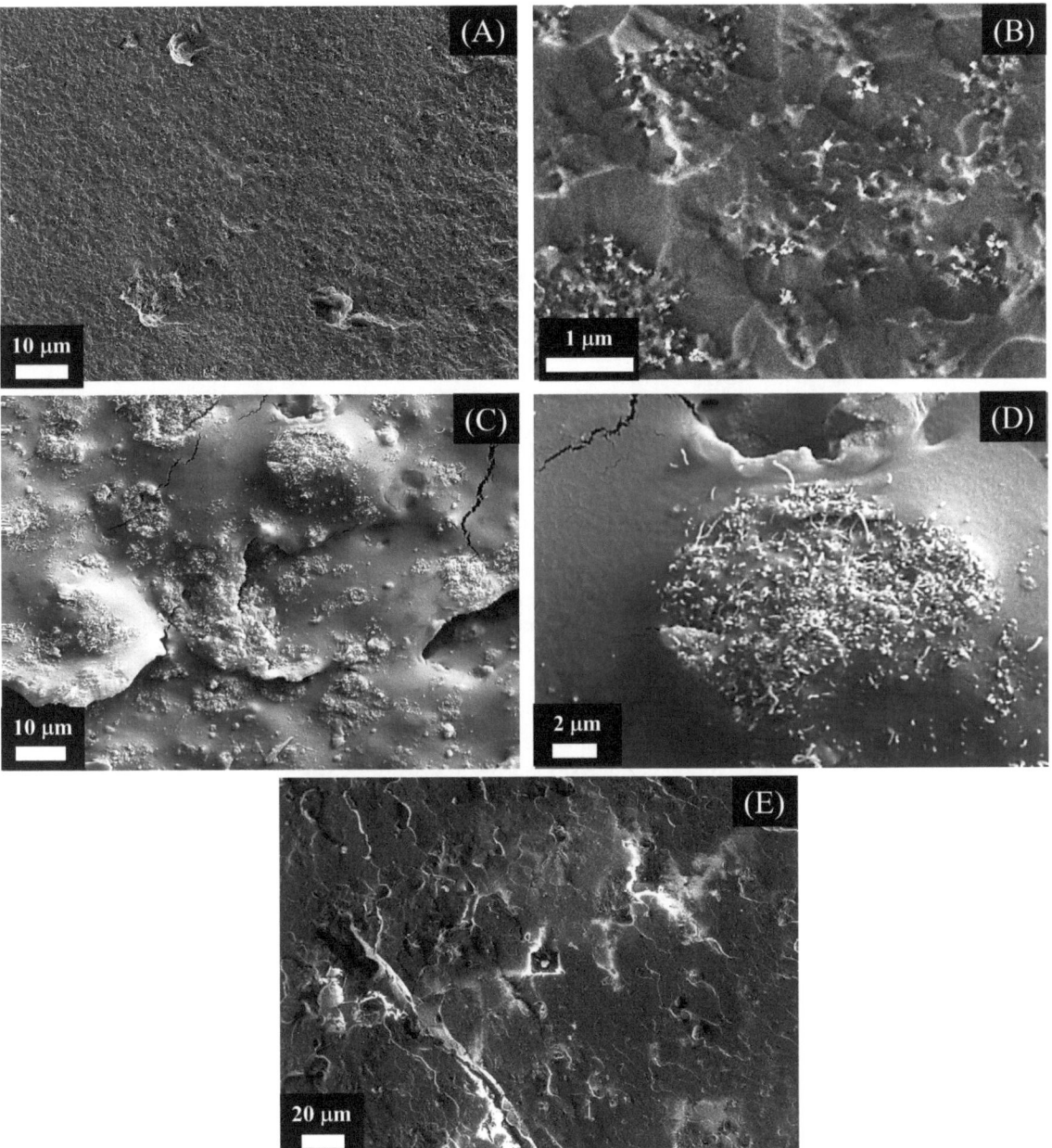

Figure 2. SEM microimages of (**A**,**B**) PUR/CB and (**C**,**D**) PUR/MWCNT nanocomposite samples with 5 mass% of conductive nanofillers and (**E**) Neat PUR.

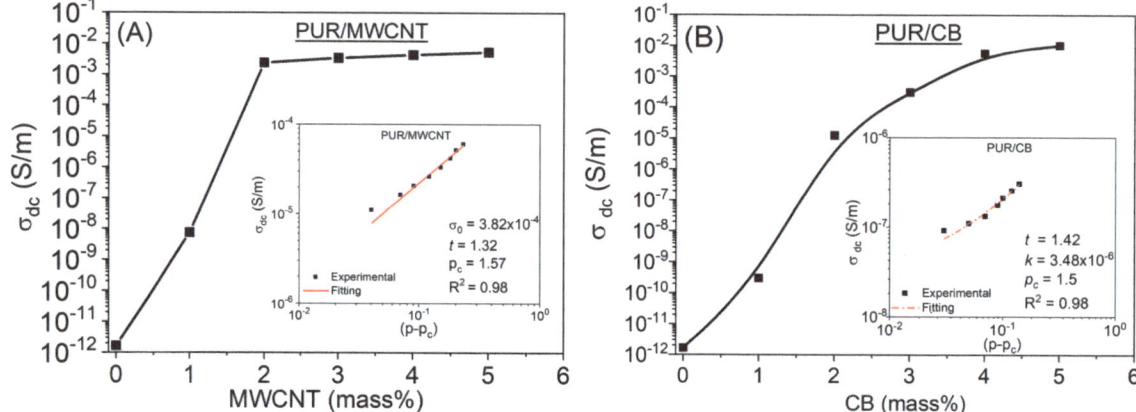

Figure 3. σ_{dc} behavior in relation to the conductive nanofiller mass fractions for (**A**) PUR/MWCNT and (**B**) PUR/CB nanocomposite. The Figure presents the double-logarithmic plot, based on Equation (2), for the PUR/MWCNT nanocomposite and for the PUR/CB nanocomposite.

The σ_{dc} behavior of PUR/MWCNT and PUR/CB nanocomposites, composed of both insulating and conductive phases, can be described by power-type equations. When the conductive fillers fraction reaches the critical volume fraction pc in the percolation threshold region, geometric connectivity begins to form throughout the system, resulting in uninterrupted and continuous conductive paths [20,42]. Thus, the σ_{dc} behavior of a nanocomposite can be expressed by:

$$\sigma_{dc} = k(p - p_c)^t \quad (2)$$

where p is the concentration of the conductive phase, k is the preexponential constant, and t is the critical conductivity exponent. For two-dimensional systems, its value is 1.3–1.5, while, for three-dimensional systems, t is 1.6 to 2.0 [43]. Using the data of Figure 3A,B, we generated the graph of log (σ_{dc}) as a function of log ($p - p_c$), while the fitting was performed using Equation (2). A linear fit was performed on the points of inside graphics of Figure 3A,B to estimate the t and k values for the PUR/MWCNT and PUR/CB nanocomposite samples, in which the t values were 1.32 and 1.42, while the k values were 3.82×10^{-3} and 3.48×10^{-6}, respectively.

The t values for the PUR/MWCNT and PUR/CB samples agree with the universal percolation theory [44]. The electrical conduction process of charge carriers occurs through the conductive two-dimensional network by geometric contact between the CB nanoparticles in the PUR/CB nanocomposite samples. Similar behavior occurs for PUR/MWCNT nanocomposite samples in which the conductive two-dimensional network is formed by geometric contact of the MWCNTs [45,46].

Based on analytical micromechanical analysis, Mazaheri et al. predicted the electrical conductivity and the percolation behavior of polymer nanocomposites containing spherical carbon black nanoparticles as fillers. In the proposed model, quantum electron tunneling is accounted for, as well as the thickness of the interphase region, the radius of the filler, the conductivity of the filler, the conductivity of the interphase region, and the conductivity of the matrix [47]. In addition to the geometrical and physical properties of the CB fillers, polymer matrix, and interphase layer, the authors also validated the analytical formulation of the model [47]. The model produces meaningful physical results by accounting for electrical conductivity, potential barrier height, and tunneling distance within the interphase region [47]. The model described the behavior of electrical conductivity in both insulating and percolation transition regions, at very low volume fractions, accurately [47]. In a study by Zare and Rhee, the electrical conductivity of CNT-filled samples was tuned using a mechanics model, assuming extended CNT. A study of the extended CNT was conducted by considering the interphase and tunneling areas and estimating the conductivity of prolonged CNT based on the resistances of conductive nanofiller, interphase, and tunneling districts [48]. In nanocomposite systems, conductivity directly depends on the network size and interphase depth, but electrical conductivity of CNT and interphase are ineffective [48]. In addition, the narrow and undersized tunnels benefit the nanocomposite's conductivity because of its low percolation onset and little tunneling resistance [48].

3.3. Impedance Spectroscopy (IS)

IS is a useful technique for examining the dielectric and electrical properties of various types of materials. IS can be used to investigate the dependence between the electrical properties of a material and frequency of the applied electric field for an extensive range of frequencies (from 10^{-4} to 10^7 Hz) [49]. This technique is relatively easy to use, and results can be related to polarization effects, dielectric properties, microstructure, defects, and electrical conduction mechanisms [49]. It is possible to determine the electrical and dielectric behaviors of a material from a set of values related to the complex impedance formalism (Z^*) expressed by:

$$Z^* = Z' + iZ'' \tag{3}$$

where Z' and Z'' are the real and imaginary parts of the complex impedance, respectively, and i is the imaginary unit ($\sqrt{-1}$). The sets of values obtained from Z^* are the (a) complex admittance (Y^*), (b) complex dielectric permittivity (ε^*), (c) complex electrical module (M^*), and (d) complex electrical conductivity (σ^*) [49]. The Z^* formalism is related to each other by the interrelation factor $\mu = j\omega C_0$, where C_0 is the vacuum capacitance, ω is the angular frequency, and j ($\sqrt{-1}$) is the imaginary factor.

Figure 4 shows the real parts ($\sigma'(f)$) of the complex electrical conductivity with respect to the AC electrical field frequency for PUR/MWCNT and PUR/CB nanocomposites. It should be noted that both nanocomposites exhibit $\sigma'(f)$ similar to disordered solids, i.e., frequency-dependent electrical conduction, primarily at high frequencies [50–52]. This behavior has been most evident in samples with nanofiller concentrations below 3 mass%.

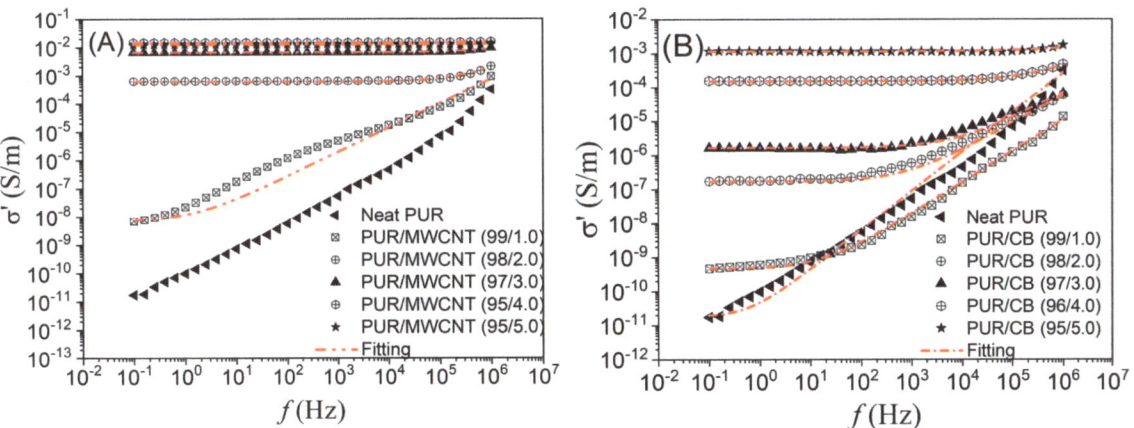

Figure 4. Real part of complex electrical conductivity as a function of frequency for (**A**) PUR/MWCNT and (**B**) PUR/CB with different mass fractions of the conductive nanofiller.

As illustrated in Figure 4A,B, the PUR/MWCNT and PUR/CB nanocomposite samples, with different quantities of conductive nanofillers, exhibited two well-defined regions of the $\sigma'(f)$ curve, namely a frequency-independent region and a frequency-dependent region. According to Kilbride and Coleman, the frequency-independent behavior of $\sigma'(f)$ can be attributed to the variation in the correlation length with the conductive nanofiller content, which has been discussed in relation to a biased random walk in a three-dimensional network [53]. In this case, the frequency-independent $\sigma'(f)$ behavior in the low-frequency region reflects a large-distance transport of charge carriers via conducting percolative paths inside the nanocomposite, which is improved by the electric-field concentration effect [54].

For the PUR/MWCNT and PUR/CB samples, a plateau is observed in the $\sigma'(f)$ curve at low frequencies, which is approximately equal to σ_{dc} [1,54]. However, the transition (critical frequency) from the frequency-independent to the frequency-dependent $\sigma'(f)$ shifted to a higher frequency with the increase in the concentration of the conductive nanofiller dispersed in the nanocomposite, which may be related to the formation of a conductive two or three-dimensional network with increasing concentrations of both MWCNT and CB nanoparticles.

The frequency-dependent behavior of PUR/MWCNT and PUR/CB nanocomposite samples occurs when the critical frequency is reached. $\sigma'(f)$ increases with frequency due to the hopping conduction of charge carriers between more closely located states with lower energy barriers, as well as the effect of spatial charge polarization at the nanofiller–PUR interface [53,55]. For samples containing concentrations above 2 mass% of MWCNT and 3 mass% of CB, the critical frequency (which normally occurs at higher frequencies) could not be obtained due to the conductive nature of the nanocomposites. This is due to limitations of the equipment needed to study at higher frequencies.

The $\sigma'(f)$ behavior in Figure 4A,B, for both nanocomposites, can be described by the Jonscher's power law:

$$\sigma'(f) = \sigma_{dc} + A\omega^n \qquad (4)$$

where σ_{dc} is the DC electrical conductivity (plateau or frequency-independent region), A is the preexponential factor, related to the strength of polarizability, and n is the fractional exponent that can vary between 0 and 1 [56]. The exponent n is related to the degree of interaction between the charge carriers and the networks around them, suggesting that electrical conduction may occur through hopping and/or interfacial polarization [57,58].

Table 2 presents the results of fitting log $\sigma'(f)$ versus log (frequency) curves (Figure 4A,B) using Equation (4) for all nanocomposite samples. According to the results, the n values ranged from 0.7 to 1.0, indicating that electrical conduction involves the hopping of charges between localized states or spatial charges trapped at the interface between the PUR matrix and the conductive nanofillers [56,57].

Table 2. Parameters obtained from fitting of the experimental data using the Jonscher's equation for PUR/MWCNT and PUR/CB nanocomposite samples and neat PUR.

Samples	σ_{dc} (S/m)	A	n	R²
Neat PUR	1.72×10^{-11}	3.27×10^{-11}	1.15	0.95
PUR/CB (99/1)	4.65×10^{-10}	3.52×10^{-11}	0.92	0.97
PUR/CB (98/2)	1.73×10^{-7}	8.49×10^{-10}	0.80	0.99
PUR/CB (97/3)	1.66×10^{-6}	1.88×10^{-9}	0.76	0.97
PUR/CB (96/4)	1.15×10^{-4}	1.40×10^{-8}	0.72	0.99
PUR/CB (95/5)	1.15×10^{-3}	3.41×10^{-7}	0.70	0.80
PUR/MWCNT (99/1)	7.01×10^{-9}	5.38×10^{-9}	0.86	0.96
PUR/MWCNT (98/2)	6.17×10^{-4}	1.39×10^{-8}	0.83	0.97
PUR/MWCNT (97/3)	6.65×10^{-3}	4.06×10^{-8}	0.79	0.95
PUR/MWCNT (96/4)	1.44×10^{-2}	2.05×10^{-8}	0.75	0.92
PUR/MWCNT (95/5)	1.21×10^{-2}	1.08×10^{-8}	077	0.90

The frequency-dependence of the real (ε') and imaginary (ε'') dielectric constants of PUR/MWCNT and PUR/CB nanocomposite samples, measured at room temperature, is shown in Figure 5. The ε' represents the ability of a material to become polarized in the presence of an electric field, whereas the ε'' represents the dielectric losses incurred by a material due to its increased conductivity. Figure 5A,C illustrates the behavior of ε' as a function of frequency for PUR/MWCNT and PUR/CB nanocomposites, respectively. At a higher concentration of conductive fillers, more charge carriers are trapped at the conductive filler–polymer matrix interfaces. In this regard, when the CPC is under the influence of electric fields, electrons move to and accumulate at one end of the conductive cluster (filler–matrix interface), while the other end is more positive, which generates a dipole moment across the clusters [59]. On the other hand, when the electric field is overturned, the reverse behavior occurs. However, as the frequency of the electric field increased, the residence time of the trapped electrons at the interface decreased, resulting in a decrease in the polarization of the system [15]. This behavior can be observed as a reduction in ε' with the increase in the frequency, particularly in the high-frequency range, for all nanocomposite samples. This decrease is principally attributed to the Maxwell–Wagner–Sillars (MWS) polarization and space charge polarization in the bulk nanocomposite samples [60,61].

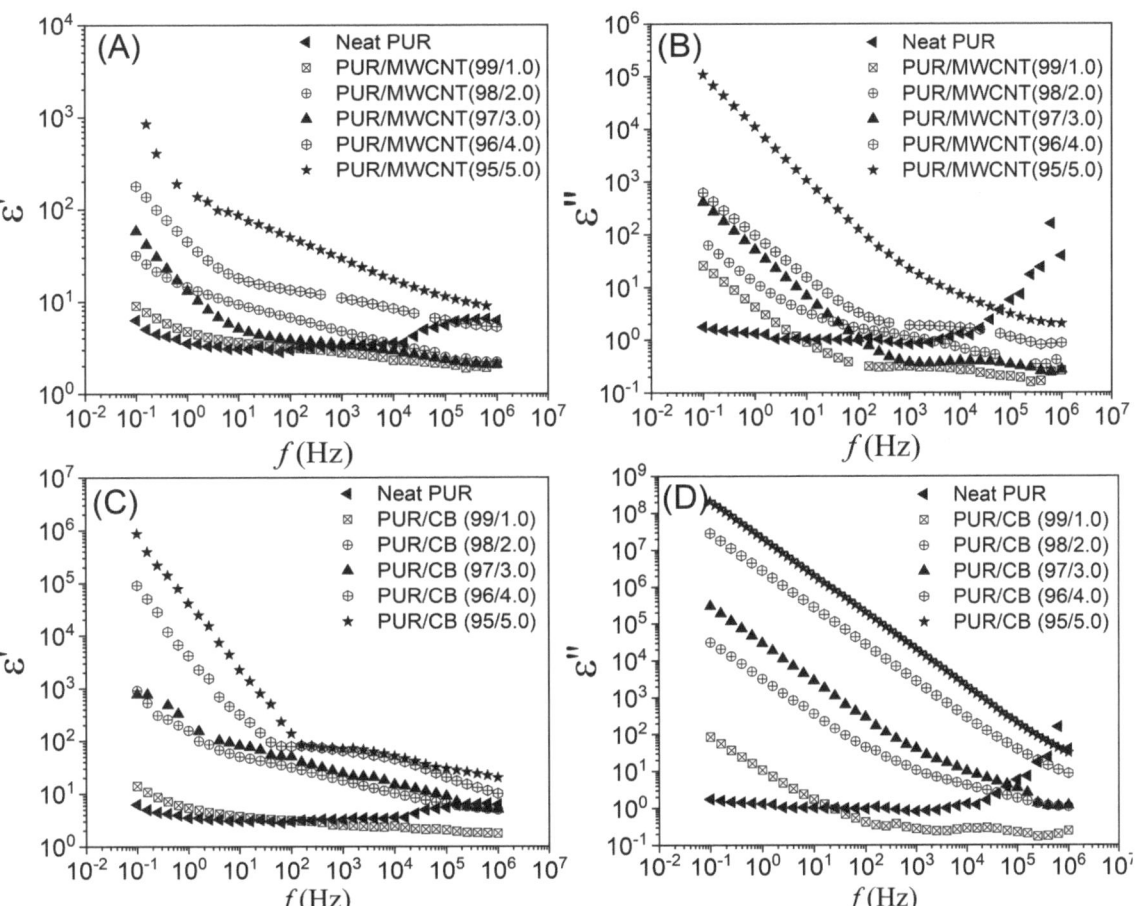

Figure 5. Complex dielectric constant, as a function of frequency for (**A**,**B**) PUR/MWCNT and (**C**,**D**) PUR/CB, with different mass fractions of the conductive nanofiller.

Figure 5B,D shows the frequency dependences of ε'' for the PUR/MWCNT and PUR/CB nanocomposite samples, respectively. In the range of low frequencies, a significant increase in ε'' can be observed due to the increasing conductivity of the nanocomposite samples [62,63]. The dielectric loss comes from the fact that the charge carriers are no longer participating in the sample polarization process and are starting to participate in the conduction process through the percolative conductive network in the composite [15].

Asandulesa et al., 2021 observed similar behavior for ε' and ε'' as a function of frequency for PTB7:PC71BM photovoltaic polymer blend. According to the authors, a high dielectric constant reduces the Coulomb attraction between electrons and holes in excitons and donor–acceptor exciplexes, thereby increasing the power conversion efficiency (PCE) [64].

In Table 3, the values of ε' and ε'' are shown at frequencies of 1 kHz and 1 MHz for both conductive nanocomposites. As seen, the values of ε' and ε'' were higher for samples with concentrations higher than the percolation threshold, mainly for samples 96/4.0 and 95/5.0. As a result of (i) having a greater amount of charge carriers trapped in the polymer-nanofiller interface, as well as (ii) the samples becoming more conductive as the nanofiller amount increased, the samples became more conductive. Asandulesa et al., 2020 observed similar behavior for ε' and ε'' as a function of frequency for the PTB7:PC71BM photovoltaic polymer blend. According to the authors, a high dielectric constant reduces the Coulomb attraction between electrons and localized states [65].

Table 3. Dielectric parameters ε' and ε'' for conductive nanocomposites samples at frequencies of 1 kHz and 1 MHz.

Samples	ε' (1 kHz)	ε' (1 MHz)	ε'' (1 kHz)	ε'' (1 MHz)
PUR	3.52	6.23	0.82	39.2
PUR/MWCNT1	2.87	2.16	0.32	0.26
PUR/MWCNT2	4.72	2.21	1.15	0.41
PUR/MWCNT3	3.45	2.11	0.36	0.27
PUR/MWCNT4	11.01	5.30	1.88	0.86
PUR/MWCNT5	29.44	8.94	21.95	2.01
PUR/CB1	2.56	1.88	0.27	0.24
PUR/CB2	17.80	5.04	10.66	1.08
PUR/CB3	24.31	5.19	41.01	1.17
PUR/CB4	65.21	10.16	2801.92	8.64
PUR/CB5	70.14	20.12	20,515.62	31.32

3.4. Piezoresistivity Analysis

In-situ piezoresistance measurements were performed while mechanically stretching the PUR/MWCNT and PUR/CB nanocomposite films across the elastic region. In Figure 6, a variation in electrical resistance is observed during the elastic regime for PUR/MWCNT and PUR/CB nanocomposite samples containing 3, 4, and 5 mass% of MWCNT and CB, over 16 cycles of loading and unloading. During each cycle, the deformation of the film increased linearly with the applied strain. It was observed that the slope was similar during unloading, as well, until it reached zero strain. The electrical resistance also exhibited a delayed response to deformation, which may be attributed to the viscoelasticity of the PUR matrix, which is related to the duration of the time needed for the polymer chains to organize as opposed to the strain applied [66].

During all applied deformations, the piezoresistive behavior exhibited a good linearity between the electrical resistivity variation and strain cycles. As the deformation of the nanocomposite is increased by an external stimulus, the electrical resistance increases, while, as the deformation decreases, the electrical resistance decreases. Consequently, the initial increase in electrical resistivity of the PUR/MWCNT and PUR/CB nanocomposites samples can be attributed to the breakdown of the conductive percolation network formed by the MWCNT and CB, which leads to a decrease in the distribution of conduction paths. As the strain decreases (to zero) and the sample deformation returns to its initial length, the percolation paths are reconstructed and the electrical resistance decreases as well. However, because of the viscoelasticity of the PUR matrix, this decrease in resistance did not occur immediately upon removal of mechanical stress. For the different MWCNT and CB fractions, maximum strength was reached at different times under constant strain, as the alignment of MWCNT and CB particles occurred later for low concentrations of conductive fillers. Figure 7 illustrates the piezoresistivity test in conductive nanocomposites specimens during stress cycles, as well as the piezoresistive effect.

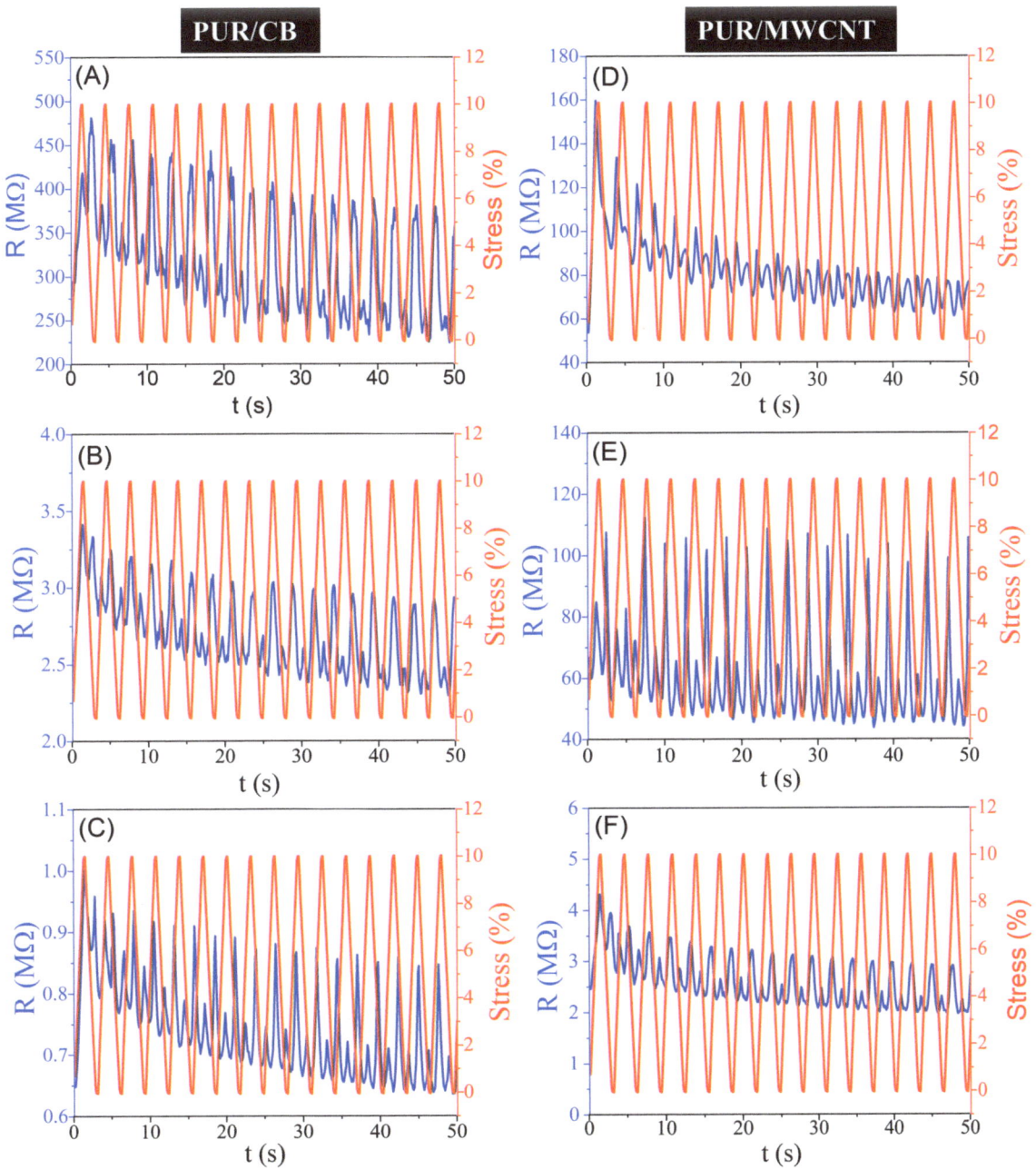

Figure 6. Piezoresistive response of (**A–C**) PUR/CB and (**D–F**) PUR/MWCNT nanocomposite with 3, 4 and 5 mass% for 10% of strain and 16 cycles.

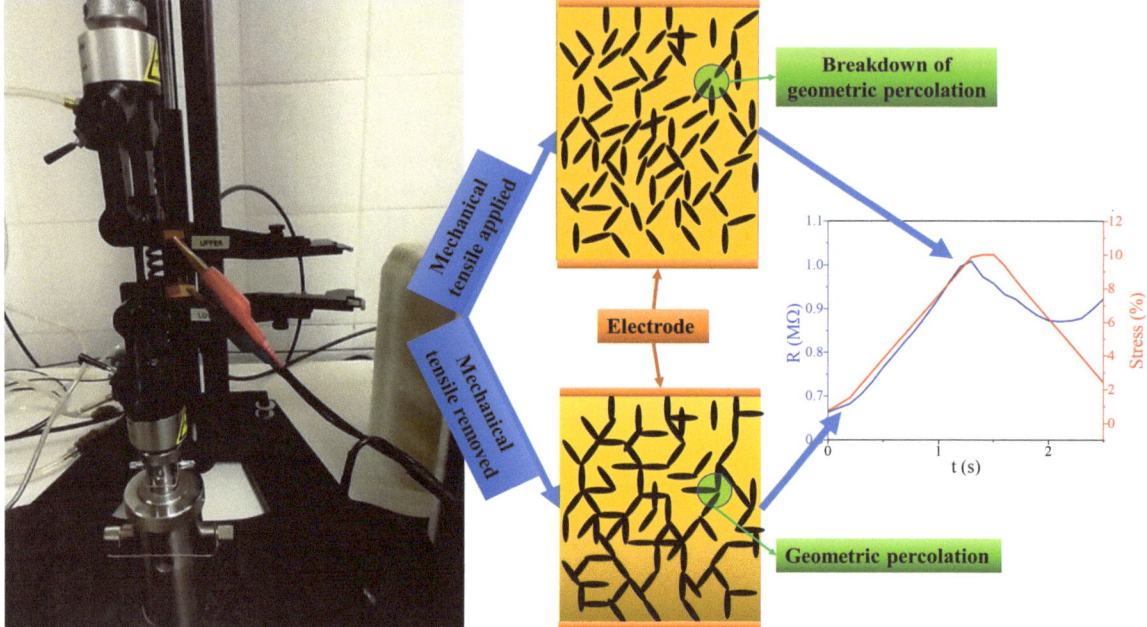

Figure 7. A schematic representation of the piezoresistivity test in conductive nanocomposites during stress cycles.

The same behavior was observed by Gonçalves et al. for a MWCNT/elastomer styrene-ethylene/butylene-styrene (SEBS) composite and by Costa et al. for polymer blends based on polyaniline (PANI) and SEBS [67,68]. Considering the outstanding piezoresistive properties observed by both groups during experimental testing, the samples have a high potential for advanced electromechanical sensor applications [67,68].

4. Conclusions

In this study, flexible films of PUR-based nanocomposites with nanoparticles of MWCNT and CB were obtained by the casting method and simple synthesis route. The SEM images showed that the MWCNT nanoparticles tended to agglomerate more than the CB nanoparticles, owing to van der Waals forces. This behavior corroborated the DC conductivity analysis, in which the PUR/CB nanocomposite exhibited a lower percolation threshold than that of the PUR/MWCNT nanocomposite (1.5 and 1.57, respectively). Both nanocomposites exhibited a nine-fold increase in σ_{dc} compared to the PUR above the percolation threshold.

The $\sigma'(f)$ spectra, as a function of the frequency, indicated that the PUR/MWCNT and PUR/CB nanocomposites with concentrations of the conductive phase above the percolation threshold had two distinct regions, with one independent and the other dependent on the frequency. The samples with concentrations below the percolation threshold exhibited a frequency-dependent behavior, which is characteristic of disordered systems in which charge carriers hop between localized states within conductive regions. Furthermore, the adjustment performed using the Jonscher's equation for $\sigma'(f)$ confirmed that the electric transport in the PUR/MWCNT and PUR/CB nanocomposites occurred via a hopping mechanism. The frequency-dependent behavior of ε' for PUR/MWCNT and PUR/CB could be attributed to the MWS polarization and space charge polarization in the bulk nanocomposite samples.

The PUR/MWCNT and PUR/CB nanocomposites with contents of 3, 4, and 5 mass% exhibited good piezoresistive properties up to a strain of 10%, which demonstrated their high stabilities and good electrical resistance responses. Thus, the PUR/MWCNT and PUR/CB nanocomposites have potential for applications as piezoresistive sensors.

Author Contributions: Conceptualization, D.S.M., I.C.R., J.C.Q. and M.J.S.; methodology, D.S.M., I.C.R., J.C.Q., C.R.C., B.O.N. and M.J.S.; software, C.R.C., M.J.S. and J.A.M.; validation, I.C.R., C.R.C., B.O.N., J.A.M. and M.J.S.; formal analysis, D.S.M., I.C.R. and J.C.Q.; investigation D.S.M., I.C.R., J.C.Q. and M.J.S.; resources, I.C.R. and M.J.S.; data curation, C.R.C., B.O.N., J.A.M. and M.J.S.; writing—original draft preparation, I.C.R., J.C.Q., C.R.C., J.A.M. and M.J.S.; writing—review and editing, I.C.R., C.R.C., B.O.N., J.A.M. and M.J.S.; visualization, I.C.R., J.A.M. and M.J.S.; supervision, I.C.R. and M.J.S.; project administration, I.C.R. and M.J.S.; funding acquisition, I.C.R. and M.J.S. All authors have read and agreed to the published version of the manuscript.

Funding: This research was funded by Sao Paulo State Funding Agency (FAPESP), grant number grant 2017/19809-5.

Institutional Review Board Statement: Not applicable.

Informed Consent Statement: Not applicable.

Data Availability Statement: The data presented in this study are available upon request from the corresponding author.

Acknowledgments: The authors acknowledge the Instituto Federal de Educação, Ciência e Tecnologia Goiano (IFGoiano), São Paulo Research Foundation (FAPESP) and the Pró-Reitoria de Pesquisa (PROPE-UNESP), for the financial support provided.

Conflicts of Interest: The authors declare no conflict of interest.

References

1. Silva, M.J.; Sanches, A.O.; Cena, C.R.; Nagashima, H.N.; Medeiros, E.S.; Malmonge, J.A. Study of the Electrical Conduction Process in Natural Rubber-based Conductive Nanocomposites Filled with Cellulose Nanowhiskers Coated by Polyaniline. *Polym. Compos.* **2021**, *42*, 1519–1529. [CrossRef]
2. Kalia, S.; Dufresne, A.; Cherian, B.M.; Kaith, B.S.; Avérous, L.; Njuguna, J.; Nassiopoulos, E. Cellulose-Based Bio- and Nanocomposites: A Review. *Int. J. Polym. Sci.* **2011**, *2011*, 837875. [CrossRef]
3. Siracusa, V.; Blanco, I. Bio-Polyethylene (Bio-PE), Bio-Polypropylene (Bio-PP) and Bio-Poly(Ethylene Terephthalate) (Bio-PET): Recent Developments in Bio-Based Polymers Analogous to Petroleum-Derived Ones for Packaging and Engineering Applications. *Polymers* **2020**, *12*, 1641. [CrossRef]
4. Costa, J.G.L.; Rodrigues, P.H.F.; Paim, L.L.; Sanches, A.O.; Malmonge, J.A.; da Silva, M.J. 1-3 Castor Oil-Based Polyurethane/PZT Piezoelectric Composite as a Possible Candidate for Structural Health Monitoring. *Mater. Res.* **2020**, *23*, 1–9. [CrossRef]
5. Kaikade, D.S.; Sabnis, A.S. Polyurethane Foams from Vegetable Oil-Based Polyols: A Review. *Polym. Bull.* **2022**, *80*, 2239–2261. [CrossRef]
6. Alves, F.C.; dos Santos, V.F.; Monticeli, F.M.; Ornaghi, H.; da Sliva Barud, H.; Mulinari, D.R. Efficiency of Castor Oil–Based Polyurethane Foams for Oil Sorption S10 and S500: Influence of Porous Size and Statistical Analysis. *Polym. Polym. Compos.* **2021**, *29*, S1063–S1074. [CrossRef]
7. Saha, P.; Khomlaem, C.; Aloui, H.; Kim, B.S. Biodegradable Polyurethanes Based on Castor Oil and Poly (3-Hydroxybutyrate). *Polymers* **2021**, *13*, 1387. [CrossRef]
8. Sardari, A.; Sabbagh Alvani, A.A.; Ghaffarian, S.R. Castor Oil-Derived Water-Based Polyurethane Coatings: Structure Manipulation for Property Enhancement. *Prog. Org. Coat.* **2019**, *133*, 198–205. [CrossRef]
9. Chen, Y.-C.; Tai, W. Castor Oil-Based Polyurethane Resin for Low-Density Composites with Bamboo Charcoal. *Polymers* **2018**, *10*, 1100. [CrossRef]
10. Gama, N.; Ferreira, A.; Barros-Timmons, A. Cure and Performance of Castor Oil Polyurethane Adhesive. *Int. J. Adhes. Adhes.* **2019**, *95*, 102413. [CrossRef]
11. Panda, S.S.; Panda, B.P.; Nayak, S.K.; Mohanty, S. A Review on Waterborne Thermosetting Polyurethane Coatings Based on Castor Oil: Synthesis, Characterization, and Application. *Polym. Plast. Technol. Eng.* **2018**, *57*, 500–522. [CrossRef]
12. Zhang, W.; Zhang, Y.; Liang, H.; Liang, D.; Cao, H.; Liu, C.; Qian, Y.; Lu, Q.; Zhang, C. High Bio-Content Castor Oil Based Waterborne Polyurethane/Sodium Lignosulfonate Composites for Environmental Friendly UV Absorption Application. *Ind. Crops Prod.* **2019**, *142*, 111836. [CrossRef]
13. Lu, J.; Zhang, Y.; Tao, Y.; Wang, B.; Cheng, W.; Jie, G.; Song, L.; Hu, Y. Self-Healable Castor Oil-Based Waterborne Polyurethane/MXene Film with Outstanding Electromagnetic Interference Shielding Effectiveness and Excellent Shape Memory Performance. *J. Colloid. Interface Sci.* **2021**, *588*, 164–174. [CrossRef] [PubMed]

14. Vieira Amorim, F.; José Ribeiro Padilha, R.; Maria Vinhas, G.; Ramos Luiz, M.; Costa de Souza, N.; Medeiros Bastos de Almeida, Y. Development of Hydrophobic Polyurethane/Castor Oil Biocomposites with Agroindustrial Residues for Sorption of Oils and Organic Solvents. *J. Colloid. Interface Sci.* **2021**, *581*, 442–454. [CrossRef] [PubMed]
15. Rebeque, P.V.; Silva, M.J.; Cena, C.R.; Nagashima, H.N.; Malmonge, J.A.; Kanda, D.H.F. Analysis of the Electrical Conduction in Percolative Nanocomposites Based on Castor-Oil Polyurethane with Carbon Black and Activated Carbon Nanopowder. *Polym. Compos.* **2019**, *40*, 7–15. [CrossRef]
16. Merlini, C.; Pegoretti, A.; Vargas, P.C.; da Cunha, T.F.; Ramôa, S.D.A.S.; Soares, B.G.; Barra, G.M.O. Electromagnetic Interference Shielding Effectiveness of Composites Based on Polyurethane Derived from Castor Oil and Nanostructured Carbon Fillers. *Polym. Compos.* **2019**, *40*, E78–E87. [CrossRef]
17. Wang, M.; Tang, X.-H.; Cai, J.-H.; Wu, H.; Shen, J.-B.; Guo, S.-Y. Construction, Mechanism and Prospective of Conductive Polymer Composites with Multiple Interfaces for Electromagnetic Interference Shielding: A Review. *Carbon. N. Y.* **2021**, *177*, 377–402. [CrossRef]
18. Bakkali, H.; Dominguez, M.; Batlle, X.; Labarta, A. Universality of the Electrical Transport in Granular Metals. *Sci. Rep.* **2016**, *6*, 29676. [CrossRef]
19. Deniz, W.D.S.; Lima, T.H.C.; Arlindo, E.P.S.; Junior, G.C.F. Prevulcanized Natural Rubber and Carbon Black: High-Deformation Piezoresistive Composites. *Mater. Lett.* **2019**, *253*, 427–429. [CrossRef]
20. Bunde, A.; Dieterich, W. Percolation in Composites. *J. Electroceram* **2000**, *5*, 81–92. [CrossRef]
21. Folorunso, O.; Hamam, Y.; Sadiku, R.; Ray, S.S.; Joseph, A.G. Parametric Analysis of Electrical Conductivity of Polymer-Composites. *Polymers* **2019**, *11*, 1250. [CrossRef] [PubMed]
22. Yang, Q.Q.; Liang, J.Z. A Percolation Model for Insulator-Metal Transition in Polymer-Conductor Composites. *Appl. Phys. Lett.* **2008**, *93*, 131918. [CrossRef]
23. Sen, A.K.; Bhattacharya, S. Variable Range Hopping Conduction in Complex Systems and a Percolation Model with Tunneling. In *Continuum Models and Discrete Systems*; Springer: Dordrecht, The Netherlands, 2004; pp. 367–373.
24. Psarras, G.C. Hopping Conductivity in Polymer Matrix–Metal Particles Composites. *Compos. Part A Appl. Sci. Manuf.* **2006**, *37*, 1545–1553. [CrossRef]
25. Huang, J.C. Carbon Black Filled Conducting Polymers and Polymer Blends. *Adv. Polym. Technol.* **2002**, *21*, 299–313. [CrossRef]
26. Gupta, A.; Goyal, R.K. Electrical Properties of Polycarbonate/Expanded Graphite Nanocomposites. *J. Appl. Polym. Sci.* **2019**, *13*, 47274. [CrossRef]
27. Mogha, A.; Kaushik, A. Functionalized Multiwall Carbon Nanotubes to Enhance Dispersion in Castor Oil-Based Polyurethane Nanocomposites. *Fuller. Nanotub. Carbon. Nanostruct.* **2021**, *29*, 907–914. [CrossRef]
28. Dai, Z.; Jiang, P.; Zhang, P.; Wai, P.T.; Bao, Y.; Gao, X.; Xia, J.; Haryono, A. Multiwalled Carbon Nanotubes/Castor-oil–Based Waterborne Polyurethane Nanocomposite Prepared Using a Solvent-free Method. *Polym. Adv. Technol.* **2021**, *32*, 1038–1048. [CrossRef]
29. Acuña, P.; Zhang, J.; Yin, G.-Z.; Liu, X.-Q.; Wang, D.-Y. Bio-Based Rigid Polyurethane Foam from Castor Oil with Excellent Flame Retardancy and High Insulation Capacity via Cooperation with Carbon-Based Materials. *J. Mater. Sci.* **2021**, *56*, 2684–2701. [CrossRef]
30. Sinh, L.H.; Luong, N.D.; Seppälä, J. Enhanced Mechanical and Thermal Properties of Polyurethane/Functionalised Graphene Oxide Composites by *in Situ* Polymerisation. *Plast. Rubber Compos.* **2019**, *48*, 466–476. [CrossRef]
31. Mohammadi, A.; Hosseinipour, M.; Abdolvand, H.; Najafabadi, S.A.A.; Sahraneshin Samani, F. Improvement in Bioavailability of Curcumin within the Castor-oil Based Polyurethane Nanocomposite through Its Conjugation on the Surface of Graphene Oxide Nanosheets. *Polym. Adv. Technol.* **2022**, *33*, 1126–1136. [CrossRef]
32. Li, J.-W.; Tsen, W.-C.; Tsou, C.-H.; Suen, M.-C.; Chiu, C.-W. Synthetic Environmentally Friendly Castor Oil Based-Polyurethane with Carbon Black as a Microphase Separation Promoter. *Polymers* **2019**, *11*, 1333. [CrossRef] [PubMed]
33. Kumar, S.; Tewatia, P.; Samota, S.; Rattan, G.; Kaushik, A. Ameliorating Properties of Castor Oil Based Polyurethane Hybrid Nanocomposites via Synergistic Addition of Graphene and Cellulose Nanofibers. *J. Ind. Eng. Chem.* **2022**, *109*, 492–509. [CrossRef]
34. Min, Y.K.; Eom, T.; Kim, H.; Kang, D.; Lee, S.-E. Independent Heating Performances in the Sub-Zero Environment of MWCNT/PDMS Composite with Low Electron-Tunneling Energy. *Polymers* **2023**, *15*, 1171. [CrossRef] [PubMed]
35. Kamedulski, P.; Lukaszewicz, J.P.; Witczak, L.; Szroeder, P.; Ziolkowski, P. The Importance of Structural Factors for the Electrochemical Performance of Graphene/Carbon Nanotube/Melamine Powders towards the Catalytic Activity of Oxygen Reduction Reaction. *Materials* **2021**, *14*, 2448. [CrossRef] [PubMed]
36. Rozhin, P.; Kralj, S.; Soula, B.; Marchesan, S.; Flahaut, E. Hydrogels from a Self-Assembling Tripeptide and Carbon Nanotubes (CNTs): Comparison between Single-Walled and Double-Walled CNTs. *Nanomaterials* **2023**, *13*, 847. [CrossRef] [PubMed]
37. Chung, D.D.L. A Critical Review of Piezoresistivity and Its Application in Electrical-Resistance-Based Strain Sensing. *J. Mater. Sci.* **2020**, *55*, 15367–15396. [CrossRef]
38. Nayak, L.; Rahaman, M.; Aldalbahi, A.; Kumar Chaki, T.; Khastgir, D. Polyimide-Carbon Nanotubes Nanocomposites: Electrical Conduction Behavior under Cryogenic Condition. *Polym. Eng. Sci.* **2017**, *57*, 291–298. [CrossRef]
39. Spahr, M.E.; Gilardi, R.; Bonacchi, D. Carbon Black for Electrically Conductive Polymer Applications. In *Fillers for Polymer Applications*; Springer: Cham, Switzerland, 2017; pp. 375–400.

40. Zhang, Q.; Wang, J.; Zhang, B.-Y.; Guo, B.-H.; Yu, J.; Guo, Z.-X. Improved Electrical Conductivity of Polymer/Carbon Black Composites by Simultaneous Dispersion and Interaction-Induced Network Assembly. *Compos. Sci. Technol.* **2019**, *179*, 106–114. [CrossRef]
41. Silva, M.J.D.; Kanda, D.H.F.; Nagashima, H.N. Mechanism of Charge Transport in Castor Oil-Based Polyurethane/Carbon Black Composite (PU/CB). *J. Non Cryst. Solids* **2012**, *358*, 270–275. [CrossRef]
42. Kirkpatrick, S. Percolation and Conduction. *Rev. Mod. Phys.* **1973**, *45.4*, 574–588. [CrossRef]
43. Shao, W.Z.; Xie, N.; Zhen, L.; Feng, L.C. Conductivity Critical Exponents Lower than the Universal Value in Continuum Percolation Systems. *J. Phys. Condens. Matter* **2008**, *20*, 395235. [CrossRef]
44. Panda, M.; Srinivas, V.; Thakur, A.K. Non-Universal Scaling Behavior of Polymer-Metal Composites across the Percolation Threshold. *Results Phys.* **2015**, *5*, 136–141. [CrossRef]
45. Marsden, A.J.; Papageorgiou, D.G.; Vallés, C.; Liscio, A.; Palermo, V.; Bissett, M.A.; Young, R.J.; Kinloch, I.A. Electrical Percolation in Graphene–Polymer Composites. *2d Mater.* **2018**, *5*, 032003. [CrossRef]
46. Shi, Y.D.; Li, J.; Tan, Y.J.; Chen, Y.F.; Wang, M. Percolation Behavior of Electromagnetic Interference Shielding in Polymer/Multi-Walled Carbon Nanotube Nanocomposites. *Compos. Sci. Technol.* **2019**, *170*, 70–76. [CrossRef]
47. Mazaheri, M.; Payandehpeyman, J.; Jamasb, S. Modeling of Effective Electrical Conductivity and Percolation Behavior in Conductive-Polymer Nanocomposites Reinforced with Spherical Carbon Black. *Appl. Compos. Mater.* **2022**, *29*, 695–710. [CrossRef]
48. Zare, Y.; Rhee, K.Y. Tuning of a Mechanics Model for the Electrical Conductivity of CNT-Filled Samples Assuming Extended CNT. *Eur. Phys. J. Plus* **2021**, *137*, 24. [CrossRef]
49. Macdonald, J.R. Impedance Spectroscopy: Models, Data Fitting, and Analysis. *Solid State Ion.* **2005**, *176*, 1961–1969. [CrossRef]
50. Dyre, J.C.; Schrøder, T.B. Universality of Ac Conduction in Disordered Solids. *Rev. Mod. Phys.* **2000**, *72*, 873–892. [CrossRef]
51. Bianchi, R.F.; Leal Ferreira, G.F.; Lepienski, C.M.; Faria, R.M. Alternating Electrical Conductivity of Polyaniline. *J. Chem. Phys.* **1999**, *110*, 4602–4607. [CrossRef]
52. Dyre, J.C. The Random Free-Energy Barrier Model for Ac Conduction in Disordered Solids. *J. Appl. Phys.* **1988**, *64*, 2456–2468. [CrossRef]
53. Kilbride, B.E.; Coleman, J.N.; Fraysse, J.; Fournet, P.; Cadek, M.; Drury, A.; Hutzler, S.; Roth, S.; Blau, W.J. Experimental Observation of Scaling Laws for Alternating Current and Direct Current Conductivity in Polymer-Carbon Nanotube Composite Thin Films. *J. Appl. Phys.* **2002**, *92*, 4024–4030. [CrossRef]
54. Ma, B.; Wang, Y.; Wang, K.; Li, X.; Liu, J.; An, L. Frequency-Dependent Conductive Behavior of Polymer-Derived Amorphous Silicon Carbonitride. *Acta Mater.* **2015**, *89*, 215–224. [CrossRef]
55. Alptekin, S.; Tataroğlu, A.; Altındal, Ş. Dielectric, Modulus and Conductivity Studies of Au/PVP/n-Si (MPS) Structure in the Wide Range of Frequency and Voltage at Room Temperature. *J. Mater. Sci. Mater. Electron.* **2019**, *30*, 6853–6859. [CrossRef]
56. Dhahri, A.; Dhahri, E.; Hlil, E.K. Electrical Conductivity and Dielectric Behaviour of Nanocrystalline $La_{0.6}$ $Gd_{0.1}$ $Sr_{0.3}$ $Mn_{0.75}$ $Si_{0.25}$ O_3. *RSC Adv.* **2018**, *8*, 9103–9111. [CrossRef] [PubMed]
57. Jouni, M.; Faure-Vincent, J.; Fedorko, P.; Djurado, D.; Boiteux, G.; Massardier, V. Charge Carrier Transport and Low Electrical Percolation Threshold in Multiwalled Carbon Nanotube Polymer Nanocomposites. *Carbon. N. Y.* **2014**, *76*, 10–18. [CrossRef]
58. Freire Filho, F.C.M.; Santos, J.A.; Sanches, A.O.; Medeiros, E.S.; Malmonge, J.A.; Silva, M.J. Dielectric, Electric, and Piezoelectric Properties of Three-phase Piezoelectric Composite Based on Castor-oil Polyurethane, Lead Zirconate Titanate Particles and Multiwall Carbon Nanotubes. *J. Appl. Polym. Sci.* **2023**, *140*, e53572. [CrossRef]
59. Li, B.; Randall, C.A.; Manias, E. Polarization Mechanism Underlying Strongly Enhanced Dielectric Permittivity in Polymer Composites with Conductive Fillers. *J. Phys. Chem. C* **2022**, *126*, 7596–7604. [CrossRef]
60. Sanches, A.O.; Kanda, D.H.F.; Malmonge, L.F.; da Silva, M.J.; Sakamoto, W.K.; Malmonge, J.A. Synergistic Effects on Polyurethane/Lead Zirconate Titanate/Carbon Black Three-Phase Composites. *Polym. Test.* **2017**, *60*, 253–259. [CrossRef]
61. Amoozegar, V.; Sherafat, Z.; Bagherzadeh, E. Enhanced Dielectric and Piezoelectric Properties in Potassium Sodium Niobate/Polyvinylidene Fluoride Composites Using Nano-Silicon Carbide as an Additive. *Ceram. Int.* **2021**, *47*, 28260–28267. [CrossRef]
62. Achour, M.E.; Brosseau, C.; Carmona, F. Dielectric Relaxation in Carbon Black-Epoxy Composite Materials. *J. Appl. Phys.* **2008**, *103*, 094103. [CrossRef]
63. Lu, H.; Zhang, X.; Zhang, H. Influence of the Relaxation of Maxwell-Wagner-Sillars Polarization and Dc Conductivity on the Dielectric Behaviors of Nylon 1010. *J. Appl. Phys.* **2006**, *100*, 054104. [CrossRef]
64. Asandulesa, M.; Kostromin, S.; Tameev, A.; Aleksandrov, A.; Bronnikov, S. Molecular Dynamics and Conductivity of a PTB7:PC71BM Photovoltaic Polymer Blend: A Dielectric Spectroscopy Study. *ACS Appl. Polym. Mater.* **2021**, *3*, 4869–4878. [CrossRef]
65. Asandulesa, M.; Kostromin, S.; Podshivalov, A.; Tameev, A.; Bronnikov, S. Relaxation Processes in a Polymer Composite for Bulk Heterojunction: A Dielectric Spectroscopy Study. *Polymer* **2020**, *203*, 122785. [CrossRef]
66. Sousa, E.A.; Lima, T.H.C.; Arlindo, E.P.S.; Sanches, A.O.; Sakamoto, W.K.; Fuzari-Junior, G. de C. Multicomponent Polyurethane–Carbon Black Composite as Piezoresistive Sensor. *Polym. Bull.* **2020**, *77*, 3017–3031. [CrossRef]

67. Costa, P.; Oliveira, J.; Horta-Romarís, L.; Abad, M.-J.; Moreira, J.A.; Zapiráin, I.; Aguado, M.; Galván, S.; Lanceros-Mendez, S. Piezoresistive Polymer Blends for Electromechanical Sensor Applications. *Compos. Sci. Technol.* **2018**, *168*, 353–362. [CrossRef]
68. Gonçalves, B.F.; Costa, P.; Oliveira, J.; Ribeiro, S.; Correia, V.; Botelho, G.; Lanceros-Mendez, S. Green Solvent Approach for Printable Large Deformation Thermoplastic Elastomer Based Piezoresistive Sensors and Their Suitability for Biomedical Applications. *J. Polym. Sci. B. Polym. Phys.* **2016**, *54*, 2092–2103. [CrossRef]

Disclaimer/Publisher's Note: The statements, opinions and data contained in all publications are solely those of the individual author(s) and contributor(s) and not of MDPI and/or the editor(s). MDPI and/or the editor(s) disclaim responsibility for any injury to people or property resulting from any ideas, methods, instructions or products referred to in the content.

Communication

Milling Parameter Optimization of Continuous-Glass-Fiber-Reinforced-Polypropylene Laminate

Hanjie Hu [1,2], Bing Du [3,4,*], Conggang Ning [3], Xiaodong Zhang [3], Zhuo Wang [3], Yangyang Xiong [3], Xianjun Zeng [2] and Liming Chen [4,*]

[1] School of Aeronautics, Chongqing Jiaotong University, Chongqing 400074, China; huhj@gatri.cn
[2] The Green Aerotechnics Research Institute of CQJTU, Chongqing 401120, China; zengxj@gatri.cn
[3] Chongqing Key Laboratory of Nano–Micro Composite Materials and Devices, School of Metallurgy and Materials Engineering, Chongqing University of Science and Technology, Chongqing 401331, China; 2018441648@cqust.edu.cn (C.N.); 2018441447@cqust.edu.cn (X.Z.); 2018444415@cqust.edu.cn (Z.W.); 2019442412@cqust.edu.cn (Y.X.)
[4] College of Aerospace Engineering, Chongqing University, Chongqing 400030, China
* Correspondence: dubing@cqust.edu.cn (B.D.); clm07@cqu.edu.cn (L.C.)

Citation: Hu, H.; Du, B.; Ning, C.; Zhang, X.; Wang, Z.; Xiong, Y.; Zeng, X.; Chen, L. Milling Parameter Optimization of Continuous-Glass-Fiber-Reinforced-Polypropylene Laminate. *Materials* **2022**, *15*, 2703. https://doi.org/10.3390/ma15072703

Academic Editor: Halina Kaczmarek

Received: 28 February 2022
Accepted: 1 April 2022
Published: 6 April 2022

Publisher's Note: MDPI stays neutral with regard to jurisdictional claims in published maps and institutional affiliations.

Copyright: © 2022 by the authors. Licensee MDPI, Basel, Switzerland. This article is an open access article distributed under the terms and conditions of the Creative Commons Attribution (CC BY) license (https://creativecommons.org/licenses/by/4.0/).

Abstract: The composite-material laminate structure will inevitably encounter connection problems in use. Among them, mechanical connections are widely used in aerospace, automotive and other fields because of their high connection efficiency and reliable connection performance. Milling parameters are important for the opening quality. In this paper, continuous-glass-fiber-reinforced-polypropylene (GFRPP) laminates were chosen to investigate the effects of different cutters and process parameters on the hole quality. The delamination factor and burr area were taken as the index to characterize the opening quality. After determining the optimal milling tool, the process window was obtained according to the appearance of the milling hole. In the selected process parameter, the maximum temperature did not reach the PP melting temperature. The best hole quality was achieved when the spindle speed was 18,000 r/min and the feed speed was 1500 mm/min with the corn milling cutter.

Keywords: continuous glass fiber; thermoplastic composite; milling; process parameter

1. Introduction

The composite-material laminate structure will inevitably encounter connection problems in use. Among them, mechanical connections are widely used in aerospace, automotive and other fields because of their high connection efficiency and reliable connection performance [1–3]. The drilling process is an indispensable process. During the processing, there are defects such as fiber pullout, tearing, peeling, delamination and burrs, which seriously affect the mechanical properties of the opening holes, further affecting the performance of composite structures. Therefore, it is of great research value to find a method that can reduce the occurrence of defects and improve the quality of holes [4].

Prasad and Chaitanya [5] used the Taguchi analysis method to study the delamination factors of laminates and optimized the processing parameters. The experimental results show that the peeling delamination is most affected by the thickness of the material, followed by the feed speed and fiber orientation, and the delamination rate is most affected by feed speed, followed by thickness, fiber orientation and spindle speed. Based on the numerical analysis of drilling images, Hrechuk et al. [6] carried out the non-destructive quantification of visible defects and proposed an overall parameter Q. The results of drilling experiments show that the drilling quality Q, the number of drillings, and the tool wear are linearly related. Liu et al. [7]. proposed a new delamination scheme in which delamination caused by thrust exerted by the chisel and cutting edges is considered. Amaro et al. [8] found that the presence of holes increased the energy absorbed by the damaged and delaminated regions. The delamination distribution is also affected due to the change in the interlayer shear-stress distribution, which is the main reason for the delamination between layers with different orientations. Tsao et al. [9] proposed a new method for the equivalent delamination factor and compared it with the adjusted delamination factor and the conventional delamination factor, which is suitable for characterizing delamination at the hole exit after drilling composites. Sugita et al. [10] proposed the design of a drilling tool that is shaped to suppress burrs and delamination during composite drilling. Loja et al. [11] used four different types of tools to conduct drilling experiments on glass-fiber laminates to analyze and evaluate the delamination effect and its maximum temperature. The experiments showed that the higher the drilling speed, the higher the temperature generated by friction, and the greater the damage to the aperture diameter. Velaga et al. [12] used a double-flute twist drill with a diameter of 10 mm to conduct drilling experiments on epoxy-resin/glass-fiber composites. The experimental results found that the rotation speed of 700 rpm and the feed rate of 0.4 mm/rev were used for the optimum drilling parameters of the selected tool. Guo [13] conducted cutting experiments on continuous-glass-fiber-reinforced-polypropylene composites at different cutting angles to analyze the characteristics of right-angle cutting and the causes of damage. The experiment found that the fiber cutting angle has a significant impact on the cutting process: at $0°$ there is groove damage; at $45°$ the cutting surface is the best; at $90°$ there is a slight burr; at $135°$ the resin interface cracks and many burrs are generated. Mustari et al. [14]. investigated the effect of glass-fiber content in glass-fiber composites on the drilling process. The results show that the drilling surface quality is closely related to the percentage of fiber content in the sample. In terms of roundness deviation, delamination, fiber pull-out and taper angle, the composite samples performed better when the glass-fiber content was lower. Patel et al. [15] aimed to examine the effect of tool geometry, spindle speed and feed on thrust and delamination in hybrid hemp-glass composites. Palanikumar et al. [16] studied the parameters affecting the thrust of glass-fiber-reinforced polypropylene (GFRPP) in boreholes and established an empirical relationship for determining the thrust of GFRTP drilling. Srinivasan et al. [17] used the "Brad and Spur" drill bit to drill the GFRPP matrix composites with the aim of testing the effect of drilling parameters on the roundness error, using the Box-Behnken design (BBD) technique to determine the cutting parameters that were evaluated and optimized. The results show that the model can be effectively used to predict the response variable and thus control the roundness error. Palanikumar et al. [18] studied the effect of various feed rates on thrust, spindle speed, and bit geometry using Brad and straight plunger bits, and the experimental results showed that delamination is entirely dependent on the feed rate.

In this paper, GFRPP laminates are taken as the research object to investigate the effects of different cutters and process parameters on the hole quality. The delamination factor and burr area were taken as the index to characterize the opening quality. After determining the optimal milling tool, the process window was obtained according to the appearance of the milling hole. Afterwards, the response surface method (RSM) was utilized to analyze the influence of the process parameters on the opening quality.

2. Materials and Methods

This experimental sample was made of GFRPP prepreg in order to prepare composite material laminate, and glass-fiber-reinforced-polypropylene prepreg with a thickness of 0.3 mm was fiber reinforced using [0/45/-45/90]s. Composite laminate, milled to 2.4 mm through holes using the hot press (Qingdao Huabo Machinery Technology Co., Ltd., Qingdao, China), according to the process parameters in [19]. Milling experiments were conducted by a CNC (Computer Numerical Control) engraving machine (JingYan Instruments & Technology CO., Ltd., Dongguan, China). As shown in Figure 1, milling tools included a single-edge spiral milling cutter, double-edged spiral milling cutter, corn milling cutter and PCB (Printed Circuit Board) alloy drill with the same size where $L_1 = 38$ mm, $L_2 = 12$ mm, $d = 3.175$ mm. First, the comparison among milling tools was made under the same spindle speed and feed speed of 6000 r/min and 800 mm/min, respectively. Five holes with a diameter of 8 mm were milled by each tool. The visual-inspection method was used to choose the optimal mill tool based on the macroscopic appearance near the hole.

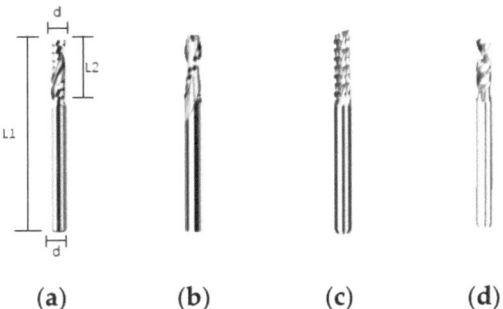

Figure 1. Schematic of cutting tools: (**a**) Single-edge milling cutter; (**b**) Double-edge milling cutter; (**c**) Corn milling cutter; (**d**) PCB alloy drill.

In the milling process of traditional composite laminates, the local thermal shock [20] caused by the presence of abrasives and fibers and the low thermal conductivity of composite materials limit the temperature dissipation during hole opening, resulting in heat accumulation at the hole exit and excessively high temperature, and even in the case of improper heat dissipation treatment, the temperature will often be higher than the glass transition temperature, resulting in a decrease in the performance of the resin matrix and a significant reduction in the quality of openings [21]. In the milling experiment, the temperature change is often mainly affected by the spindle speed and feed rate [22]. Due to the wide selection of parameters in this experiment, the effect of temperature on the quality of the opening cannot be excluded from the experimental results.

Therefore, a real-time temperature-observation platform was built, and special parameter-processing points were selected to study the effect of temperature changes on the quality of the opening during milling, as depicted in Figure 2. The infrared thermal imager was supported and fixed to the main shaft of the cutting machine by a movable bracket and was able to move on the same trajectory with the main shaft of the cutting machine. The thermal-imaging camera directly captures the milling hole, and is connected to an external receiving software device, which can display the milling temperature change on the screen in real time. A Canon digital camera (EOS 200DII, Canon (CHINA) CO., Ltd., Beijing, China) was used to take pictures of the openings, and ImageJ and Photoshop software were used to process the pictures to calculate the delamination factor and burr area during milling, as shown in Figure 3. The delamination factor F_d was defined as the actual diameter divided by the designed diameter, referring to Equation (1). The delamination factor and burr area were taken as the index to characterize the opening quality. After determining the optimal milling tool, the process window was obtained according to the appearance of the milling hole. Afterwards, the response surface method (RSM) was utilized to analyze the influence of the process parameters on the opening quality.

$$F_d = \frac{D_{max}}{D} \quad (1)$$

where D is the designed diameter of the hole and D_{max} is the actual diameter.

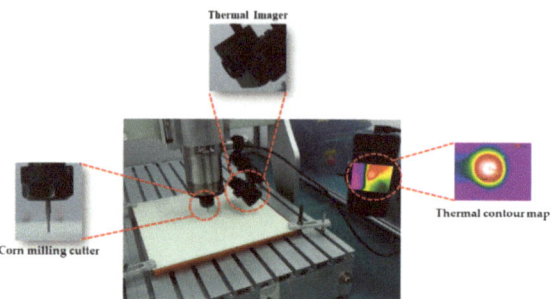

Figure 2. Thermal-imaging setup.

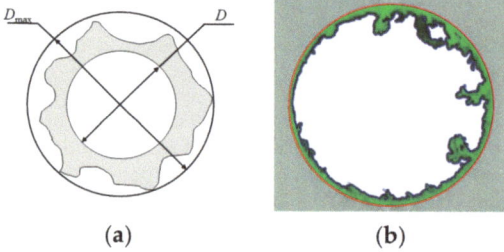

Figure 3. Schematic diagram of evaluation parameters: (**a**) Delamination factor; (**b**) Burr area.

3. Results

3.1. Determining the Process Window

As seen in Figure 4, under the spindle speed of 6000 r/min and feed speed 800 mm/min, the four milling tools had different defects. The single-edged milling cutters, double-edged milling cutters, and PCB alloy drill had many burrs and serious fiber delamination. For the corn milling cutter, no visual severe delamination was found and relatively less burrs appeared compared with the other three tools. Loja et al. [11] used four drilling tools to evaluate the delamination and maximum temperature caused by drilling in fiberglass laminates. For milling cutters and twist drills, the higher the drilling speed, the higher the temperature generated by friction, and the greater the damage to the aperture. The experimental results and literature data have certain commonality. The quality of the milling effect of the corn milling cutter was better than that of the spiral milling cutter and twist drill. The subsequent experiment was conducted by the corn milling cutter.

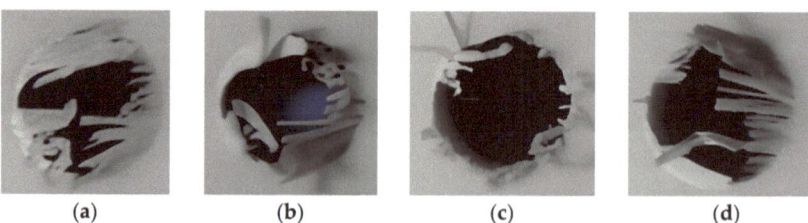

Figure 4. Morphology of milling holes: (**a**) Single-edge milling cutter; (**b**) Double-edge milling cutter; (**c**) Corn milling cutter; (**d**) PCB alloy drill.

In this step, spindle speed and feed speed were varied from 700 r/min to 24,000 r/min and 300 mm/min to 1500 mm/min, respectively, as listed in Table 1. A total of 27 groups of experiments were made and two holes were milled under each group of parameters to ensure the results. The influence of the process parameters on milling property is shown in Figure 5, where three typical situations known as tool fracture, delamination and burr were found. When the spindle speed was too low, 700 r/min, tool fracture appeared. When the spindle speed was lower than 3000 r/min, the delamination of fibers near the hole was found. On the contrary, the burr became dominant when the spindle speed was higher than 6000 r/min. Between the range of 3000 r/min to 6000 r/min was the transition zone where delamination and burr coexisted. From Figure 6, three combinations of processing parameters named F1–S2, F3–S5, and F6–S9 were selected as the thermal-imaging objects. It was shown in Figure 6 that with the increase in the spindle speed and feed speed, the maximum temperature increases, but the maximum temperature was 49.3 °C below the melting temperature of PP.

Table 1. Number of process parameters.

Feeding Speed mm/min	No.	Spindle Speed r/min	No.
300	F1	700	S1
600	F2	1400	S2
900	F3	2000	S3
1200	F4	3000	S4
1500	F5	6000	S5
1800	F6	10,000	S6
		12,000	S7
		18,000	S8
		24,000	S9

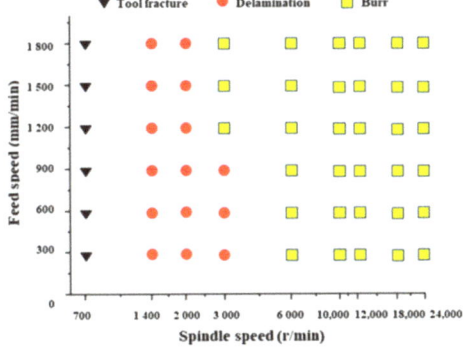

Figure 5. Influence of process parameters on milling property.

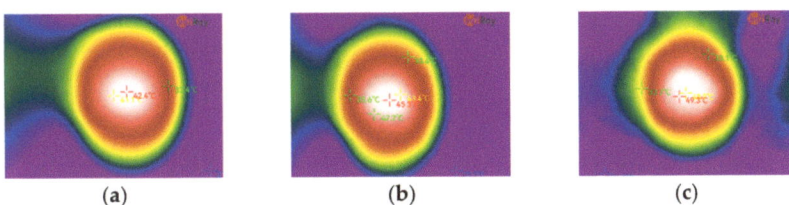

Figure 6. Thermal contour map of milling hole: (a) F1-S2; (b) F3-S5; (c) F6-S9.

3.2. Optimizing the Process Parameters

Based on the above process window, the process parameters were refined. Feed speed was chosen as 1200 mm/min, 1500 mm/min, and 1800 mm/min, and the spindle speed was chosen as 12,000 r/min, 18,000 r/min, and 24,000 r/min. The response-surface-method analysis was carried out to determine the optimal process parameters. The factor-level design table is shown in Table 2 with two factors and three levels. The experimental trials were design as in Table 3 with the corresponding characterization variables named delamination factor and burr area. It can be seen that with the change in the processing parameters, the delamination factor is the lowest at 1.0046 and the highest at 1.0374. Taking the delamination factor as the dependent variable and the specific level of each factor as the independent variable, the quadratic regression equation was established to obtain the following results as Equation (2):

$$F_d = 0.003887A^2 + 0.002667B^2 - 0.005302AB - 0.009877A - 0.005397B + 1.02 \quad (2)$$

Table 2. Factor-level design table of response surface method.

Level	A: Feeding Speed mm/min	B: Spindle Speed r/min
−1	2000	300
0	10,000	900
1	18,000	1500

Table 3. Design and test results of response surface method.

No.	A: Spindle Speed mm/min	B: Feeding Speed r/min	Delamination Factor	Burr Area mm^2
1	2000	300	1.0352	15.17
2	18,000	300	1.0237	23.67
3	2000	1500	1.0374	25.07
4	18,000	1500	1.0046	8.94
5	2000	900	1.0322	30.94
6	18,000	900	1.0173	17.54
7	10,000	300	1.0313	20.67
8	10,000	1500	1.0158	12.99
9	10,000	900	1.0179	21.11
10	10,000	900	1.0179	24.49

The significance analysis of each item in the quadratic regression equation is shown in Table 4. It shows that under the selected experimental interval, the F value of this model is 15.21628 and the probability of data error is 1.04%. The analysis of variance for the model is shown in Table 5, showing that 95% of the experimental data can be explained by this model. Adj R^2 and Pred R^2 had higher values and the difference was within 0.2, indicating that the regression model can fully fit the test results. However, Table 6 shows that the relationship between burr area and the model was not significant, although the burr area decreased with the increase in the processing parameters.

Table 4. Significance analysis of delamination factor.

Source	Sum of Squares	d_f	Mean Square	F-Value	p-Value Prob > F	
Model	0.000934	5	0.000187	15.21628	0.0104	significant
A–Spindle speed	0.000585	1	0.000585	47.66917	0.0023	
B–Feeding speed	0.000175	1	0.000175	14.23207	0.0196	
AB	0.000112	1	0.000112	9.159819	0.0389	
A^2	3.53×10^{-5}	1	3.53×10^{-5}	2.871886	0.1654	
B^2	1.66×10^{-5}	1	1.66×10^{-5}	1.352159	0.3095	
Residual	4.91×10^{-5}	4	1.23×10^{-5}			
Lack of Fit	4.91×10^{-5}	3	1.64×10^{-5}			

Table 5. ANOVA results of delamination factor.

R^2	Adj R^2	Pred R^2	Adeq Precisior
0.9501	0.8876	0.5152	11.254

Table 6. Significance analysis of burr area.

Source	Sum of Squares	d_f	Mean Square	F-Value	p-Value Prob > F	
Model	335.7867	5	67.15734	5.901067	0.0551	not significant
A–Spindle speed	73.70933	1	73.70933	6.476785	0.0636	
B–Feeding speed	26.07555	1	26.07555	2.29124	0.2047	
AB	151.5951	1	151.5951	13.32055	0.0218	
A^2	4.637278	1	4.637278	0.407474	0.5580	
B^2	84.00382	1	84.00382	7.381355	0.0532	
Residual	45.52217	4	11.38054			
Lack of Fit	39.81506	3	13.27169	2.325465	0.4412	not significant

From the above, it can be concluded that when the machining parameters were low, the hole quality was not ideal. With the increase in the spindle speed and feed speed, the hole quality was gradually improved. When the machining parameters were increased to a certain range, the effect of improving the quality of the opening became weaker. The results of the response surface method show that when the spindle speed was 18,000 r/min and the feed speed was 1500 mm/min, the hole quality was the best, and the delamination factor and the burr area were both the minimum.

4. Conclusions

In this paper, GFRPP laminates were taken as the research object to investigate the effects of different cutters and process parameters on the hole quality. The specific conclusions are as follows:

(1) Different tools have a great influence on the hole quality, and the corn milling cutter used for GFRPP laminates has the best opening quality.
(2) In the selected region, with the increase in spindle speed and feed speed, the change in temperature distribution in the entry surface near the hole is not significant, and the maximum temperature does not reach the resin-melting temperature.
(3) Using the delamination factor and burr area to represent the hole quality is feasible. The quadratic regression model is significant for delamination factor while insignificant for burr area. The reason is because that the calculation of the burr area is greatly affected by the experimental error.
(4) In the milling of presented GFRPP laminate, the best hole quality is achieved when the spindle speed is 18,000 r/min and the feed speed is 1500 mm/min with the corn milling cutter.

Author Contributions: Conceptualization, H.H., B.D. and L.C.; Data curation, X.Z. (Xiaodong Zhang) and Z.W.; Funding acquisition, B.D.; Investigation, C.N. and X.Z. (Xiaodong Zhang); Methodology, H.H. and X.Z. (Xianjun Zeng); Resources, Z.W. and Y.X.; Software, C.N. and Y.X.; Validation, C.N.; Visualization, X.Z. (Xianjun Zeng); Writing—original draft, H.H.; Writing—review & editing, B.D. and L.C. All authors have read and agreed to the published version of the manuscript.

Funding: This research was funded by Chongqing Natural Science Foundation (cstc2020jcyj-msxmX0559), Science and Technology Research Program of Chongqing Municipal Education Commission (KJQN202101531), Research Foundation (ckrc2019024) and College Students Science and Technology Innovation Training Program (2021194) of Chongqing University of Science and Technology.

Institutional Review Board Statement: Not applicable.

Informed Consent Statement: Not applicable.

Data Availability Statement: The data presented in this study are available on request from the corresponding author.

Conflicts of Interest: The authors declare no conflict of interest.

References

1. Xing, L.Y.; Bao, J.W.; Li, S.M.; Chen, X.B. Development status and facing challenge of advanced polymer matrix composites. *Acta Mater. Compos. Sin.* **2016**, *33*, 1327–1338.
2. Wu, L.H.; Yuan, Y.H. Applications of advanced fiber-reinforced composite materials in large commercial aircraft. *Ordnance Mater. Sci. Eng.* **2018**, *41*, 100–103.
3. Thoppul, S.D.; Finegan, J.; Gibson, R.F. Mechanics of mechanically fastened joints in polymer–matrix composite structures—A review. *Compos. Sci. Technol.* **2009**, *69*, 301–329. [CrossRef]
4. Jia, Z.Y.; Bi, G.J.; Wand, F.J.; Wang, X.N.; Zhang, B.Y. The research of machining mechanism of carbon fiber reinforced plastic. *J. Mech. Eng.* **2018**, *54*, 199–208. [CrossRef]
5. Prasad, K.S.; Chaitanya, G. Analysis of delamination in drilling of GFRP composites using Taguchi Technique. *Mater. Today: Proc.* **2019**, *18*, 3252–3261. [CrossRef]
6. Hrechuk, A.; Bushlya, V.; Ståhl, J.-E. Hole-quality evaluation in drilling fiber-reinforced composites. *Compos. Struct.* **2018**, *204*, 378–387. [CrossRef]
7. Liu, S.; Yang, T.; Liu, C.; Jin, Y.; Sun, D.; Shen, Y. Modelling and experimental validation on drilling delamination of aramid fiber reinforced plastic composites. *Compos. Struct.* **2020**, *236*, 111907. [CrossRef]
8. Amaro, A.M.; Reis, P.N.B.; de Moura, M.; Neto, M.A. Influence of open holes on composites delamination induced by low velocity impact loads. *Compos. Struct.* **2013**, *97*, 239–244. [CrossRef]
9. Tsao, C.C.; Kuo, K.L.; Hsu, I.C. Evaluation of novel approach on delamination factor after drilling composite laminates. In *Key Engineering Materials*; Trans Tech Publications Ltd.: Zurich, Switzerland, 2010; Volume 443, pp. 626–630.
10. Sugita, N.; Shu, L.; Kimura, K.; Arai, G.; Arai, K. Dedicated drill design for reduction in burr and delamination during the drilling of composite materials. *CIRP Ann.* **2019**, *68*, 89–92. [CrossRef]
11. Loja, M.A.R.; Alves, M.S.F.; Bragança, I.M.F.; Rosa, R.S.B.; Barbosa, I.C.J.; Barbosa, J.I. An assessment of thermally influenced and delamination-induced regions by composites drilling. *Compos. Struct.* **2018**, *202*, 413–423. [CrossRef]
12. Velaga, M.; Cadambi, R.M. Drilling of GFRP composites for minimising delamination effect. *Mater. Today Proc.* **2017**, *4*, 11229–11236. [CrossRef]
13. Guo, H.B. Analysis of Formation Mechanism and Restraining Methods of GFRPP Composites Drilling Damage. Master's Dissertation, Dalian University of Technology, Dalian, China, 2018.
14. Mustari, A.; Chakma, P.; Sobhan, R.; Dhar, N. Investigation on roundness and taper of holes in drilling GFRP composites with variable weight percentages of glass fiber. *Mater. Today Proc.* **2020**, *38*, 2578–2583. [CrossRef]
15. Patel, K.; Gohil, P.P.; Chaudhary, V. Investigations on drilling of hemp/glass hybrid composites. *Mater. Manuf. Process.* **2018**, *33*, 1714–1725. [CrossRef]
16. Palanikumar, K.; Srinivasan, T.; Rajagopal, K.; Latha, B. Thrust Force Analysis in Drilling Glass Fiber Reinforced/Polypropylene (GFR/PP) Composites. *Mater. Manuf. Process.* **2014**, *31*, 581–586. [CrossRef]
17. Srinivasan, T.; Palanikumar, K.; Rajagopal, K. Roundness Error Evaluation in Drilling of Glass Fiber Reinforced Polypropylene (GFR/PP) Composites Using Box Behnken Design (BBD). In *Applied Mechanics and Materials*; Trans Tech Publications Ltd.: Zurich, Switzerland, 2015; Volume 766–767, pp. 844–851. [CrossRef]
18. Srinivasan, T.; Palanikumar, K.; Rajagopal, K. Influence of Process Parameters on Delamination of Drilling of (GF/PC) Glass Fiber Reinforced Polycarbonate Matrix Composites. *Adv. Mater. Res.* **2014**, *984–985*, 355–359. [CrossRef]
19. Du, B.; Chen, L.M.; Liu, H.C.; He, Q.H.; Li, W.G. Resistance Welding of Glass Fiber Reinforced Thermoplastic Composite: Ex-perimental Investigation and Process Parameter Optimization. *Chin. J. Aeronaut.* **2020**, *33*, 3469–3478. [CrossRef]
20. Sorrentino, L.; Turchetta, S.; Bellini, C. In process monitoring of cutting temperature during the drilling of FRP laminate. *Compos. Struct.* **2017**, *168*, 549–561. [CrossRef]
21. Hou, G.; Luo, B.; Zhang, K.; Luo, Y.; Cheng, H.; Cao, S.; Li, Y. Investigation of high temperature effect on CFRP cutting mechanism based on a temperature controlled orthogonal cutting experiment. *Compos. Struct.* **2021**, *268*, 113967. [CrossRef]
22. Giasin, K.; Ayvar-Soberanis, S. Evaluation of Workpiece Temperature during Drilling of GLARE Fiber Metal Laminates Using Infrared Techniques: Effect of Cutting Parameters, Fiber Orientation and Spray Mist Application. *Materials* **2016**, *9*, 622. [CrossRef] [PubMed]

Article

Conversion of Polyethylene to Low-Molecular-Weight Oil Products at Moderate Temperatures Using Nickel/Zeolite Nanocatalysts

Hyungjin Cho [1], Ahyeon Jin [1], Sun Ju Kim [1], Youngmin Kwon [1], Eunseo Lee [1], Jaeman J. Shin [1,2] and Byung Hyo Kim [1,2,*]

[1] Department of Materials Science and Engineering, Soongsil University, Seoul 06978, Republic of Korea; hyungjincho@soongsil.ac.kr (H.C.); jshin@ssu.ac.kr (J.J.S.)
[2] Department of Green Chemistry and Materials Engineering, Soongsil University, Seoul 06978, Republic of Korea
* Correspondence: byunghyokim@ssu.ac.kr; Tel.: +82-2-829-8218

Abstract: Polyethylene (PE) is the most widely used plastic, known for its high mechanical strength and affordability, rendering it responsible for ~70% of packaging waste and contributing to microplastic pollution. The cleavage of the carbon chain can induce the conversion of PE wastes into low-molecular-weight hydrocarbons, such as petroleum oils, waxes, and natural gases, but the thermal degradation of PE is challenging and requires high temperatures exceeding 400 °C due to its lack of specific chemical groups. Herein, we prepare metal/zeolite nanocatalysts by incorporating small-sized nickel nanoparticles into zeolite to lower the degradation temperature of PE. With the use of nanocatalysts, the degradation temperature can be lowered to 350 °C under hydrogen conditions, compared to the 400 °C required for non-catalytic pyrolysis. The metal components of the catalysts facilitate hydrogen adsorption, while the zeolite components stabilize the intermediate radicals or carbocations formed during the degradation process. The successful pyrolysis of PE at low temperatures yields valuable low-molecular-weight oil products, offering a promising pathway for the upcycling of PE into higher value-added products.

Keywords: nickel nanocatalyst; upcycling; hydrogen; polyethylene

1. Introduction

Polyethylene (PE) is the polymer produced in the largest quantity, with an annual production of 100 million tons. PE is used in packaging materials such as plastic bags, beverage caps, and straws due to its low cost and high chemical stability [1–6]. The usage of PE as a disposable item results in a massive amount of PE waste, which accounts for approximately 34% of total plastic waste [7], leading to significant environmental and social issues [8]. Most PE waste is either landfilled or incinerated, resulting in problems such as landfill capacity limitations, resource loss, and additional environmental pollution [6,9–11]. Mechanical recycling through melt reprocessing allows for the reutilization of about 16% of plastics but suffers from property degradation [9,12–14]. To overcome these drawbacks, chemical recycling methods, known as upcycling, have gained much attention [10,14–17].

Upcycling primarily involves the thermal degradation of PE, cracking the hydrocarbon chains into low-molecular-weight products such as oil, naphtha, and gas, and thus creating higher-value materials [8,18–20]. Since PE is composed of C-C bonds without specific functional groups, the thermal degradation temperature is relatively high at around 400 °C, necessitating a considerable energy input [21]. To mitigate the high energy consumption in the degradation of PE, recent studies have aimed to lower the reaction temperature [8] by using hydrogenation catalysts [6,22–25].

The addition of hydrogen during the thermal degradation of PE facilitates C-C bond cleavage, enhancing upcycling efficiency [12,25–27]. The hydrogenation catalysts with a high efficiency can be selected by a volcano plot. The Sabatier principle posits that the ideal catalyst ought to establish interactions of moderate affinity with both reactant species and resultant products. The catalyst's binding strength should be sufficiently robust to engage with reactants effectively, yet suitably feeble to facilitate the dissociation and subsequent formation of products. The volcano plot, rooted in this principle, delineates the correlation between a catalyst's catalytic activity and its binding affinity in a systematic manner [28]. In the volcano plot, noble metals such as Pt, Rh, and Ru can be the most effective catalysts for hydrogenation reactions [1,27,29–31]. However, considering the high cost and limited reserves of these catalytic elements, the development of cost-effective metal-based nanocatalysts is in high demand [27]. Nickel catalysts exhibit a good catalytic activity among non-noble metals, with an abundant availability and low cost, making them suitable for industrial-scale reactions [31]. Several groups have investigated nickel-based thermal degradation catalysts for PE upcycling using supported nickel precursors via impregnation or deposition methods. Dionisios and coworkers utilized Ni/SiO_2 catalysts for PE degradation, achieving diesel yields of 40–70%, which is comparable to the diesel yield achieved when using Pt- and Ru-based catalysts [27]. Rui Cai and coworkers obtained gas-phase hydrocarbon yields of up to 89.1% through C-C bond cleavage when using Ni- and Ru-based catalysts [32]. Pioneering studies have shown the applicability of Ni-based catalysts for PE upcycling but still require a high reaction temperature or produce a low oil yield.

Herein, we synthesized uniform-sized nickel nanoparticles and incorporated the particles into zeolite supports for PE-upcycling catalysts (Figure 1). The nanometer-sized grain size of the Ni nanoparticles of the catalysts offers advantages in terms of a high surface area to increase the catalytic efficiency [33]. Large-sized nanoparticles among a non-uniform nanoparticle ensemble have a low surface atom ratio, and thus nanoparticle-based catalysts with a broad size distribution usually show weak catalytic properties. Nickel nanoparticles are supported on porous zeolites [30,34,35], which serve not only as a support but also as a Lewis acid site to enhance C-C bond cleavage by initiating and stabilizing radicals [1,29,36–38]. We conducted PE thermal degradation at a relatively low temperature of 350 °C by adjusting the hydrogen pressure and catalyst quantity.

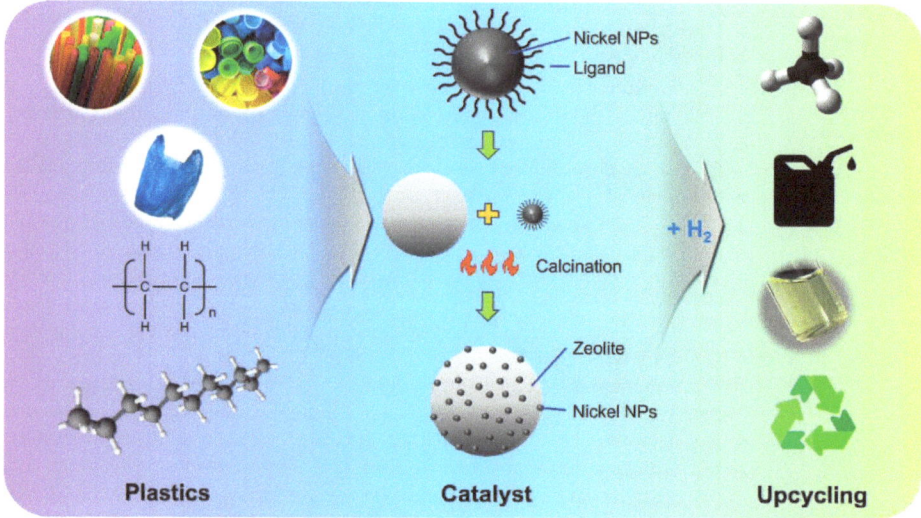

Figure 1. Schematic diagram of upcycling of PE using nickel/zeolite nanocatalysts.

2. Materials and Methods

2.1. Materials

Nickel(II) acetylacetonate (Ni(acac)$_2$, 90%), Dioctyl ether (99%), trioctylphosphine (90%), oleylamine (70%) and low-density polyethylene were purchased from Sigma-Aldrich (St. Louis, MO, USA). Zeolite HY (zeolite Y modified with hydrogen) was purchased from Thermo Scientific (Waltham, MA, USA).

2.2. Methods

2.2.1. Synthesis of 10 nm Nickel Nanoparticles

The monodisperse Ni nanoparticles were synthesized by following a previous report [39]. The synthetic procedure for the preparation of the nickel–oleylamine complex was commenced with a reaction between 1.55 g of oleylamine and 0.257 g of Ni(acac)$_2$ at 100 °C for a duration of 30 min under an inert atmosphere. Subsequently, 5 mL of dioctyl ether was injected into the complex solution, which underwent degassing for 1 h at 100 °C under vacuum conditions. Following this step, 0.3 mL of trioctylphosphine was carefully added to the solution under an argon atmosphere. The temperature was gradually elevated to 250 °C at a rate of 2 °C/min and maintained at this level for 30 min. The color of the solution transitioned from bluish-green to black at approximately 200 °C, indicative of the formation of nickel nanoparticles. Upon completion of the reaction, the product was swiftly cooled to room temperature. Nanoparticles were then precipitated by adding a mixture of toluene and excess ethanol in a 1:3 ratio. Finally, the obtained nanoparticles were dispersed in hydrophobic solvents such as hexane or toluene for further analysis and applications.

2.2.2. Preparation of Polyethylene Degradation Catalyst

Synthesized nanoparticles were impregnated onto a zeolite support. Specifically, nickel nanoparticles were synthesized and dispersed onto Y-zeolite, with a weight ratio of 1:10, where 0.5 g of nickel nanoparticles was impregnated onto 5 g of zeolite, which had an approximate size of 1 μm. The nanoparticles, bearing phosphine ligands, were dispersed in hexane, a non-polar solvent. The zeolite support was then immersed in the dispersion, followed by sonication for 30 min and stirring at 1000 rpm. The sonication and vigorous stirring were repeated several times to ensure uniform impregnation of the nanocatalyst onto the zeolite surface. Subsequently, the solvent was removed using a rotary evaporator, and the ligands attached to the particles were eliminated by calcination in air atmosphere at 450 °C for 3 h using a muffle furnace.

2.2.3. Polyethylene Upcycling

PE upcycling was conducted using a batch reactor with mechanical stirring, wherein the pressure and catalyst amount were controlled. To ensure the reaction took place under inert condition, the reactor was purged with N_2 gas for 10 min to remove oxygen from the interior. Subsequently, H_2 gas was introduced and evacuated up to 15 bar, repeating this process five times to transition the interior to a hydrogen atmosphere. Pressure and temperature were measured using pressure gauges and temperature sensors built into the batch reactor. The reaction was carried out at 350 °C for 2 h under H_2 atmosphere. Pressure-related experiments were conducted by varying the H_2 pressure to 1 bar, 10 bar, and 20 bar. The catalytic efficiency was measured by varying the catalyst amount to 1%, 5%, and 10% by weight while fixing the H_2 pressure at 10 bar and the amount of PE at 5 g.

2.2.4. Characterization of Materials

Transmission electron microscopy (TEM) and energy dispersive spectrometer (EDS) analyses were conducted using a JEOL JEM-2100F instrument (Tokyo, Japan) operating at an accelerating voltage of 200 kV. Fourier transform infrared spectroscopy (ATR FT-IR) measurements were performed using a VERTEX 70 FT-IR spectrometer (Bruker, Ettlingen, Germany). Gel Permeation Chromatography (GPC) was carried out using a Waters GPC system with tetrahydrofuran (THF) as the solvent and a flow rate of 1.0 mL/min. Gas chromatography mass spectroscopy (GC-MS) analysis was performed using the Chromatec Crystal 9000 instrument from CHROMATEC (Yoshkar-Ola, Russia) with a VF-5MS column. Thermogravimetry analysis (TGA) was performed using a Metler Toledo H8-1430KR (Columbus, OH, USA), under N_2 atmosphere with a heating rate of 10 °C/min. Brunauer–Emmett–Teller analysis (BET) was performed using a Quantachrome Instruments ASIQM00002200-7X (Boynton Beach, FL, USA). X-ray diffraction (XRD) was performed using a Bruker D2 phase.

3. Results and Discussion

3.1. Analysis of Nanoparticles and Catalysts

The Ni nanoparticles were synthesized by thermal decomposition of $Ni(acac)_2$–oleylamine complexes in dioctyl ether solvent at 250 °C. Dioctyl ether used as a solvent is degassed before reaction to remove moisture. The formation of $Ni(acac)_2$–oleylamine complexes enables the thermal decomposition of Ni precursors at a moderate temperature [39]. The color change from light green (for the $Ni(acac)_2$) to bluish-green (for the $Ni(acac)_2$–oleylamine) may be attributed to the coordination number changes of Ni from 4 to 6 [40].

The synthesized uniform nickel nanoparticles have a size of 10 nm (10.3 nm ± 0.8 nm) based on TEM analysis using ImageJ program (Figure 2a). TEM images of Ni nanoparticles synthesized using trioctylphosphine as a surfactant are depicted in Figure 2a. The TEM image and corresponding size distribution histogram reveal the high monodispersity of the particles (Figure 2b).

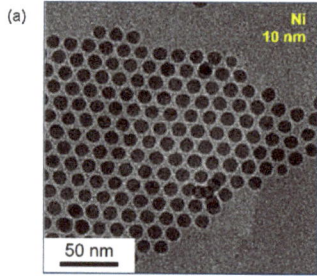
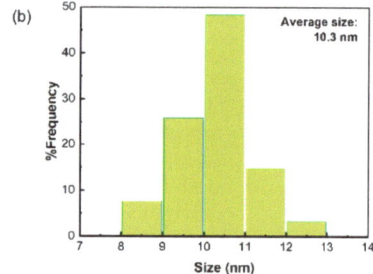

Figure 2. (**a**) TEM image of 10 nm nickel nanoparticles and (**b**) corresponding size distribution.

We mixed the nanoparticles with zeolite before conducting a 450 °C calcination process. Since the surface of the nanoparticles is surrounded by long-chain hydrocarbon ligands, which make it difficult for reactants and catalyst materials to gain access, we conducted a calcination process to remove the ligands.

Scanning transmission electron microscopy (STEM) imaging and energy dispersive analysis (EDS) of the calcined catalysts showed that the Ni nanoparticles were successfully impregnated onto the zeolite support (Figure 3). The presence of nickel nanoparticles on the zeolite surface was confirmed through EDS analysis. It is presumed that the relatively high amount of phosphorus of the EDS data occurs because the phosphorus of trioctylphosphine used as ligands for the nanoparticles remains after calcination. The size of the nickel nanoparticles on the zeolite surface ranged from approximately 10 to 30 nm, indicating that some slight agglomeration occurred during the calcination process, but they still exhibited a smaller size distribution compared to conventional catalysts [41,42].

The X-ray diffraction (XRD) pattern exhibits a Y-zeolite crystal structure with small and broad peaks at 43°, presumably due to the fact that Ni nanoparticles have a polycrystalline structure with a grain size of ~3 nm (Figure S2) [39]. The small and broad XRD peaks of Ni nanoparticles resulting from the polycrystalline nature are enshrouded by the strong XRD peaks from Y-zeolite with a grain size of about 1 μm.

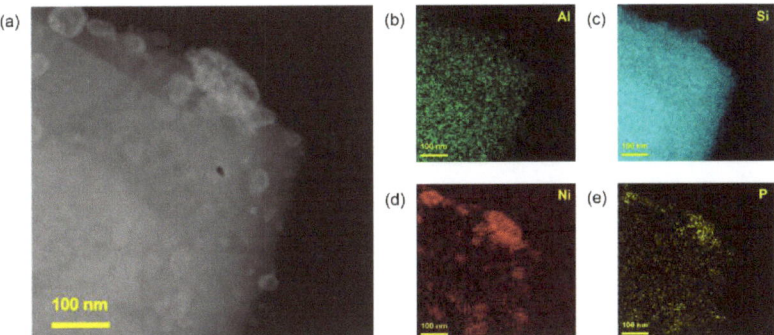

Figure 3. STEM image of Ni on Y-zeolite; weight ratio 1:10. (**a**) Ni on Y-zeolite. EDS data for each element: (**b**) Al, (**c**) Si, (**d**) Ni and (**e**) P.

Through Brunauer–Emmett–Teller (BET) analysis, we measured the surface area of the Y-zeolite before and after nanoparticle loading. The surface area of the pure Y-zeolite was 1283 m^2/g, and decreased to 976 m^2/g after loading 10 nm Ni nanoparticles onto the zeolites, blocking the pores with the incorporated nanoparticles (Figure S1). Both pure Y-zeolite and Ni/zeolite catalysts show a similar average pore size diameter, 3.83 nm, indicating that the structure of the zeolite remained intact even after calcination at 450 °C.

The dispersion (*D*) of catalysts is calculated by the following equation [43–45],

$$D = \frac{6V}{A \times d} \times 100 (\%)$$

where *V* is the atomic volume of a Ni atom, *A* is the surface density of a Ni atom on the Ni(111) plane, and *d* is the average particle size. The dispersion of our Ni-based nanocatalysts is 8.53%.

3.2. Analysis of PE Upcycling
3.2.1. Effect of Hydrogen Pressure on PE Upcycling

To investigate the impact of hydrogen pressure on the PE pyrolysis when using nickel on Y-zeolite catalysts, we carried out the process with different hydrogen pressures of 1 bar, 10 bar, and 20 bar. The obtained products consisted of gas, liquid, and wax phases. Under the 1 bar hydrogen pressure, PE catalytic pyrolysis resulted in 34% wax-phase, 5% liquid-phase, and 61% gas-phase products. Upon increasing the hydrogen pressure to 10 bar, the percentage of gas-phase product increased to 43%, while that of wax-phase product decreased to 50%. We achieved a 25% yield of liquid-phase oil product under the condition of a hydrogen pressure of 20 bar (Figure 4a). The results suggest that increasing pressure enhances the C-C bond cleavage reaction of PE. As hydrogen pressure increases, the molecular weight of PE decreases, leading to a higher yield of liquid-phase and gas-phase hydrocarbons.

FT-IR analysis has the capability of differentiating between PE polymers and pyrolysis products with a low molecular weight. The C-H stretch peaks in the range of 2900–3000 cm^{-1} for hydrocarbons exhibit slightly different peak positions depending on the surrounding environment. The C-H stretch peak for -CH$_2$- in the middle of the hydrocarbon chain appears at 2920 cm^{-1}, while the peak for -CH$_3$ at the chain end is located at 2950 cm^{-1} [46,47]. The difference originates from the weakening of the C-H bond strength when increasing the number of electron-donating alkyl groups. By comparing the peaks at 2950 cm^{-1} and 2920 cm^{-1}, it is possible to determine the qualitative ratio of carbons at end sites of the hydrocarbon chain. In Figure 4b, the FT-IR spectra of PE polymers with a very low number of chain ends show marginal peaks at 2950 cm^{-1} corresponding to -CH$_3$. The intensity of the -CH$_3$ stretch peaks becomes more prominent for the PE upcycling product, indicating the degradation of long polymer chains and an increase in chain ends. Similarly, the -CH$_3$ bend peaks at 1376 cm^{-1} are more distinct in the products after hydrogenation compared to PE, and the intensity of the peak increases at a higher hydrogen pressure. The FT-IR analysis indicates that higher hydrogen concentrations lead to the production of shorter hydrocarbons with lower molecular weights. FT-IR spectra also show the presence of aromatic hydrocarbons.

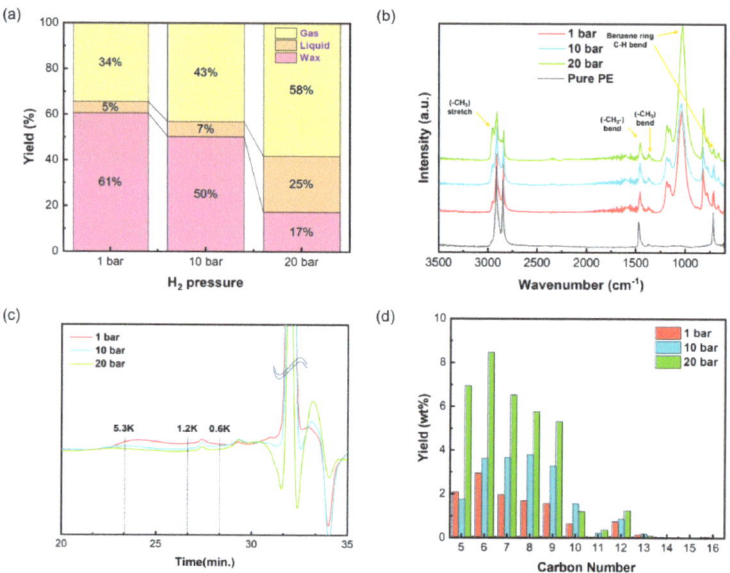

Figure 4. Effect of hydrogen pressure on the PE upcycling. (**a**) Percentage yield of each phase, (**b**) FT-IR spectra, (**c**) GPC data, and (**d**) GC-MS carbon number data of PE upcycling product under different hydrogen pressures of 1 bar, 10 bar, and 20 bar. The PE upcycling was conducted at 350 °C for 2 h, and the catalyst loading amount, when compared to the polymer, was 10 wt%.

We also conducted gel permeation chromatography (GPC) of the obtained liquid and solid products to identify the molecular weight range of the pyrolyzed product under different hydrogen pressures. When the catalytic pyrolysis was conducted under 1 bar hydrogen pressure, the resulting product showed a GPC peak starting at 22 min, which is expected to contain molecules with a molecular weight ranging from 1000 Da to 5000 Da. In contrast, when the degradation was carried out under 10 bar and 20 bar pressure, almost no peak could be identified at the same range, suggesting that degradation occurred into even lower molecular weight hydrocarbons at higher pressures. The system GPC peak at around 32 min for the pyrolysis product at the reaction condition of 20 bar hydrogen pressure indicates the generation of a significant amount of low-molecular-weight hydrocarbons (Figure 4c).

The obtained liquid products were analyzed using gas chromatography-mass spectrometry (GC-MS). The results of the analysis of the products obtained at 1 bar, 10 bar, and 20 bar pressures are shown in Figure 4d. The liquid products contained hydrocarbons ranging from C5 to C16, with a predominance of products closer to C9. Considering that the hydrocarbons from C5 to C12 are utilized as gasoline, the liquid products obtained at 1 bar, 10 bar, and 20 bar pressures primarily consist of gasoline-range hydrocarbons with percentages of 97.75%, 98.97%, and 99.69%, respectively [9,48]. The quantitative analysis indicates that our catalyst exhibits a high selectivity for hydrocarbons in the gasoline range. Additionally, the negligible amount of hydrocarbons with carbon numbers higher than C13 suggests that the catalyst effectively breaks down long-chain PE into hydrocarbons within the fuel range.

The FT-IR, GPC, and GC-MS analyses indicate that a higher degradation rate is observed under the condition of a high hydrogen pressure, suggesting that the hydrogen-involved process may play a key role in the pyrolysis mechanism. The PE degradation mechanism occurs in accordance with Scheme 1. The degradation process of PE using the nickel/zeolite catalysts involves the formation of radicals on the PE polymer chain through Lewis-acidic Y-zeolite [49], the adsorption of hydrogen onto the nickel surfaces, and the generation of short chains by reacting adsorbed hydrogen and radicals. When inferring that more degradation occurs at a higher hydrogen pressure, one sees that the adsorption of hydrogen or the reaction between adsorbed hydrogen and radicals may be the rate-determining step. The hydrogen binding energy of nickel and platinum is 3.07 eV and 0.094 eV, respectively. The higher binding energy of nickel compared to platinum suggests that the chemisorption process could contribute to differences in reaction rates [50,51]. The effect of hydrogen pressure varies with different catalyst systems, potentially due to differences in adsorption and desorption energy. Further studies are required to gain a deeper understanding of the catalytic mechanism [52].

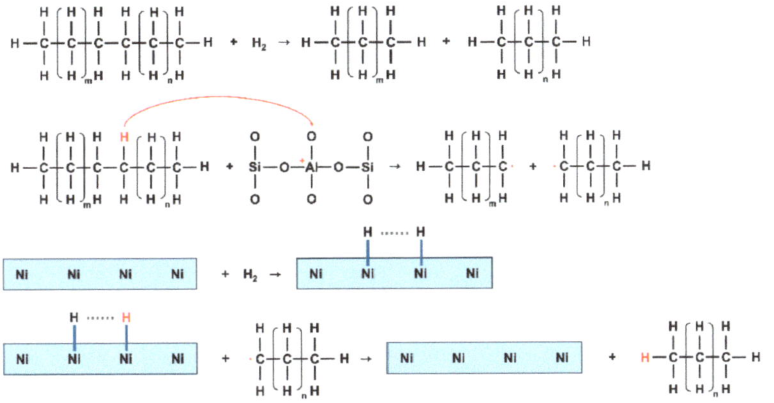

Scheme 1. Suggested PE degradation mechanism using zeolite and nickel nanoparticles. Symbols highlighted in red indicate atoms or electrons involved in reactions.

3.2.2. Effect of Catalyst Amount on PE Upcycling

To investigate how the quantity of the catalysts affects the degradation of PE, PE pyrolysis experiments were conducted by changing the Ni/Y-zeolite catalyst to PE ratio, as follows: 0 wt%, 1 wt%, 5 wt%, and 10 wt%. The reactions were carried out at 350 °C and 10 bar hydrogen pressure for 2 h.

The pyrolysis experiment involving a control of the catalyst amount elucidated the effect of catalysts on the pyrolysis temperature. Without catalyst, the pyrolysis temperature of 350 °C proves inadequate for breaking down the resilient C-C bond, as shown in Figure 5a. After the reaction at 350 °C for 2 h, 93.4% of PE polymers remain in solid form, with only 6.6% of polymer chains undergoing cleavage and converting to gas form. Conversely, in the presence of catalysts under the same reaction conditions, most polymer molecules are converted to gas, liquid, and wax forms. The high degradation temperature of PE is confirmed by TGA (Figure S3). TGA data of PE polymers under N_2 atmosphere indicate that PE degradation initiates at 400 °C, which is 50 °C higher than the degradation temperature observed in the presence of the Ni/Y-zeolite catalysts. These findings underscore the significant role of the Ni-based catalysts in reducing the activation energy required for C-C bond cleavage.

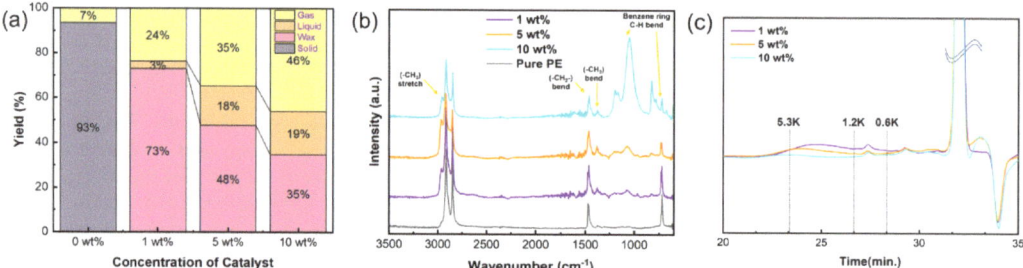

Figure 5. (**a**) Percentage yield of each phase, (**b**) FT-IR spectra, and (**c**) GPC data of PE upcycling product at different catalyst wt%: 1 wt%, 5 wt%, and 10 wt%.

The portion of oily liquid-phase product among the products obtained after pyrolysis of PE with 1 wt% of catalysts was only 2% (Figure 5a). This indicates that with low amounts of catalysts, the C-C bond cleavage reaction does not occur sufficiently to produce oil. With a higher catalyst loading of 10 wt%, the oil-phase liquid fraction escalated to 7% of the products resulting from the pyrolysis reaction, alongside a noticeable increase in gas-phase products, presumably due to the catalyst-induced end-cleavage reactions. Considering that the oil-phase product is the most valuable out of the three phases, the nickel/Y-zeolite catalyst impregnated with 10 wt% seems to offer economic benefits. Adjusting catalyst quantities proves conducive to specific product outcomes: a large wt% of the catalyst favors the natural gas production of PE, while a low wt% of the catalyst ratio would be preferable for obtaining wax-phase products.

The extent of the hydrocarbon chain cleavage of PE was assessed via the IR spectrum of the product. An increase in the proportion of carbon at end sites compared to pure PE was confirmed through a comparison of peaks at 2950 cm^{-1} and 2930 cm^{-1} (Figure 5b). Interestingly, a distinctive peak positioned at 1035 cm^{-1} was observed for the IR spectrum of catalytically pyrolyzed product with a high wt% of the nickel/zeolite catalysts. The catalytic pyrolysis of PE also yielded aromatic compounds, evident from the appearance of the benzene ring, as indicated by the 1035 cm^{-1} peak in the IR spectra.

The degradation of PE to low-molecular-weight carbon is also confirmed by the solubility change and GPC analysis. While the reactants, high-molecular-weight PE, are insoluble in any solvent, such as tetrahydrofuran (THF), the products, hydrocarbon molecules resulting after the catalytic pyrolysis reaction, are well dissolved in the THF solution, indicating the cleavage of PE polymer chains. The GPC data of upcycled PE using 1 wt% nickel/zeolite catalysts show a strong peak in the 1000 to 5000 Da range. The highest molecular weight of pyrolyzed products is 5300 Da. Based on the GPC analysis, we conclude that a significant portion of the PE polymer molecules has been cleaved into low-molecular-weight hydrocarbons. Furthermore, the wax form of the solid product implies a significant reduction in mechanical strength compared to the initial PE. The phenomenon is corroborated by the GPC results, which show the low molecular weights of the product.

4. Conclusions

In this study, nickel-based catalysts for PE upcycling were prepared by loading nickel nanoparticles onto zeolite supports. PE polymer chains were successfully decomposed into gas, liquid, and wax phases at 350 °C by using the nickel/zeolite catalysts. As the hydrogen pressure and catalyst amount increased, the proportion of gas- and liquid-phase products increased. PE upcycling using the nickel/zeolite nanocatalysts enabled the conversion of plastic waste into high-value liquid products at a relatively moderate pyrolysis temperature. Overall, our research highlights the potential of nickel-nanoparticle-based catalysts supported on zeolites to efficiently upcycle PE at lower temperatures, thereby contributing to the advancement of sustainable strategies for plastic waste management.

Supplementary Materials: The following supporting information can be downloaded at: https://www.mdpi.com/article/10.3390/ma17081863/s1, Figure S1. Absorption (black)-desorption (red) nitrogen isotherms of the (a) Pure Y-zeolite (surface area 1307.960 m^2/g), (b) Nickel on Y-zeolite (surface area 1079.580 m^2/g). Figure S2. X-ray diffraction spectra of nickel on Y-zeolite (red) and Y-zeolite (black). Due to nickel's polycrystalline structure, its XRD peak appears smaller and less prominent compared to that of zeolite, which forms a crystalline structure. Figure S3. Thermogravimetric curve of pure LDPE at a heating rate of 10 °C/min under N$_2$ atmosphere.

Author Contributions: Conceptualization, B.H.K.; methodology, H.C., A.J., Y.K., E.L. and B.H.K.; formal analysis, H.C., S.J.K., J.J.S. and B.H.K.; investigation, B.H.K.; resources, B.H.K.; data curation, H.C. and B.H.K.; writing—original draft preparation, H.C. and B.H.K.; writing—review and editing, H.C. and B.H.K.; visualization, B.H.K.; supervision, B.H.K.; project administration, B.H.K.; funding acquisition, B.H.K. All authors have read and agreed to the published version of the manuscript.

Funding: This research was funded by National Research Foundation of Korea (NRF), grant number NRF-2021R1C1C1014339.

Institutional Review Board Statement: Not applicable.

Informed Consent Statement: Not applicable.

Data Availability Statement: Data are contained within the article and Supplementary Materials.

Conflicts of Interest: The authors declare no conflict of interest.

References

1. Jia, C.; Xie, S.; Zhang, W.; Intan, N.N.; Sampath, J.; Pfaendtner, J.; Lin, H. Deconstruction of High-Density Polyethylene into Liquid Hydrocarbon Fuels and Lubricants by Hydrogenolysis over Ru Catalyst. *Chem. Catal.* **2021**, *1*, 437–455. [CrossRef]
2. Barbosa, A.S.; Biancolli, A.L.; Lanfredi, A.J.C.; Rodrigues, O.; Fonseca, F.C.; Santiago, E.I. Enhancing the Durability and Performance of Radiation-Induced Grafted Low-Density Polyethylene-Based Anion-Exchange Membranes by Controlling Irradiation Conditions. *J. Membr. Sci.* **2022**, *659*, 120804. [CrossRef]
3. Geyer, R.; Jambeck, J.R.; Law, K.L. Production, Use, and Fate of All Plastics Ever Made. *Sci. Adv.* **2017**, *3*, e1700782. [CrossRef] [PubMed]
4. Kamal, R.S.; Shaban, M.M.; Raju, G.; Farag, R.K. High-Density Polyethylene Waste (Hdpe)-Waste-Modified Lube Oil Nanocomposites as Pour Point Depressants. *ACS Omega* **2021**, *6*, 31926–31934. [CrossRef] [PubMed]
5. Shah, A.A.; Hasan, F.; Hameed, A.; Ahmed, S. Biological Degradation of Plastics: A Comprehensive Review. *Biotechnol. Adv.* **2008**, *26*, 246–265. [CrossRef] [PubMed]
6. Kunwar, B.; Cheng, H.N.; Chandrashekaran, S.R.; Sharma, B.K. Plastics to Fuel: A Review. *Renew. Sustain. Energy Rev.* **2016**, *54*, 421–428. [CrossRef]
7. Diaz-Silvarrey, L.S.; Zhang, K.; Phan, A.N. Monomer Recovery through Advanced Pyrolysis of Waste High Density Polyethylene (HDPE). *Green Chem.* **2018**, *20*, 1813–1823. [CrossRef]
8. Bagri, R.; Williams, P.T. Catalytic Pyrolysis of Polyethylene. *J. Anal. Appl. Pyrolysis* **2002**, *63*, 29–41. [CrossRef]
9. Panda, A.K.; Singh, R.K.; Mishra, D.K. Thermolysis of Waste Plastics to Liquid Fuel a Suitable Method for Plastic Waste Management and Manufacture of Value Added Products—A World Prospective. *Renew. Sustain. Energy Rev.* **2010**, *14*, 233–248. [CrossRef]
10. Li, L.; Luo, H.; Shao, Z.; Zhou, H.; Lu, J.; Chen, J.; Huang, C.; Zhang, S.; Liu, X.; Xia, L.; et al. Converting Plastic Wastes to Naphtha for Closing the Plastic Loop. *J. Am. Chem. Soc.* **2023**, *145*, 1847–1854. [CrossRef]
11. Hussein, Z.A.; Shakor, Z.M.; Alzuhairi, M.; Al-Sheikh, F. The Yield of Gasoline Range Hydrocarbons from Plastic Waste Pyrolysis. *Energy Sources Part A Recovery Util. Environ. Eff.* **2022**, *44*, 718–731. [CrossRef]
12. Vance, B.C.; Najmi, S.; Kots, P.A.; Wang, C.; Jeon, S.; Stach, E.A.; Zakharov, D.N.; Marinkovic, N.; Ehrlich, S.N.; Ma, L.; et al. Structure–Property Relationships for Nickel Aluminate Catalysts in Polyethylene Hydrogenolysis with Low Methane Selectivity. *JACS Au* **2023**, *3*, 2156–2165. [CrossRef] [PubMed]
13. Schyns, Z.O.; Shaver, M.P. Mechanical Recycling of Packaging Plastics: A Review. *Macromol. Rapid Commun.* **2020**, *42*, e2000415. [CrossRef] [PubMed]
14. Kim, R.; Delva, L.; Van Geem, K. Mechanical and Chemical Recycling of Solid Plastic Waste. *Waste Manag.* **2017**, *69*, 24–58.
15. Rahimi, A.; García, J.M. Chemical Recycling of Waste Plastics for New Materials Production. *Nat. Rev. Chem.* **2017**, *1*, 0046. [CrossRef]
16. Jehanno, C.; Alty, J.W.; Roosen, M.; De Meester, S.; Dove, A.P.; Chen, E.Y.-X.; Leibfarth, F.A.; Sardon, H. Critical Advances and Future Opportunities in Upcycling Commodity Polymers. *Nature* **2022**, *603*, 803–814. [CrossRef]
17. Jin, H.; Gonzalez-Gutierrez, J.; Oblak, P.; Zupančič, B.; Emri, I. The Effect of Extensive Mechanical Recycling on the Properties of Low Density Polyethylene. *Polym. Degrad. Stab.* **2012**, *97*, 2262–2272. [CrossRef]

18. Onwudili, J.A.; Insura, N.; Williams, P.T. Composition of Products from the Pyrolysis of Polyethylene and Polystyrene in a Closed Batch Reactor: Effects of Temperature and Residence Time. *J. Anal. Appl. Pyrolysis* **2009**, *86*, 293–303. [CrossRef]
19. Kiran, N.; Ekinci, E.; Snape, C.E. Recyling of Plastic Wastes via Pyrolysis. *Resour. Conserv. Recycl.* **2000**, *29*, 273–283. [CrossRef]
20. Al-Salem, S.M.; Lettieri, P.; Baeyens, J. Recycling and Recovery Routes of Plastic Solid Waste (PSW): A Review. *Waste Manag.* **2009**, *29*, 2625–2643. [CrossRef]
21. Rauert, C.; Pan, Y.; Okoffo, E.D.; O'Brien, J.W.; Thomas, K.V. Extraction and Pyrolysis-GC-MS Analysis of Polyethylene in Samples with Medium to High Lipid Content. *J. Environ. Expo. Assess.* **2022**, *1*, 13. [CrossRef]
22. Zhang, W.; Kim, S.; Wahl, L.; Khare, R.; Hale, L.; Hu, J.; Camaioni, D.M.; Gutiérrez, O.Y.; Liu, Y.; Lercher, J.A. Low-Temperature Upcycling of Polyolefins into Liquid Alkanes via Tandem Cracking-Alkylation. *Science* **2023**, *379*, 807–811. [CrossRef] [PubMed]
23. Zhang, F.; Zeng, M.; Yappert, R.D.; Sun, J.; Lee, Y.-H.; LaPointe, A.M.; Peters, B.; Abu-Omar, M.M.; Scott, S.L. Polyethylene Upcycling to Long-Chain Alkylaromatics by Tandem Hydrogenolysis/Aromatization. *Science* **2020**, *370*, 437–441. [CrossRef] [PubMed]
24. Wang, C.; Xie, T.; Kots, P.A.; Vance, B.C.; Yu, K.; Kumar, P.; Fu, J.; Liu, S.; Tsilomelekis, G.; Stach, E.A.; et al. Polyethylene Hydrogenolysis at Mild Conditions over Ruthenium on Tungstated Zirconia. *JACS Au* **2021**, *1*, 1422–1434. [CrossRef] [PubMed]
25. Celik, G.; Kennedy, R.M.; Hackler, R.A.; Ferrandon, M.; Tennakoon, A.; Patnaik, S.; LaPointe, A.M.; Ammal, S.C.; Heyden, A.; Perras, F.A.; et al. Upcycling Single-Use Polyethylene into High-Quality Liquid Products. *ACS Cent. Sci.* **2019**, *5*, 1795–1803. [CrossRef] [PubMed]
26. Wang, Y.-Y.; Tennakoon, A.; Wu, X.; Sahasrabudhe, C.; Qi, L.; Peters, B.G.; Sadow, A.D.; Huang, W. Catalytic Hydrogenolysis of Polyethylene under Reactive Separation. *ACS Catal.* **2024**, *14*, 2084–2094. [CrossRef]
27. Vance, B.C.; Kots, P.A.; Wang, C.; Granite, J.E.; Vlachos, D.G. Ni/SiO$_2$ Catalysts for Polyolefin Deconstruction via the Divergent Hydrogenolysis Mechanism. *Appl. Catal. B Environ.* **2023**, *322*, 122138. [CrossRef]
28. Medford, A.J.; Vojvodic, A.; Hummelshøj, J.S.; Voss, J.; Abild-Pedersen, F.; Studt, F.; Bligaard, T.; Nilsson, A.; Nørskov, J.K. From the Sabatier Principle to a Predictive Theory of Transition-Metal Heterogeneous Catalysis. *J. Catal.* **2015**, *328*, 36–42. [CrossRef]
29. Arroyave, A.; Cui, S.; Lopez, J.C.; Kocen, A.L.; LaPointe, A.M.; Delferro, M.; Coates, G.W. Catalytic Chemical Recycling of Post-Consumer Polyethylene. *J. Am. Chem. Soc.* **2022**, *144*, 23280–23285. [CrossRef]
30. Ding, W.B.; Tuntawiroon, W.; Liang, J.; Anderson, L.L. Depolymerization of Waste Plastics with Coal over Metal-Loaded Silica-Alumina Catalysts. *Fuel Process. Technol.* **1996**, *49*, 49–63. [CrossRef]
31. Zhao, Z.; Li, Z.; Zhang, X.; Li, T.; Li, Y.; Chen, X.; Wang, K. Catalytic Hydrogenolysis of Plastic to Liquid Hydrocarbons over a Nickel-Based Catalyst. *Environ. Pollut.* **2022**, *313*, 120154. [CrossRef]
32. Si, X.; Chen, J.; Wang, Z.; Hu, Y.; Ren, Z.; Lu, R.; Liu, L.; Zhang, J.; Pan, L.; Cai, R.; et al. Ni-Catalyzed Carbon–Carbon Bonds Cleavage of Mixed Polyolefin Plastics Waste. *J. Energy Chem.* **2023**, *85*, 562–569. [CrossRef]
33. Singh, S.B.; Tandon, P.K. Catalysis: A brief review on nano-catalyst. *J. Energy Chem. Eng.* **2014**, *2*, 106–115.
34. Akubo, K.; Nahil, M.A.; Williams, P.T. Aromatic Fuel Oils Produced from the Pyrolysis-Catalysis of Polyethylene Plastic with Metal-Impregnated Zeolite Catalysts. *J. Energy Inst.* **2019**, *92*, 195–202. [CrossRef]
35. Susastriawan, A.A.P.; Purnomo; Sandria, A. Experimental Study the Influence of Zeolite Size on Low-Temperature Pyrolysis of Low-Density Polyethylene Plastic Waste. *Therm. Sci. Eng. Prog.* **2020**, *17*, 100497.
36. Jiang, C.; Wang, Y.; Luong, T.; Robinson, B.; Liu, W.; Hu, J. Low Temperature Upcycling of Polyethylene to Gasoline Range Chemicals: Hydrogen Transfer and Heat Compensation to Endothermic Pyrolysis Reaction over Zeolites. *J. Environ. Chem. Eng.* **2022**, *10*, 107492. [CrossRef]
37. Somorjai, G.A.; Blakely, D.W. Mechanism of Catalysis of Hydrocarbon Reactions by Platinum Surfaces. *Nature* **1975**, *258*, 580–583. [CrossRef]
38. Tennakoon, A.; Wu, X.; Paterson, A.L.; Patnaik, S.; Pei, Y.; LaPointe, A.M.; Ammal, S.C.; Hackler, R.A.; Heyden, A.; Slowing, I.I.; et al. Catalytic Upcycling of High-Density Polyethylene via a Processive Mechanism. *Nat. Catal.* **2020**, *3*, 893–901. [CrossRef]
39. Cho, H.; Lee, N.; Kim, B.H. Synthesis of Highly Monodisperse Nickel and Nickel Phosphide Nanoparticles. *Nanomaterials* **2022**, *12*, 3198. [CrossRef]
40. Park, J.; Kang, E.; Son, U.; Park, M.; Lee, K.; Kim, J.; Kim, W.; Noh, H.-J.; Park, J.-H.; Bae, C.J.; et al. Monodisperse Nanoparticles of Ni and NiO: Synthesis, Characterization, Self-assembled Superlattices, and Catalytic Applications in the Suzuki Coupling Reaction. *Adv. Mater.* **2005**, *17*, 429–434. [CrossRef]
41. Pang, H.; Yang, G.; Li, L.; Yu, J. Toward the Production of Renewable Diesel over Robust Ni Nanoclusters Highly Dispersed on a Two-Dimensional Zeolite. *NPG Asia Mater.* **2023**, *15*, 24. [CrossRef]
42. Barton, R.R.; Carrier, M.; Segura, C.; Fierro, J.L.G.; Escalona, N.; Peretti, S.W. Ni/HZSM-5 Catalyst Preparation by Deposition-Precipitation. Part 1. Effect of Nickel Loading and Preparation Conditions on Catalyst Properties. *Appl. Catal. A Gen.* **2017**, *540*, 7–20. [CrossRef]
43. Escobar, J.; Núñez, S.; Montesinos-Castellanos, A.; de los Reyes, J.A.; Rodríguez, Y.; González, O.A. Dibenzothiophene Hydrodesulfurization over PdPt/Al$_2$O$_3$–TiO$_2$. Influence of Ti-Addition on Hydrogenating Properties. *Mater. Chem. Phys.* **2016**, *171*, 185–194.
44. Rios-Escobedo, R.; Ortiz-Santos, E.; Colín-Luna, J.A.; de León, J.N.D.; del Angel, P.; Escobar, J.; de los Reyes, J.A. Anisole Hydrodeoxygenation: A Comparative Study of Ni/TiO$_2$-ZrO$_2$ and Commercial TiO$_2$ Supported Ni and NiRu Catalysts. *Top. Catal.* **2022**, *65*, 1448–1461. [CrossRef]

45. Beck, I.E.; Bukhtiyarov, V.I.; Pakharukov, I.Y.; Zaikovsky, V.I.; Kriventsov, V.V.; Parmon, V.N. Platinum Nanoparticles on Al_2O_3: Correlation between the Particle Size and Activity in Total Methane Oxidation. *J. Catal.* **2009**, *268*, 60–67. [CrossRef]
46. Marcilla, A.; Beltrán, M.I.; Navarro, R. TG/FT-IR Analysis of HZSM5 and HUSY Deactivation during the Catalytic Pyrolysis of Polyethylene. *J. Anal. Appl. Pyrolysis* **2006**, *76*, 222–229. [CrossRef]
47. Wang, S.; Tang, Y.; Schobert, H.H.; Guo, Y.; Gao, W.; Lu, X. FTIR and Simultaneous TG/MS/FTIR Study of Late Permian Coals from Southern China. *J. Anal. Appl. Pyrolysis* **2013**, *100*, 75–80. [CrossRef]
48. Sarıkoç, S. Fuels of the Diesel-Gasoline Engines and Their Properties. *Diesel Gasol. Engines* **2020**, 31–45. [CrossRef]
49. Palčić, A.; Valtchev, V. Analysis and Control of Acid Sites in Zeolites. *Appl. Catal. A Gen.* **2020**, *606*, 117795. [CrossRef]
50. Hideshi, O.; Wintzer, M.E.; Nakamura, R. Non-Zero Binding Enhances Kinetics of Catalysis: Machine Learning Analysis on the Experimental Hydrogen Binding Energy of Platinum. *ACS Catal.* **2021**, *11*, 6298–6303.
51. Niu, J.; Rao, B.K.; Jena, P. Binding of Hydrogen Molecules by a Transition-Metal Ion. *Phys. Rev. Lett.* **1992**, *68*, 2277–2280. [CrossRef]
52. Laursen, A.B.; Varela, A.S.; Dionigi, F.; Fanchiu, H.; Miller, C.; Trinhammer, O.L.; Rossmeisl, J.; Dahl, S. Electrochemical Hydrogen Evolution: Sabatier's Principle and the Volcano Plot. *J. Chem. Educ.* **2012**, *89*, 1595–1599. [CrossRef]

Disclaimer/Publisher's Note: The statements, opinions and data contained in all publications are solely those of the individual author(s) and contributor(s) and not of MDPI and/or the editor(s). MDPI and/or the editor(s) disclaim responsibility for any injury to people or property resulting from any ideas, methods, instructions or products referred to in the content.

Article

Electrical and Electro-Thermal Characteristics of (Carbon Black-Graphite)/LLDPE Composites with PTC Effect

Eduard-Marius Lungulescu [1,*], Cristina Stancu [2,*], Radu Setnescu [1,3], Petru V. Notingher [2] and Teodor-Adrian Badea [4]

[1] National Institute for Research and Development in Electrical Engineering ICPE-CA, 313 Splaiul Unirii, 030138 Bucharest, Romania; radu.setnescu@icpe-ca.ro
[2] Faculty of Electrical Engineering, University POLITEHNICA of Bucharest, 313 Splaiul Independentei, 060042 Bucharest, Romania; petrunot2000@yahoo.fr
[3] Department of Advanced Technologies, Faculty of Sciences and Arts, Valahia University of Târgoviște, 13 Aleea Sinaia, 130004 Targoviste, Romania
[4] Romanian Research and Development Institute for Gas Turbines COMOTI, 220D Iuliu Maniu Av., 061126 Bucharest, Romania; teodor.badea@comoti.ro
* Correspondence: marius.lungulescu@icpe-ca.ro (E.-M.L.); stcris2003@yahoo.co.uk (C.S.)

Citation: Lungulescu, E.-M.; Stancu, C.; Setnescu, R.; Notingher, P.V.; Badea, T.-A. Electrical and Electro-Thermal Characteristics of (Carbon Black-Graphite)/LLDPE Composites with PTC Effect. *Materials* **2024**, *17*, 1224. https://doi.org/ 10.3390/ma17051224

Academic Editor: Andrea Sorrentino

Received: 29 January 2024
Revised: 29 February 2024
Accepted: 4 March 2024
Published: 6 March 2024

Copyright: © 2024 by the authors. Licensee MDPI, Basel, Switzerland. This article is an open access article distributed under the terms and conditions of the Creative Commons Attribution (CC BY) license (https:// creativecommons.org/licenses/by/ 4.0/).

Abstract: Electrical properties and electro-thermal behavior were studied in composites with carbon black (CB) or hybrid filler (CB and graphite) and a matrix of linear low-density polyethylene (LLDPE). LLDPE, a (co)polymer with low crystallinity but with high structural regularity, was less studied for *Positive Temperature Coefficient* (PTC) applications, but it would be of interest due to its higher flexibility as compared to HDPE. Structural characterization by scanning electron microscopy (SEM) confirmed a segregated structure resulted from preparation by solid state powder mixing followed by hot molding. Direct current (DC) conductivity measurements resulted in a percolation threshold of around 8% (w) for CB/LLDPE composites. Increased filler concentrations resulted in increased alternating current (AC) conductivity, electrical permittivity and loss factor. Resistivity-temperature curves indicate the dependence of the temperature at which the maximum of resistivity is reached ($T_{max(R)}$) on the filler concentration, as well as a differentiation in the $T_{max(R)}$ from the crystalline transition temperatures determined by DSC. These results suggest that crystallinity is not the only determining factor of the PTC mechanism in this case. This behavior is different from similar high-crystallinity composites, and suggests a specific interaction between the conductive filler and the polymeric matrix. A strong dependence of the PTC effect on filler concentration and an optimal concentration range between 14 and 19% were also found. Graphite has a beneficial effect not only on conductivity, but also on PTC behavior. *Temperature* vs. *time* experiments, revealed good temperature self-regulation properties and current and voltage limitation, and irrespective of the applied voltage and composite type, the equilibrium superficial temperature did not exceed 80 °C, while the equilibrium current traversing the sample dropped from 22 mA at 35 V to 5 mA at 150 V, proving the limitation capacities of these materials. The concentration effects revealed in this work could open new perspectives for the compositional control of both the self-limiting and interrupting properties for various low-temperature applications.

Keywords: LLDPE composite; PTC; conductivity; self-regulating temperature; carbon filler

1. Introduction

It is widely known that by introducing a quantity of a conductive material in the form of powder (such as carbon black, graphite, carbon nanotubes, graphene, metallic powders) into a polymeric matrix, the electrical conductivity of the composite material can be approximately 10 orders of magnitude higher than that of the pure polymer [1–4]. While retaining the valuable properties of polymers—mechanical strength, flexibility, chemical inertness and easy processability—conductive polymer composites (CPCs) currently have numerous applications, with those related to electrical engineering arguably being the most significant [5]. These applications include electrodes for fuel cells and electrochemical sources [6], high-capacitance capacitors [2], high voltage power cable shields [1], screens for the electromagnetic protection of electronic equipment and human beings [7,8], radar wave absorption [2], electronic packaging, antistatic fabrics [2,6,9], temperature sensors, current and voltage protection [10], heating elements, etc. [1,2,11–14].

As for the polymeric matrix used to obtain CPCs, a wide range of thermoplastic, thermosetting, or elastomeric polymers, either singularly or in blends, is employed [11,15]. Among all polymers, polyethylene, which holds the largest share in synthetic polymer production, and especially HDPE, is most frequently mentioned as the matrix for CPCs. Various materials, metallic, oxide, or carbon-based, are mentioned as conductive fillers [16]. Among them, carbon materials are of greatest interest because they have a lower density than metals or oxides and comparable electrical conductivity, are more stable to corrosion than metals, and are less susceptible to initiating the oxidation of the polymer matrix. Commonly used carbonaceous conductive fillers include carbon black (CB [10,14]), graphite (Gr), carbon fibers (CF), and carbon nanotubes (CNT, MWCNT) [6]. Carbon black and graphite are widely used to prepare conductive composites because they present high electrical conductivity and chemical stability, are relatively inexpensive and have a good commercial availability. Carbon black has a primary structure composed of spherical nanoparticles of around 30 nm size and imparts easy processability to composites [11,17,18]. However, CB particles are often agglomerated into large aggregates due to strong van der Waals forces, which, combined with the unfavorable influence of aspect ratio [19], result in relatively high percolation threshold values (15–20% w/w). Nevertheless, considerably lower values of the percolation threshold have been reported, even for CB, by creating segregated structures (with conductive material particles unevenly distributed within the polymer matrix), through the use of hybrid fillers or modified particles [20], as well as by using of high-viscosity polymer matrices [11,21].

In general, fillers with high aspect ratio values, such as carbon fibers or CNTs, enable the achievement of low percolation threshold values [19,20]. However, instances of non-percolation are also known, typically caused by weak dispersion of the filler, as seen with acid-treated MWCNTs/PMMA [19].

A more recent trend involves the use of hybrid fillers, consisting of mixtures of particles with different characteristics, especially with different aspect ratio values. Essentially, combining particles with high aspect ratios and other particles (conductive or not) with low aspect ratios and microscopic dimensions creates so-called excluded volumes in the polymeric matrix, with a favorable effect on increasing electrical conductivity and achieving lower percolation thresholds [2]. Such structures are also considered segregated, meaning that conductive particles are unevenly distributed on a microscopic scale around islands of low aspect ratio particles, and electrical conduction can be described by combining percolation theory and the Voronoi geometric model [2].

The methods for dispersing fillers In the polymeric matrix applicable to thermoplastic matrices include melt mixing [19,21,22], solution mixing [9], dry mixing of components [23], in situ polymerization [11] and others. Subsequently, items with the desired shape are usually formed by hot press molding.

A distinct category of CPCs is composed of CPCs that present an increase in electrical resistivity with increasing temperature, i.e., the PTC effect (positive temperature coefficient) and, in particular, an important jump in resistivity (of several orders of magnitude), which results in the transition of the material from a conductor to an insulator. If the temperature exceeds the value corresponding to the maximum resistivity, a decrease in resistivity can be observed (NTC effect). Beyond the temperature corresponding to the maximum resistivity, a decrease in resistivity (*Negative Temperature Coefficient*—NTC effect) can be observed upon further heating.

A plausible explanation of the PTC behavior of these materials is based on depercolation near a structural transition point of the polymer matrix. This effect is very clear in the case of polymer matrices with high crystallinity, such as HDPE, for which the temperature of reaching the maximum resistivity of the composite is very close to or coincides with the peak temperature of the melting endotherm of the polymer [22]. For other systems, the resistivity jump can be correlated with the glass transition, as for example in the case of epoxy matrices, or with the activation of the motion of some polymer chain segments, as in the case of amorphous matrix systems [22]. The amplitude of the resistivity jump depends on the concentration and distribution of the filler material [11]. Such materials, especially those with no or a low NTC effect, are interesting for applications where self-limiting electrical power is involved, such as self-regulating heating elements, current limiters, overcurrent protection, micro-switches, sensors, etc. [18,23].

Other CPCs often exhibit a monotonous decrease in resistivity with temperature. This effect, called NTC (negative temperature coefficient) does not enable the use of materials in self-limiting applications, but makes them useful for other applications, such as resistive temperature sensors [24].

In the design of CPC and PTC materials, the study of the dependence of the electrical conductivity on the concentration of the conductive charge is of particular importance, as it allows the rational dosage of the components as well as the control of the properties of the resulting composites. Typically, the graphical representation of the dependence of ρ as a function of the conductive filler concentration ϕ (expressed as either volume or mass fraction) is presented as a sigmoidal curve on which three important regions are distinguished (see Figure S1 in the Supplementary Materials and [19] for typical conductivity or resistivity vs. ϕ curves, respectively), namely: (i) an initial region, in which although the content of conductive charge added to the polymer increases, the resistivity decreases very little, so that the material practically remains an insulator; (ii) upon reaching a certain concentration, called the critical concentration (ϕ_{cr}), the resistivity drops suddenly, by several orders of magnitude, and the material becomes electrically conductive; and (iii) for $\phi > \phi_{cr}$, increasing in conductive filler concentration results in a slow increase in conductivity. The specific shape of the σ (or ρ) vs. ϕ curves may suggest the mechanism of electrical conduction in the composites. Thus, it is considered that when electric percolation is reached, a sufficiently large number of conductive paths are formed that ensure the passage of the electric current through the sample. A theory [25] considers that all the particles of a conductive path must be in direct physical contact to ensure the continuity of the path, while another theory [26,27] admits that between any two consecutive particles of the conductive phase, a gap (whose maximum width has been estimated at 2 nm) that electrons can traverse through a tunneling effect could exist. In both theories, the conductor–insulator transition specific to PTC materials is due to the appearance of, respectively, an increase in the gap between the conductive particles that leads to the interruption of direct contact, or an increase in the gap width over the tunelation limit. It was demonstrated that either one of the mechanisms can dominate depending on the composition of the material: for high values of the filler concentration $\phi > \phi_{cr}$, the electrical conductivity satisfies the law of electrical conduction and it can be assumed that the conductive particles are in direct contact, while for values of $\phi < \phi_{cr}$, electron tunneling predominates [28].

In this study, the influence of the composition on the DC and AC electrical conductivity properties, the variation of electrical resistivity with temperature, and the electro-thermal effect of a composite with a hybrid conductive filler (CB + Gr) and LLDPE matrix were investigated. To our knowledge, LLDPE is less studied as a matrix for PTC materials, although it could be of interest due to its low crystallinity, high structural regularity, high flexibility and elevated melting temperature, close to that of HDPE [24,29]. We also demonstrated the strong effects of filler concentration on both the PTC intensity and transition temperature (from PTC to NTC) of these materials. The combination of CB and graphite used to impart the electrical conductivity of these materials would be of practical interest due to the economic effectiveness and wide availability of these materials.

2. Materials and Methods

2.1. Materials

Linear low-density polyethylene (LLDPE) powder with density 0.935 g/cm^3 and MFI 5.0 m/10 min at 190 °C/2.16 kg (producer data), type RX 806 Natural from Resinex (Bucharest, Romania), was used, in the as-received state, as a polymer matrix. Carbon black (Fast Extruder Furnace—FEF type) and natural graphite flakes (CR10) were used as conductive fillers (see more details in [30]). Irganox 1010 (pentaerythritol tetrakis(3-(3,5-di-tert-butyl-4-hydroxyphenyl) propionate) was used in concentration of 0.05% (w/w) for matrix stabilization against oxidation.

The list of abbreviations used in this manuscript is presented in Table 1.

Table 1. List of abbreviations and symbols.

Abbreviation	Full Name/Description
AC	Alternating Current
CB	Carbon Black
(CB, Gr)/LLD	Composites with LLDPE matrix and CB and Gr
CF	Carbon Fibers
CPC	Conductive Polymer Composite
DC	Direct Current
DSC	Differential Scanning Calorimetry
ΔH	Transition enthalpy (from DSC)
ε', ε''	complex relative permitivities (AC)
FTIR	Fourier Transform InfraRed spectroscopy
ϕ	Fraction (either mass or volume) of conductive filler within the composite
ϕ_c	Critical concentration of the filler
Gr	Graphite
HDPE	High Density Polyethylene
LDPE	Low Density Polyethylene
LLDPE	Linear Low-Density Polyethylene
NTC	Negative Temperature Coefficient
PTC	Positive Temperature Coefficient
R	Electrical resistance
ρ	Electrical resistivity
ρ_{DC}	Direct current resistivity
ρ_V	Volume resistivity
RMS voltage	Root Mean Square voltage (effective voltage = 0.707 of peak voltage, in AC measurements)
σ', σ''	complex conductivities (AC)
σ_{dc}	Direct current conductivity
SEM	Scanning Electron Microscopy
T_c, $T_{c(DSC)}$ (see the footnote for T_m)	peak temperature of crystallization endotherm (in DSC)
Teq	Equilibrium temperature (T_{eq} denotes the practically constant value of the surface temperature reached after few minutes of sample exposure to electric field, in T vs. t, U measurements)
T_m; $T_{m(DSC)}$ (The notation $T_{m(DSC)}$ is used for better distinguish between the maximum of temperatures in either DSC and R vs. T measurements)	Peak temperature of crystallinity melting in DSC
T_{offset}	Offset temperature in DSC or R vs. T-heating measurements
T'_{offset}	Offset temperature in R vs. T-cooling measurements
T_{onset}	Onset temperature in DSC or R vs. T-heating measurements
UHMWPE	Ultra-High Molecular Weight Polyethylene

2.2. Preparation of Composites and Samples for Measurements

Composites were obtained through hot pressing of physical mixtures of polymer particles (LLDPE) and carbonaceous particles of carbon black and graphite, following a previously described procedure [31]. The dispersion of conductive particles was initially achieved through physical pre-mixing, followed by high-shear solid-state mixing applied to the pre-mixed system. Composite sample formation was conducted by hot pressing (at 160 °C, 6.5 bar) in a rectangular mold (190 × 120 × 0.7 mm) using a hydraulic press. Cooling was performed under pressure at a slow rate of approximately 1 °C/min until reaching 60 °C, followed by pressure release and natural cooling to ambient temperature.

The code names of the prepared composite samples are shown in Table 2. The composition of these materials is also indicated within the table.

Table 2. Code names and composition of the studied samples.

| Sample Code | Polymeric Matrix | Filler | | Total C (%, w) |
	LLDPE (%, w)	Carbon Black (%, w)	Graphite (%, w)	
LLD 0	100 (neat)	0	0	0
LLD 44	92	4	4	8
LLD 80	92	8	0	8
LLD 82	90	8	2	10
LLD 100	90	10	0	10
LLD 120	88	12	0	12
LLD 122	86	12	2	14
LLD 140	86	14	0	14
LLD 142	84	14	2	16
LLD 160	84	16	0	16
LLD 162	82	16	2	18
LLD 190	81	19	0	19
LLD 192	79	19	2	21

The measurement of electrical properties was conducted on plate-shaped samples (with approximate dimensions of 100 × 100 × 0.7 mm) obtained from the plates resulting from molding. The measurements of electro-thermal properties and the temperature dependence of resistivity were performed on specimens with dimensions of 30 × 25 × 0.7 mm, obtained from plates similar to those used for electrical property measurements. Prior to measurements, the extremities of each sample were covered on both faces by conductive silver paste. The width of these conductive traces was 5 mm. Finally, a copper foil was tightly fixed on the silver conductive traces, forming the electrodes of each measured specimen.

2.3. Sample Conditioning

Before the electrical measurements, all samples were thermally preconditioned in an oven for 72 h at a temperature of 50 °C. After preconditioning, the thickness of each sample was measured at 5 points (in the center and in the 4 corners). The average value of the thickness of the samples was 0.713 ± 0.028 mm.

A similar pre-conditioning treatment, i.e., 72 h at 30 ± 1 °C, 50 ± 5% r.h., was applied before other measurements described below (namely FTIR, DSC, R vs. T and T vs. t, U).

3. Instruments and Methods

3.1. Scanning Electron Microscopy

The microstructural and morphological analyses of specimens were performed on secondary electron images (ETD detector—Everhart Thornley Detector, FEI Company, Hillsboro, OR, USA) by scanning electron microscopy using a FEI F50 Inspect instrument. The analysis was performed on the specimen's cross-section under the following conditions: acceleration voltage 10 kV, acquisition time 50 s, spot size 3 nm and working distance 6.0–6.9 mm.

3.2. Differential Scanning Calorimetry

DSC thermograms of the blank LLDPE sample and (CB-Gr)/LLDPE composites were recorded using a Setaram 131 evo instrument (Setaram, Caluire-et-Cuire, France), employing 30 µL aluminum crucibles with pierced lids. The mass of each sample was 4.5 ± 0.1 mg, except for the pure LLDPE, which were 3.6 ± 0.1 mg. The samples were heated at a constant rate (10 °C/min) from 35 °C to 200 °C under a nitrogen flow (50 mL/min), after maintaining 5 min at 200 °C, the sample was cooled back under nitrogen at a rate of 10 °C/min till 35 °C. Two such cycles were applied for each sample, and measurements were conducted in duplicate.

On the heating curve the onset temperature (T_{onset}), peak maximum temperature (T_m), and offset temperature (T_{offset}) of the endothermic peak were determined, along with the melting enthalpy (ΔH_m). Similarly, on the cooling curve, the parameters of the solidification endotherm were determined, namely the onset temperature (T'_{onset}), peak maximum temperature (T_c), offset temperature (T'_{offset}) and the crystallization enthalpy of (ΔH_c). Throughout the work, especially when comparing R vs. T data, the terms $T_{m(DSC)}$ and $T_{c(DSC)}$ were used instead of T_m and T_c, respectively, in order to emphasize the DSC origin of these data.

The degree of crystallinity (Cr (%)) was calculated using a simple formula [32]:

$$Cr\ (\%) = 100 \cdot \Delta H_{cor} / \Delta H_{100\%} \tag{1}$$

where ΔH_{cor} is the latent heat of fusion ΔH corrected with the polymer percentage (p) in the composite.

$$\Delta H_{cor} = \Delta H \cdot p / 100 \tag{2}$$

The $\Delta H_{100\%}$ = 279 (J/g) represents the melting enthalpy of 100% crystalline polyethylene [32]. The data processing procedure for DSC was described in more detail in previous works [30,31,33].

3.3. FTIR Spectroscopy

The infrared spectra were recorded in ATR for both composites and the LLDPE base polymer. The spectral range was 4000–400 cm^{-1}, resolution 2 cm^{-1}. For each measurement, 48 scans were performed. For the carbon-containing samples, a baseline correction before the main peaks was necessary.

The calculation on the spectra, including peak wavelength, peak absorbance and spectra comparison were performed by specific applications Spectra Manager (Jasco, Tokyo, Japan) and Essential FTIR (Biorad, Tokyo, Japan).

3.4. DC Measurements

To determine the direct current (DC) conductivity, an apparatus described earlier [34] was employed. The setup comprised a Keithley model 6517 B electrometer (Keithley Instruments Inc., Cleveland, OH, USA), a Keithley 8009 measurement cell and a PC, supplemented with a Trade Raypa oven with forced air circulation (and adjustable temperature between 30 and 250 °C) in which the measurement cell was placed for measurements at different temperatures.

The values of the absorption currents (I_a) were measured for 600 s, for two different applied voltage values (U_0), namely 1 V and 100 V. The DC electrical conductivity was calculated with the following equation:

$$\sigma_{DC} = \frac{I_a}{U_0} \times \frac{g}{A} \tag{3}$$

where: g is the thickness of the planar sample and $A = \pi D^2/4$ is the area of the upper cylindrical electrode with diameter D.

3.5. AC Measurements

The experimental determinations of the components of the complex relative permittivites (ε_r' and ε_r''), complex conductivities (σ' and σ'') and the loss factor (tgδ) were performed with an Alpha-A Novocontrol impedance analyzer from Novocontrol Technologies GmbH & Co. KG, Montabaur, Germany. The real part of the complex conductivity (σ_{ca})

was determined for all samples at the RMS voltage value $U_{RMS} = U = 1$ V and frequencies ranging from 1 Hz to 500 MHz.

The study of temperature dependence (T) of the electrical resistivity (ρ) was conducted by measuring the electrical resistance of the specimens (R) at varying temperatures, using the experimental setup already presented [32], on specimens removed from the plate, with a 20 mm space between electrodes. Since, in this case, the electrodes were applied across the whole surface at the ends of the specimen, the experimentally measured value was volume resistance. As $\rho_V = RA/g$, where A is the specimen surface and g its thickness, the curves ρ_V vs. T and R vs. T are identical. Hence, below, the curves R vs. T will be analyzed.

For obtaining the R vs. T curves, the specimens were placed in a Memmert oven, and the temperature was programmed to increase at a heating rate of 1 °C/min. The electrical resistance was measured using a multimeter positioned outside the oven and connected to the specimen through conductors. For high resistivity specimens, an insulation tester of type UNI-T UT512 (Uni-trend Technology Co., Ltd., Dongguan, China) was used. Parameters characterizing the variation of resistivity during the heating of the samples are shown in Figure 1.

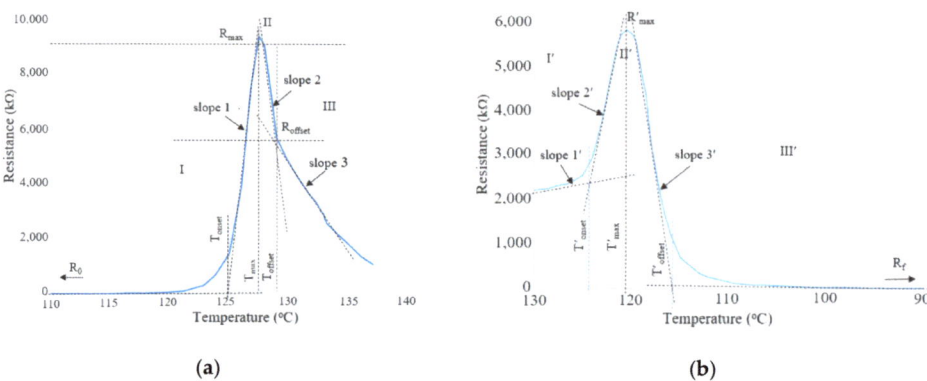

Figure 1. Typical R vs. T heating (**a**) and T cooling (**b**) curve and kinetic parameters.

3.6. Measurement of the Electro-Thermal Effect

The measurement of the dependence of the temperature at the surface of the sample on time and applied voltage (either AC or DC), the so called electro-thermal effect, was carried out using a setup similar to the one described earlier [31]. The current temperature (T), the equilibrium temperature (T_{eq}), the intensity of the electric current flowing through the sample (I), as well as the intensity upon reaching the temperature equilibrium (I_{eq}), were determined.

4. Results and Discussion
4.1. Structural Characterization
4.1.1. SEM (Scanning Electron Microscopy)

The morphology of the composites under study is depicted in Figure 2a,b, representing the LLD 190 and LLD 192 composites, respectively. These images can be qualitatively analyzed on the basis of particle shape, dimensions and gray nuance intensities. First of all, the materials show a pronounced micro-heterogeneous character in which the presence of shiny filamentous particles, with dimensions in the order of microns, as well as spheroidal particles with dimensions of 30–100 nm, are observed. The filamentous particles can be assigned to the insulating polymer, while the small spheroidal particles are primary carbon black particles (30 nm) or associations of primary particles (30–100 nm), resulting from the breakdown of the initial aggregates (see a typical image of CB used within this work in reference [30]). If these CB particles are part of a conductive path that allows electrical charges to flow, they appear darker (see for example [35]). Note that during measurement, the sample heats up, so a significant part of the conductive paths could break, resulting in the increased brightness of the CB particles. The graphite particles are not visible in this image, but their presence can be easily observed in lower magnification images (for example, see Figure 2c,d), with dimensions in the order of ten microns, typical for graphite flakes (see a typical SEM image of these particles in reference [30]). Hence, the darker areas in the vicinity of the filiform polymer particles as well as the spheroidal particles are related to conductive or potentially conductive areas belonging to conductive paths within the material.

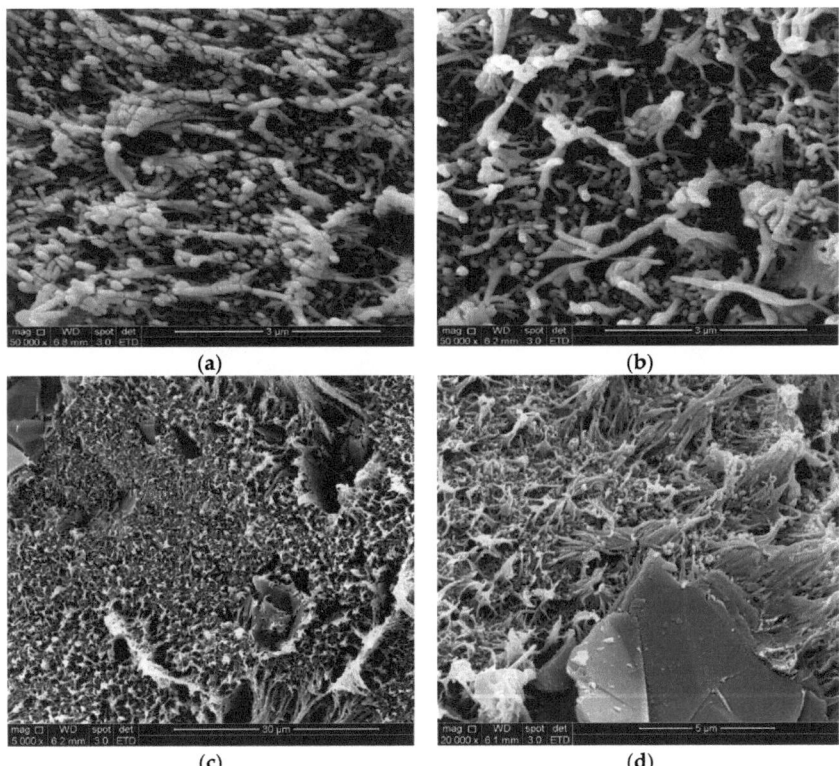

Figure 2. SEM images of some studied composites in a fresh state (unconditioned samples) at different magnifications: (**a**) LLD 190 (50,000×); (**b**) LLD 192 (50,000×); (**c**) LLD 192 (5000×); (**d**) LLD 122 (20,000×).

4.1.2. Differential Scanning Calorimetry

The blank polymer sample (LLD 0) presents an endotherm with a peak at 125.8 °C in the heating curves at the first cycle and 125.0 °C at the second cycle, with the difference possibly arising from the sample's history (mainly due to different solidification conditions during molding and cooling in the DSC instrument, respectively). Similarly, the composites show slightly higher values of T_m, ΔH_{cor}, and crystallinity during the first scan as compared to the second. Besides the previously mentioned cooling conditions, slow crystallization at room temperature within the time span between sample preparation and DSC measurement is also possible. With the second scan, the differences between the parameters of different samples are small, as indicated by the curves shown in Figure 3 and the data in Table S1 from the Supplementary Materials. Thus, the Tm values of the composites have an average of 125.5 °C, with a standard deviation of 0.17. A slight decreasing trend with carbon content can be observed (Figure 4), with lower values for composites with CB as compared to CB + Gr composites. Compared to unmodified LLDPE, all Tm values are slightly higher, suggesting that carbon particles may, to some extent, favor the crystallization of LLDPE, possibly by improving the heat transfer, but increasing carbon content may negatively affect the crystallization process. A similar increase in the crystallinity of LLDPE-based composites was previously reported for different fillers [36–38] and was assigned to a possible nucleation effect of the filler [38].

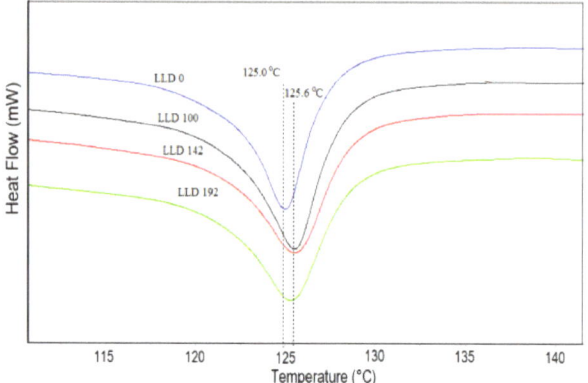

Figure 3. DSC curves of LLDPE and two different (CB-Gr)/LLDPE composites.

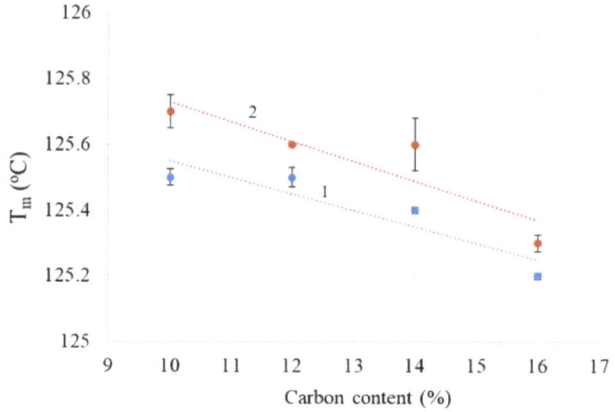

Figure 4. T_m decrease with total carbon content: (1) composites with CB; (2) composites with CB and Gr.

The cooling curves (Figure 5, Table S2) lead to similar conclusions, indicating that the crystallization temperature of the composites is higher than that of the pure polymer. Additionally, the CB and graphite-containing samples present slightly higher crystallization temperatures than those containing CB only. In all cases, the variation of T_c with the concentration of conductive filler was very small (the standard deviation of T_c values was 0.42 for CB samples, and 0.58 for CB and graphite-containing samples.

The vertical lines indicate the melting temperature of LLDPE (125.0 °C) and the average melting temperature of the studied CB and CB, Gr composites.

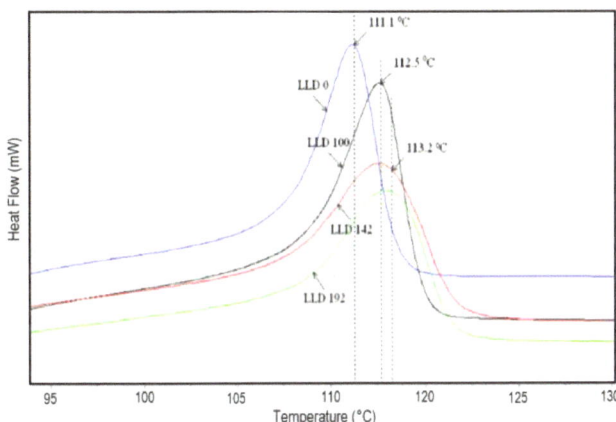

Figure 5. DSC cooling curves of LLDPE (blank) and three different (CB, Gr)/LLDPE composites.

4.1.3. FTIR

The spectrum of the blank LLDPE sample is shown in Figure 6. It is typical for polyethylenes, showing two main bands at 2915 cm^{-1} and 2848 cm^{-1}.

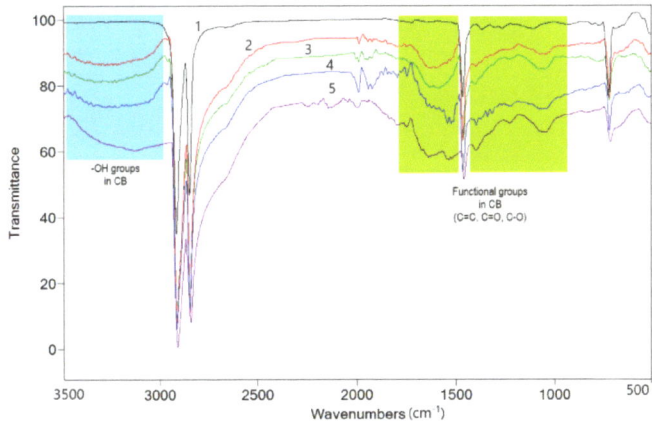

Figure 6. FTIR spectra of polymer matrix and different composites: 1—LLD 0; 2—LLD 120; 3—LLD 190; 4—LLD 122; 5—LLD 192.

As mentioned by Nikishida and Coates [39] regarding the differentiation between low-density polyethylene (LDPE) and linear low-density polyethylene (LLDPE), the bands at 890 cm^{-1} (vinylidene groups) and 910 cm^{-1} (terminal vinyl groups) present very weak intensities in LLDPE, while for LDPE, the band at 890 cm^{-1} is dominant [40]. Indeed, the intensity of these bands in the blank sample is very weak, confirming the LLDPE nature of the polymer. Another band, attributed to CH$_3$, specifically occurring at 1378 cm^{-1} (proportional to the number of branches), is also weak in intensity in our case.

The incorporation of CB within the polymer matrix increased optical absorption in the range 3500–3000 cm^{-1}, in the form of a broad band with a maximum at ca. 3300 cm^{-1}. This wide band shows an increasing trend with the CB content of the samples and is attributed to the -OH groups on the CB surface (oxidized -COOH and -OH groups as well as adsorbed water molecules [41–44]). The weak band at 2962 cm^{-1} is attributed to C-H groups of the raw material residues from CB synthesis [41], while the bands at 1796, 1740 cm^{-1} (also weak and absent from the matrix spectrum), to oxidized groups on the CB surface. The broad band between 1700 and 1470 cm^{-1} corresponds to different absorptions of the structural elements of CB, as well as to the oxygenated groups occurring on its surface, such as C=C (1640 cm^{-1}–graphitization, 1600 cm^{-1}), C=O (1680 cm^{-1}), C=O chelated with phenolic hydroxyls (1600 cm^{-1}), etc. [41,45]). The spectral range between 1420 and 760 cm^{-1} also contains bands that can be assigned to oxygenated groups on the CB surface, such as 1225 cm^{-1} and 1074 cm^{-1} (C-O-C in highly stable cyclic ethers [43,46]) or 1398 cm^{-1} (C=O in carbonyl and carboxylic compounds [44,45]). The addition of graphite in CB composites did not lead to rise of new bands, the most important effect appearing to be a partial splitting of the band between 1700 cm^{-1} and 1420 cm^{-1}, an effect that is more clearly observed in samples with a higher CB content.

4.2. Electrical Properties

4.2.1. DC Conductivity

It was observed that the DC conductivity (σ_{DC}) values depend on both the values and durations of the voltage application (Figure 7). In addition, the σ_{DC} values depend on filler concentration (see Figure 8 for CB) and temperature (see Figure S2). The influence of another important factor, namely the molding pressure, will be the subject of a future paper. This paper refers only to the results at 6 bars, a value experimentally found as the minimum required to impart adequate conduction and PTC properties.

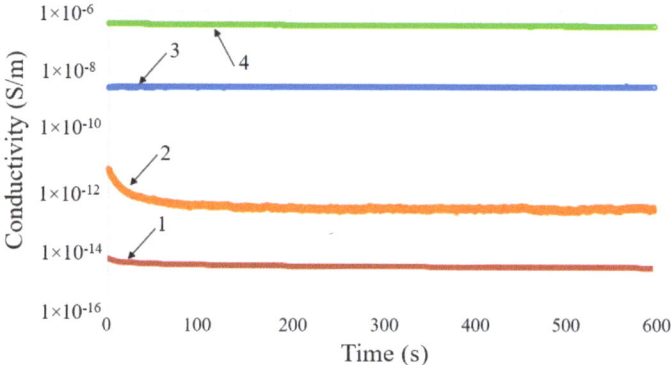

Figure 7. σ_{DC} vs. time, measured at 1 V for LLD neat samples (1) and LLD 80 (2), and at 100 V for LLD 44 (3) and LLD 82 (4).

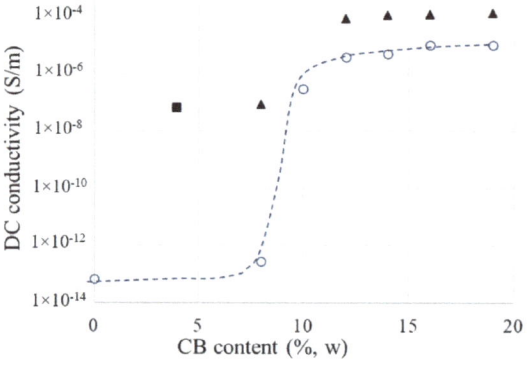

Figure 8. DC conductivity vs. carbon black and graphite content of the studied composites: (○) CB-containing samples; (▲) CB + 2% graphite; (■) CB + 4% graphite.

In Figure 8, which illustrates the influence of the CB content (ϕ_{CB}) on the DC conductivity of CB/LLD composites, three regions can be observed, namely: (i) for low concentrations between 0 and 8%, when σ_{dc} increases slowly with ϕ_{CB}, (ii) between 8 and 10%, σ_{dc} increases sharply with a slight change in ϕ_{CB}, and (iii) the relatively slow growth domain of σ_{dc} after the jump. The critical concentration, i.e., the minimum concentration at which the composite becomes a conductor (ϕ_c), corresponds to the jump (ii) and is, in our case, approximately 8%.

The partial substitution of CB with graphite, or the addition of Gr to a certain composition, resulted in the increased conductivity of the composites. Therefore, considering samples LLD 44 and LLD 80, both having a same total content of carbon materials, i.e., 8% (w), the conductivity of the sample with Gr was 68% higher than that of the sample with CB, when is measured at 1 min from the application of the voltage (1 V), and 35% higher when is measured at 10 min. The higher σ_{DC} values observed with the samples containing graphite are related to the considerably higher electrical conductivity of graphite as compared to CB [33], as well as to the morphological characteristics of Gr (high aspect ratio values), which make the transport of charge carriers through the sample easier.

A slight influence of temperature on conductivity was also observed in the range of 30–50 °C. In essence, both the blank sample and the composites with low filler content (i.e., with low values of conductivity) showed a slight decrease in conductivity with time and temperature, which could be due to the extinction of charge carriers generated within the materials during their processing. The materials with higher carbon filler content showed a slight increase in conductivity over time and with temperature (see Figure S2, Table S3), possibly due to a thermally and/or electrically activated local alignment of conductive particles, which could result in a slight increase in the number of conductive paths. Such effects could prevail at some point over the matrix dilatation effect, leading to fluctuations in the monotonous increase in resistivity with temperature, especially at lower temperatures (close to r.t.) and at moderate filler concentrations. The intensity of matrix dilatation effects tends to increase with both the temperature and the conductivity of the sample, as indicated by the R vs. T curves, which become smoother as the resistivity of the material decreases. The fluctuation trend in resistance values with increasing temperature is particularly visible with the LLD 100 and LLD 120 samples.

4.2.2. AC Conductivity

Similarly, to DC conductivity, AC conductivity increases with the content of conductive filler ϕ. This increase is mainly attributed to the growth in the number and dimensions of conductive paths.

As depicted in Figure 9, for all samples, the values of σ_{ca} rise with the increase in the frequency of the measuring voltage. A similar behavior was reported for other polymer (nano)composites [34,47]. For example, an increase in frequency from 50 Hz to 1 kHz leads to a 29.5-fold increase in σ_{ca} for the LLD 80 sample and a 7.5-fold increase for LLD 122. In the case of the LLD 120 samples, which had the highest CB content among the analyzed samples, σ_{ca} practically did not vary with the frequency (Figure 9, curve 5). The increase in AC conductivity with frequency can be assigned to charge carrier hopping within the symmetric hopping model in solids with microscopic disorder [34].

The measurement of complex relative permittivity (ε_r') as a function of frequency revealed (Figure 10) that, in the presence of a conductive filler, the values of ε_r' are higher compared to the unfilled sample (neat), a result consistent with previous reports for other polymers [28]. More significant differences between the filled and the neat samples were observed in the 1 Hz–1 kHz range, while at frequencies close to 10 MHz, these differences become less pronounced. This increase in ε_r' is primarily attributed to the new interfaces created by the filler particles, leading to the emergence of inhomogeneity polarization. This phenomenon, in turn, intensifies orientation, electronic and ionic polarizations in the polymeric matrix, resulting in an increase in ε_r' values. On the other hand, a high filler content also increases the number of clusters and reduces the number of individual carbon black particles dispersed in the polymeric matrix. Consequently, for ϕ values greater than 10%, the matrix/filler interface areas decrease, leading to a reduction in ε_r' values [1,34]. This phenomenon may explain the position of the LLD122 curve on the diagram in Figure 10.

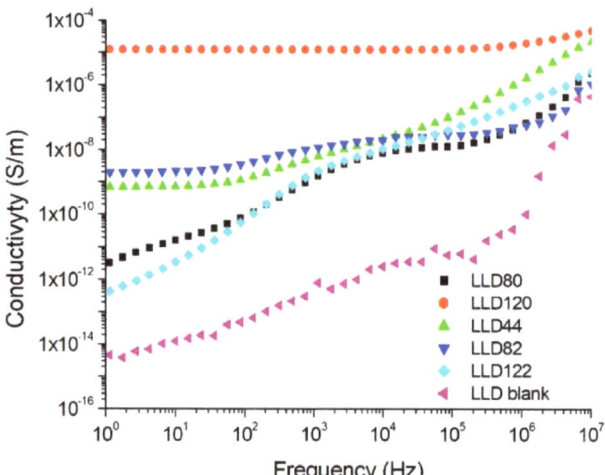

Figure 9. Variation of AC conductivity with the frequency of the measuring voltage for different composites with an LLDPE matrix (U = 1 V).

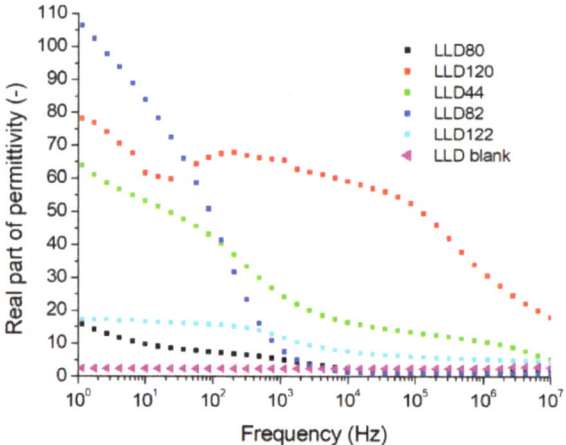

Figure 10. Variation with frequency of the real part of the relative complex permittivity for different composites with an LLDPE matrix (U = 1 V).

The values of the loss tangent (tan δ) depended on frequency, as shown in Figure 11, exhibiting a trend to increase with the rise in the content of conductive particles, driven by losses due to inhomogeneity polarization.

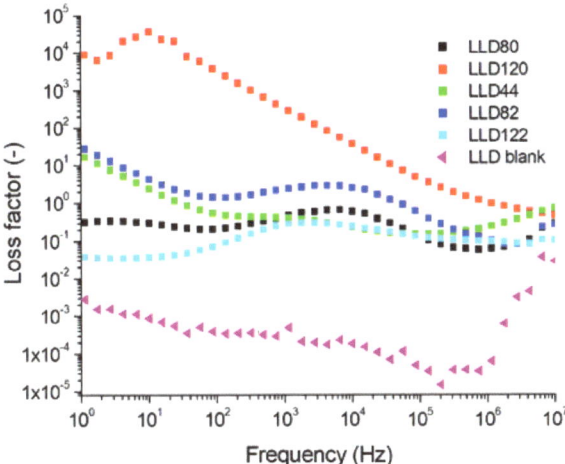

Figure 11. The variation with frequency of the loss factor for LLD neat samples (U = 1 V).

4.3. Resistivity (Resistance) vs. Temperature (PTC Behavior)

The R vs. T curves of samples with carbon contents higher than the critical concentration of 8% (as resulted from the DC conductivity measurements, see above), are similar regardless of the carbon material concentration. These curves present the three typical regions described in the literature (see for example [16]) for PTC composites, namely: (i) a slow increase in resistivity at the beginning of heating from room temperature; (ii) sudden increase of resistivity at elevated temperatures near the polymer melting temperature and reaching a maximum in the proximity of the crystallinity melting temperature (T_m) of the polymer; and (iii) a gradual decrease in resistivity after reaching the resistivity maximum (NTC effect in the molten state). The interpretation of the phenomena involved in processes (i) and (ii) is usually based on the concept of the thermal expansion of the polymer matrix, which results in increased space between the conductive particles and a gradual disruption of the conductive paths (depercolation). Near a crystalline transition point, the thermally induced volume expansion abruptly increases, causing a resistivity jump, an effect that enables various practical applications, as mentioned above. The decrease in resistivity after reaching the resistivity maximum (NTC effect) is attributed to the formation of conductive aggregates that enable charge carriers to move through the molten polymer [19].

Although the shape of the R vs. T curves of (CB, Gr)/LLDPE composites follows the general pattern described above, a strong influence of the conductive filler content on the parameters of these curves has been observed (Figure 12). Thus, the onset temperatures (T_{onset}) and those of reaching the resistivity maximum (T_{max}) decreased as the conductive filler content decreased. Also, the R_{max} values increased as the filler content decreased (see Table S4).

This strong dependence of the R vs. T curves on filler concentration is different from that observed with similar composites with an HDPE-based matrix and seems to be determined by the low crystallinity content of LLDPE. Hence, only for the samples with high conductive filler content, the T_{max} values are close to those corresponding to the melting of the crystalline phase of LLDPE (T_m) as measured by DSC. Unlike $T_{max(R)}$, $T_{m(DSC)}$ varies very little with the conductive filler concentration, as already mentioned (see Figures 3 and 4 and Table S1). This effect of resistivity on filler concentration was not observed or was insignificant in the case of (CB, Gr)/(LLDPE + HDPE) composites [31]. However, Zhang P et al. [18] reported a dependence of the ρ vs. T curves on the filler concentration for the Gr/(LLDPE + HDPE) system, for filler concentrations of 35–50% (mass), but the ρ vs. T of these materials approach the ideal curve from the literature only for high graphite contents $\geq 40\%$ (for HDPE: LLDPE 1:1) or for LLDPE: HDPE ratios $\geq 60:40$. Furthermore, the variation of $T_{max(R)}$ with the filler concentration seems to be rather small. The large difference observed between the values of $T_{max(DSC)}$ and $T_{max(R)}$ was explained by the complex contribution of the total crystallinity of the polymer matrix and the specific volume dilatation [14]. However, the PTC intensity seems to be rather low, with the highest values of the $\log_{10}(\rho_{max}/\rho_0)$ ratio being slightly below 3, despite the large graphite concentrations used. The study presented by Zhang R. et al. [12] concerning the influence of carbon fiber (CF) concentration on the ρ_{DC} vs. T curves of CF/(UHMWPE+LDPE) composites revealed a behavior similar to that highlighted by us: the onset temperature of the PTC effect decreased considerably (50–123 °C) with the filler content (2.5–10% vol), showing an optimal value (maximum PTC) for 5% CF (vol). The peak temperature of the melting endotherm practically did not change with the CF concentration, but the SEM morphological information suggests rather a homogeneous structure of the matrix and a homogeneous dispersion of CF between the two polymers.

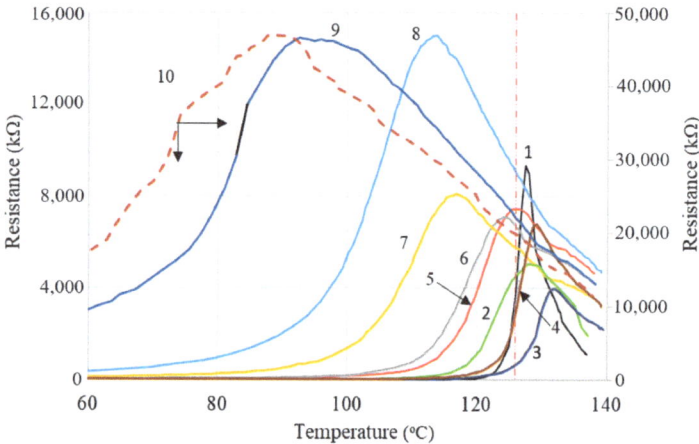

Figure 12. R vs. T curves from (CB, Gr)/LLDPE composites at first heating cycle: 1—LLD 192; 2—LLD 190; 3—LLD 162; 4—LLD 160; 5—LLD 142; 6—LLD 140; 7—LLD 122; 8—LLD 82; 9—LLD 120; 10—LLD 100. The vertical red line correspond to the average $T_{m(DSC)}$ of the composites (125.5 °C).

The optimal values of the concentration of conductive fillers, for which the highest intensities of the PTC effect were obtained (Table S4), correspond to samples LLD 140–LLD 190, i.e., those for which the T_{max} values approach the value of the crystallinity melting temperature determined by DSC ($T_{m(DSC)}$). However, it should be noted that all conductive samples showed significant jumps in resistivity with increasing temperature. The samples with high r.t resistivity (LLD 100, LLD 122 and LLD 82) exhibit broad peaks of resistivity, which occur at considerably lower temperatures than the melting temperature of the polymer. Combining the experimental observations and the data from the literature, it can be concluded that the behavior of the studied composites in R vs. T measurements could be described more properly by the Ohe and Natio model [34,35], based on tunneling through the interparticle gaps, rather than the model based on sudden expansion at the melting temperature. Thus, for samples with a lower filler content, depercolation occurs as a result of tunneling interruption due to the thermally induced modification of the space gap distribution between the conductive particles (a contribution of molecular movements or the melting of pseudo-crystalline domains is to be also considered). On the other hand, at high concentrations of the conductive filler, a more significant volumetric expansion is necessary (like what is produced near the crystallinity melting temperature), to sufficiently increase the distances between conductive particles; hence, a mechanism based on thermal expansion predominates in such a case.

If we compare the values of $T_{m(DSC)}$ and $T_{max(R)}$ for a same sample, it can be observed that excluding the samples with a high carbon contents (LLD 192, LLD 190, LLD 162 and LLD 160), the reaching of the R_{max} value and the subsequent decrease upon heating occur in the "solid" phase, at lower temperatures than $T_{m(DSC)}$. This fact suggests that the aggregates of conductive particles, considered responsible for the decrease in resistivity after R_{max}, are formed even before complete polymer melting and that the mobility of these aggregates within the polymer matrix is sufficiently high to ensure electrical conduction. Therefore, the enhancement of electrical conduction at $T_{max(DSC)} > T > T_{max(R)}$ could be provided by a thermally activated mechanism of jumping/tunneling between neighboring conductive (fibrous) aggregates, which could result from the coagulation (re-arrangement) of the neighboring conductive particles. Note that the viscosity of the medium at elevated temperatures (below the melting point) is high enough to enable only the limited movement of the conductive particles around their equilibrium positions in the conductive paths. This could also explain the observed broadening of the resistivity peak (which equals to a decrease in the intensity of the NTC effect) with decreasing conductive filler concentrations.

If we compare the values of $T_{m(DSC)}$ and $T_{max(R)}$ for the same sample, it can be observed that excluding the samples with high carbon contents (LLD 192, LLD 190, LLD 162 and LLD 160), the R_{max} value is reached well before the matrix melts and the subsequent decrease occurs in the "solid" phase until a temperature equal to $T_{m(DSC)}$ is attained. This fact suggests that aggregates of the conductive particles, considered to be responsible for the decrease in resistivity after R_{max}, are formed well before complete polymer melting, and that the mobility of these aggregates within the polymer matrix is sufficiently high to ensure electrical conduction. Therefore, the decrease in electrical resistivity at $T_{m(DSC)} > T > T_{max(R)}$ could be provided by a thermally activated mechanism of jumping/tunneling between neighboring conductive (fibrous) aggregates, which may result from the coagulation (re-arrangement) of the neighboring conductive particles. Note that the viscosity of the medium at elevated temperatures (below the melting point) is high enough to enable only limited movement of the conductive particles around their equilibrium positions in the conductive paths. This fact could also explain the observed broadening of the resistivity peak (which equals to a decrease in the intensity of the NTC effect) with decreasing conductive filler concentrations.

Additionally, it is observed from the comparison of R vs. T and DSC data (Figures 12–14) that even for samples with a high carbon content, there is no perfect correspondence between the $T_{max(R)}$ and $T_{max(DSC)}$ values. For some of the samples mentioned above as having a high carbon content, the $T_{max(R)}$ values are slightly higher than the $T_{m(DSC)}$, although the heating rates in the R vs. T measurements were lower. This behavior suggests that the same distribution of particles that exists in the solid state persists in the highly viscous fluid material that resulted from the crystallinity melting. On the other hand, the attainment of R_{max} before the crystallinity melting for the samples with a lower carbon content suggests that the smaller volume expansion of the amorphous phase at $T < T_{m(DSC)}$ would be large enough to increase the interparticle distances interrupting so the conductive paths in those samples.

The decrease in resistivity after $T_{max(R)}$ could be explained by the formation of aggregates in the amorphous zone of the not yet melted polymer, through a continuous process apparently unaffected by the melting of the polymer (the shape of the R vs. T curves did not change dramatically in the proximity of $T_{m(DSC)}$). In understanding the behavior of the composites below $T_{m(DSC)}$, we should take into account that the conductive material is distributed in the amorphous region of the polymer (see, e.g., Zhang P et al. [19]) and that the composite preparation method provides a certain degree of inhomogeneity in the conductive particles distribution. Hence it seems that it can be expected that the changes in volume and viscosity in either the amorphous or pseudo-crystalline phases and/or the activation of some thermally induced molecular movements [19] influence the electrical conductivity of the material more than the melting crystallinity, which is present in a relatively small proportion, according to the DSC data (see the Supplementary Materials). In particular, in the case of LLDPE, which exhibits a complicated behavior during melting and solidification [48], such processes can influence the distribution of conductive particles and, consequently, the resistivity of the material during heating.

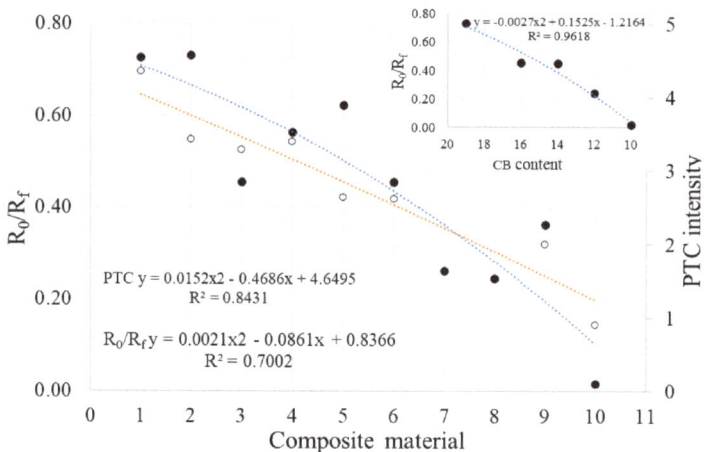

Figure 13. R_0/R_f (●) and *PTC* intensity (○) of different composites (see numbers on the abscise): 1—LLD 192; 2—LLD 190; 3—LLD 162; 4—LLD 160; 5—LLD 142; 6—LLD 140; 7—LLD 122; 8—LLD 120; 9—LLD 82; 10—LLD 100. In the inset, decreasing R_0/R_f values for CB/LLDPE composites as CB content decreases.

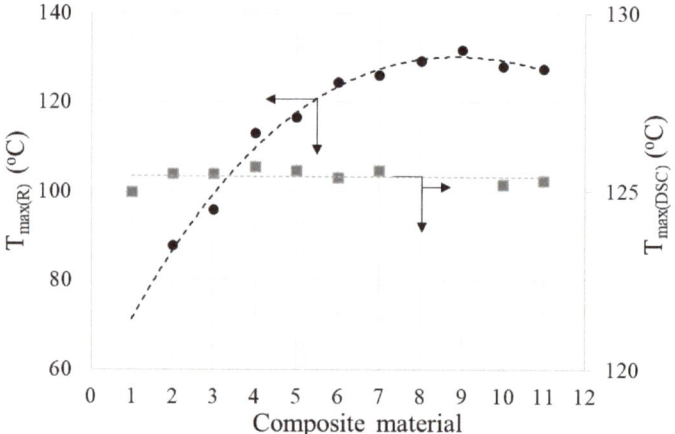

Figure 14. $T_{max(R)}$ and $T_{max(DSC)}$ vs. carbon content of composites (see numbers on the abscise): 1—LLD 0; 2—LLD 100; 3—LLD 82; 4—LLD 120; 5—LLD 122; 6—LLD 140; 7—LLD 142; 8—LLD 160; 9—LLD 162; 10—LLD 190; 11—LLD 192.

The analysis of the data in Tables S1 and S4 and in Figures 13 and 14 suggests a relatively complex influence of the carbon content on the R-T behavior: according to the Tmax(R) values, the composites can be divided into two groups, namely (i) samples with a high total carbon content ($\geq 14\%$ w/w) which present $T_{max(R)}$ values close to $T_{m(DSC)}$ and (ii) samples with a low total carbon content ($\leq 12\%$ w/w), for which the difference in $T_{mDSC} - T_{max(R)}$ increases as the carbon content decreases. The intensities of the PTC effect have high values in the first group and decrease as the carbon content of the sample decreases. The presence of graphite generally improved the parameters of the R-T curves parameters, but the most noticeable effect was observed at low total carbon concentrations, as seen in the case of sample LLD 82, which exhibits comparable parameters to sample LLD 122, despite having a lower carbon content. Additionally, the differences in behavior induced by the presence of graphite are clear if we compare samples LLD 122 vs. LLD 120, or LLD 82 vs. LLD 100. In addition, it shall be remarked that these results are in good agreement with the AC and DC conductivity data presented above.

In the case of cooling curves (Figure 15), similar trends were observed, but the values of R'_{max} are higher than the corresponding R_{max} values from the heating curves (excluding sample LLD 192), suggesting a slower reformation of conductive paths upon solidification.

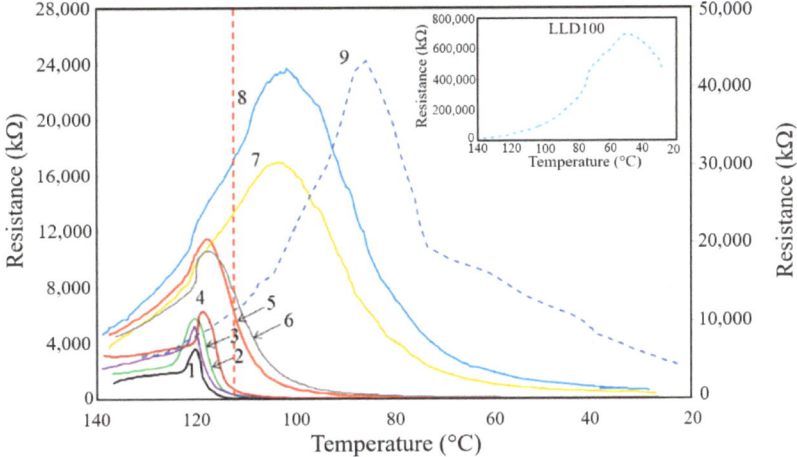

Figure 15. R vs. T curves from (CB, Gr)/LLDPE composites at first cooling cycle: 1—LLD 192; 2—LLD 190; 3—LLD 162; 4—LLD 160; 5—LLD 142; 6—LLD 140; 7—LLD 122; 8—LLD 82; 9—LLD 120. In inset, LLD 100 sample (note the max. of y scale is 800,000 kΩ). The vertical red line corresponds to average T_c of the composites (112.8 °C).

Overall, all samples showed higher resistivity values at room temperature after a heating–cooling cycle, indicating that some of the conductive pathways do not reform. Considering a proportional relationship between the number of conductive pathways and conductivity (inverse of resistivity), Figure 13 shows that the fraction of channels that do not reform (described by R_0/R_f ratio) is smaller in samples with a high carbonaceous conductive filler content (>14 w/w) and increases with the decreasing carbon content of the samples. The scattering of the R_0/R_f values seems to be influenced by the graphite-containing samples, where the variation trend is weaker than for the samples containing CB only (see Figure 13 inset).

The conductive samples with lower filler concentrations (at the end of the conductive range, i.e., LLD 82–LLD 192) exhibit a different behavior during cooling as compared to other samples in the mentioned range. Thus, the LLD 120 and LLD 100 samples, with a relatively low concentration of carbon black and no graphite, return to high resistance values after cooling, in the order of MΩ and hundreds of MΩ, respectively, categorizing them as insulators; hence, the R_0/R_f values are low. The behavior of the LLD 192 sample, which contains a high concentration of fillers, is remarkable too: the cooling curves (see Figure 15 and Table S5) show a slow increase in melt resistivity between 133 and 123 °C, after which the resistivity suddenly increases until ~120 °C ($T'_{max(R)}$). This value is far from $T_{c(DSC)}$ (see Figure 15 and Table S2). However, the maximum value (R'_{max}) is considerably lower than the corresponding value from the heating curves, representing a different behavior not only compared to the rest of the studied composites, but also in comparison to the (Gr, CB)/HDPE and (Gr, CB)/(HDPE + LLDPE) composites reported earlier [31]. This behavior indicates that the material conserves relatively high conductivity at the crystallization temperature, i.e., a significant number of the conductive pathways remain functional at $T'_{max(R)}$.

In general, it is observed (Figure 15 and Tables S2 and S5) that all samples with high carbon content (LLD 140–LLD 192) have $T'_{max(R)}$ values higher than $T_{c(DSC)}$, with little dependence on filler concentration, while the samples with lower conductive filler content present lower $T'_{max(R)}$ values (and are strongly dependent on the filler concentration) than $T_{c(DSC)}$.

4.4. Electrothermal Behavior

For all samples with a higher carbon content than the critical concentration, regardless of the carbon content and applied voltage, the temperature–time variation (T vs. t) curves on the sample surface exhibit a typical self-limiting behavior of temperature and current, enabling the use of these materials in self-regulating thermal applications (Figures 16 and 17).

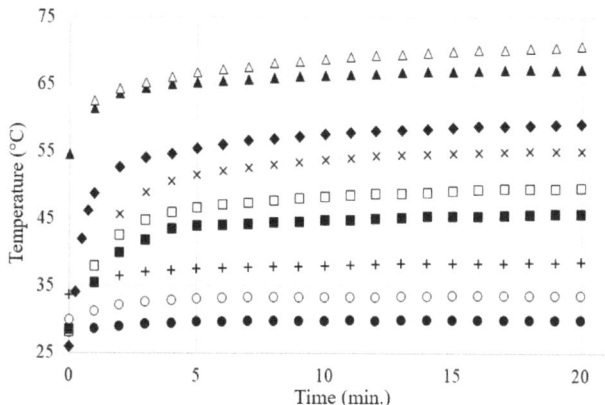

Figure 16. The T vs. t curves for different applied voltages (AC) on the same composite sample LLD 192 (s10): (●) 3 V; (○) 5 V; (+) 8 V; (■) 10 V; (□) 12 V; (×) 15 V; (▲) 20 V; (△) 25 V; (◆) 35 V Levelling of temperature increase at 20 and 25 V can be clearly observed.

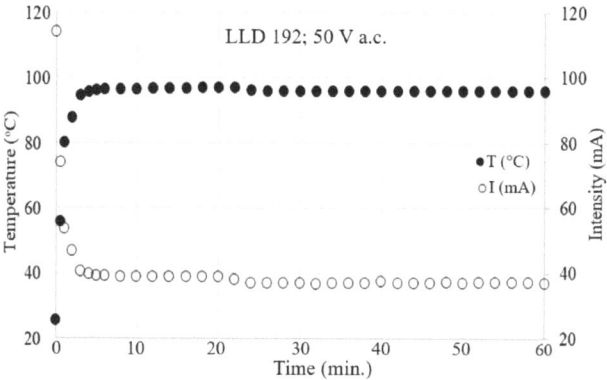

Figure 17. The T vs. t and I vs. t curves for LLD 192 sample (U = 50 V).

No significant differences were observed between the T vs. t curves resulting from the DC and AC measurements for the same applied voltages (effective voltage in AC), as can be observed in Figure 18 for the LLD 192 sample at applied voltages of 20 V.

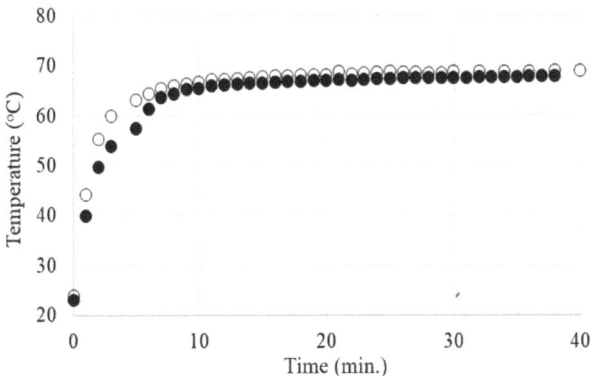

Figure 18. The T vs. t curves for LLD 192 at 20 V: (○) DC; (●) AC.

The voltage-limiting effect is illustrated in Figure 19 for the LLD 190 sample. In the low voltage range, up to 50 V in this case, the temperature increased with increasing voltage, reaching a maximum. Subsequently, the equilibrium temperature values decreased with increasing voltage. This trend was accentuated as the applied voltage increased. Such a behavior is of particular practical importance because it describes overvoltage protection.

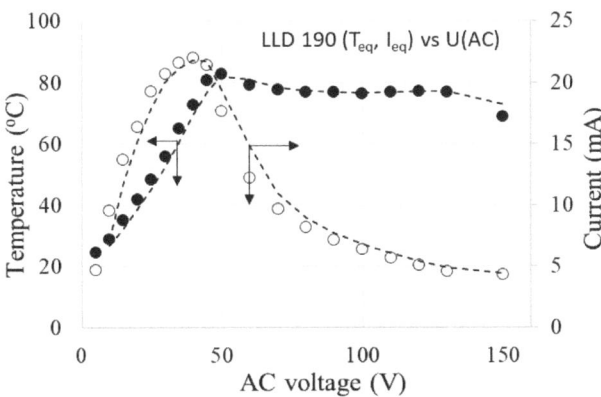

Figure 19. T_{eq} vs. applied voltage for LLD 190 sample.

It is noteworthy that the Teq value decreased with the decreasing of the conductive filler concentration (or, for achieving the same temperature, the necessary voltage was higher as the concentration of the conductive filler was lower).

Also, for any of the sample combinations (filler concentration)—the applied voltage, the maximum value of T_{eq} was far enough from $T_{m(DSC)}$, meaning that the temperature limitation occurs at considerably lower T_{eq} values. This fact is also of practical importance for the operational safety of the material as a heater, preventing its destruction by melting during use. Additionally, because the Teq values differ significantly from the $T_{m(DSC)}$ values, our previous observations concerning the limited involvement of the crystalline–amorphous transition in the PTC effect mechanism in the case of this type of polymer matrix are confirmed (see R vs. T curves).

The dependency of the thermal–voltage effect described above, observed in all of the studied samples, differs significantly from the behavior of an ohmic resistor and is determined by the self-limiting characteristics of the material. If the self-limiting effect did not occur, then the current intensity would increase proportionally with the applied voltage, following Ohm's law (U = RI, with R practically constant). However, in our case, it is clearly observed (see Figure 19) that the equilibrium current intensity through the sample decreases with the applied voltage. This effect directly affects the temperature on the surface of the sample, which remains unchanged over time for a given voltage (Figure 17) and changes very little over a wide range of voltages (Figure 19). At the same time, the decrease in I_{eq} with applied voltage suggests that as the voltage tends toward infinity, the material would tend to become an insulator (Figure 19) due to the self-limitation of the current passing through the sample.

It was also observed, especially for samples exposed to high AC voltages (>150 °C), that a slight increase in r.t. resistivity (measured as resistance) occurred after conducting long term tests. Although, after exposure to moderate voltages (25–50 V), the resistivity tends to decrease slightly, suggesting the reversibility of the process, and the possible aging of the polymeric material (or the matrix–filler assembly) at elevated temperatures and high voltages cannot be excluded. Hence, the extension of the work in the direction of degradation diagnosis is considered.

5. Conclusions

This study reaffirms the significant influence of filler content (ϕ) on the electrical properties of polyethylene/carbon black/graphite composites. A notable increase in DC conductivity was observed between 8–10% filler content due to percolation. Higher ϕ values also increased AC conductivity, permittivity and the loss factor, leading to enhanced dielectric losses and subsequent heating in electric fields.

The use of an LLDPE matrix with its low crystallinity offers greater flexibility compared to HDPE. A strong dependence of the PTC effect on carbon content was demonstrated, particularly on the temperature at which maximum resistivity occurs ($T_{max(R)}$). This effect is underexplored in the literature and highlights a unique aspect of this study.

The results revealed how the composition of a composite significantly impacts self-limiting properties and the PTC effect. Above the critical percolation concentration, there exists an optimal range where the PTC effect's intensity and stability hold practical value for (CB-Gr)/LLDPE composites. Unlike highly crystalline matrices (HDPE or HDPE/LLDPE), $T_{max(R)}$ and the DSC-determined melting temperature ($T_{m(DSC)}$) differ significantly. This indicates that factors beyond crystallinity melting influence the PTC mechanism. Further studies are needed to explore the specific interaction between the conductive filler and the LLDPE matrix.

The distinct behavior of LLDPE-matrix composites allows for the tailored control of self-limiting and interrupting properties, potentially leading to innovative low-temperature applications.

Supplementary Materials: The following supporting information can be downloaded at: https://www.mdpi.com/article/10.3390/ma17051224/s1, Figure S1: Theoretical curve of DC conductivity vs. conductive filler concentration; Figure S2: Electrical conductivity (σ_{DC}) vs. time for LLD 122 sample, at different temperatures: (1) 30 °C; (2) 40 °C; (3) 50 °C. The measurement voltage, U0 = 1V; Table S1: Kinetic parameters of melting process studied by DSC (ramp experiment, heating rate, 10 °C/min., N2 flow, 50 ml/min); Table S2: Kinetic parameters of crystallization process studied by DSC (ramp experiment, cooling rate, 10 °C/min., N2 flow, 50 ml/min); Table S3: σ_{DC} values measured at different temperatures (T) after 1 min. (σ_{dc1}) and 10 min. (σ_{dc10}) from voltage (U_0) application; Table S4: Kinetic parameters of the R vs. T heating curves of the studied composites; Table S5: Kinetic parameters of the R vs. T cooling curves of the studied composites.

Author Contributions: Conceptualization, R.S., E.-M.L. and P.V.N.; methodology, R.S., E.-M.L., P.V.N. and C.S.; investigation, R.S., E.-M.L., P.V.N., C.S. and T.-A.B.; data curation, R.S.; writing—original draft preparation, R.S. and E.-M.L.; writing—review and editing, R.S., E.-M.L., P.V.N. and C.S.; funding acquisition, E.-M.L. All authors have read and agreed to the published version of the manuscript.

Funding: This research was funded by by the Ministry of Research, Innovation and Digitization through contracts: 612PED/2022, PN23140201-42N/2023 and project number 25PFE/30.12.2021—Increasing R-D-I capacity for electrical engineering-specific materials and equipment regarding electromobility and "green" technologies within PNCDI III, Programme 1. T.B. also acknowledge the support provided by the Ministry of Research, Innovation and Digitization through the "Nucleu" Program, Grant no. 31N/2023, Project PN 23120401.

Institutional Review Board Statement: Not applicable.

Informed Consent Statement: Not applicable.

Data Availability Statement: Data are contained within the article and Supplementary Materials.

Conflicts of Interest: The authors declare no conflicts of interest.

References

1. Huang, X.; Sun, B.; Yu, C.; Wu, J.; Zhang, J.; Jiang, P. Highly conductive polymer nanocomposites for emerging high voltage power cable shields: Experiment, simulation and applications. *High Volt.* **2020**, *5*, 387–396. [CrossRef]
2. Park, S.-H.; Hwang, J.; Park, G.-S.; Ha, J.-H.; Zhang, M.; Kim, D.; Yun, D.-J.; Lee, S.; Lee, S.H. Modeling the electrical resistivity of polymer composites with segregated structures. *Nat. Commun.* **2019**, *10*, 2537. [CrossRef] [PubMed]
3. Kuan, C.-F.; Kuan, H.-C.; Ma, C.-C.M.; Chen, C.-H. Mechanical and electrical properties of multi-wall carbon nanotube/poly(lactic acid) composites. *J. Phys. Chem. Solids* **2008**, *69*, 1395–1398. [CrossRef]
4. Miyasaka, K.; Watanabe, K.; Jojima, E.; Aida, H.; Sumita, M.; Ishikawa, K. Electrical conductivity of carbon-polymer composites as a function of carbon content. *J. Mater. Sci.* **1982**, *17*, 1610–1616. [CrossRef]
5. Chen, X.; Zheng, Y.; Han, X.; Jing, Y.; Du, M.; Lu, C.; Zhang, K. Low-dimensional Thermoelectric Materials. In *Flexible Thermoelectric Polymers and Systems*; Wiley: Hoboken, NJ, USA, 2022; pp. 209–238.
6. Xi, Y.; Yamanaka, A.; Bin, Y.; Matsuo, M. Electrical properties of segregated ultrahigh molecular weight polyethylene/multiwalled carbon nanotube composites. *J. Appl. Polym. Sci.* **2007**, *105*, 2868–2876. [CrossRef]
7. Gümüş, E.; Yağımlı, M.; Arca, E. Investigation of the Dielectric Properties of Graphite and Carbon Black-Filled Composites as Electromagnetic Interference Shielding Coatings. *Appl. Sci.* **2023**, *13*, 8893. [CrossRef]
8. Wu, M.; Wu, F.; Ren, Q.; Jia, X.; Luo, H.; Shen, B.; Wang, L.; Zheng, W. Tunable electromagnetic interference shielding performance of polypropylene/carbon black composites via introducing microcellular structure. *Compos. Commun.* **2022**, *36*, 101363. [CrossRef]
9. Choi, H.-J.; Kim, M.S.; Ahn, D.; Yeo, S.Y.; Lee, S. Electrical percolation threshold of carbon black in a polymer matrix and its application to antistatic fibre. *Sci. Rep.* **2019**, *9*, 6338. [CrossRef] [PubMed]
10. Qi, Y.; Dai, C.; Gao, J.; Gou, B.; Bi, S.; Yu, P.; Chen, X. Reinforcement effects of graphite fluoride on breakdown voltage rating and Pyro-Resistive properties of carbon Black/Poly (vinylidene fluoride) composites. *Compos. Part A Appl. Sci. Manuf.* **2024**, *177*, 107947. [CrossRef]
11. Ding, X.; Wang, J.; Zhang, S.; Wang, J.; Li, S. Carbon black-filled polypropylene as a positive temperature coefficient material: Effect of filler treatment and heat treatment. *Polym. Bull.* **2016**, *73*, 369–383. [CrossRef]
12. Zhang, R.; Tang, P.; Li, J.; Xu, D.; Bin, Y. Study on filler content dependence of the onset of positive temperature coefficient (PTC) effect of electrical resistivity for UHMWPE/LDPE/CF composites based on their DC and AC electrical behaviors. *Polymer* **2014**, *55*, 2103–2112. [CrossRef]
13. Mamunya, Y.; Maruzhenko, O.; Kolisnyk, R.; Iurzhenko, M.; Pylypenko, A.; Masiuchok, O.; Godzierz, M.; Krivtsun, I.; Trzebicka, B.; Pruvost, S. Pyroresistive Properties of Composites Based on HDPE and Carbon Fillers. *Polymers* **2023**, *15*, 2105. [CrossRef] [PubMed]
14. Nagel, J.; Hanemann, T.; Rapp, B.E.; Finnah, G. Enhanced PTC Effect in Polyamide/Carbon Black Composites. *Materials* **2022**, *15*, 5400. [CrossRef] [PubMed]
15. Lai, F.; Zhao, L.; Zou, J.; Zhang, P. High positive temperature coefficient effect of resistivity in conductive polystyrene/polyurethane composites with ultralow percolation threshold of MWCNTs via interpenetrating structure. *React. Funct. Polym.* **2020**, *151*, 104562. [CrossRef]
16. Deng, H.; Lin, L.; Ji, M.; Zhang, S.; Yang, M.; Fu, Q. Progress on the morphological control of conductive network in conductive polymer composites and the use as electroactive multifunctional materials. *Prog. Polym. Sci.* **2014**, *39*, 627–655. [CrossRef]
17. Li, Y.; Huang, X.; Zeng, L.; Li, R.; Tian, H.; Fu, X.; Wang, Y.; Zhong, W.-H. A review of the electrical and mechanical properties of carbon nanofiller-reinforced polymer composites. *J. Mater. Sci.* **2019**, *54*, 1036–1076. [CrossRef]
18. Zhang, P.; Wang, B.-b. Positive temperature coefficient effect and mechanism of compatible LLDPE/HDPE composites doping conductive graphite powders. *J. Appl. Polym. Sci.* **2018**, *135*, 46453. [CrossRef]
19. Makuuchi, K.; Cheng, S. *Radiation Processing of Polymer Materials and Its Industrial Applications*; John Wiley & Sons: Hoboken, NJ, USA, 2012.
20. Xu, H.-P.; Dang, Z.-M.; Shi, D.-H.; Bai, J.-B. Remarkable selective localization of modified nanoscaled carbon black and positive temperature coefficient effect in binary-polymer matrix composites. *J. Mater. Chem.* **2008**, *18*, 2685–2690. [CrossRef]

21. Bao, Y.; Xu, L.; Pang, H.; Yan, D.-X.; Chen, C.; Zhang, W.-Q.; Tang, J.-H.; Li, Z.-M. Preparation and properties of carbon black/polymer composites with segregated and double-percolated network structures. *J. Mater. Sci.* **2013**, *48*, 4892–4898. [CrossRef]
22. Setnescu, R.; Lungulescu, E.M. Novel PTC Composites for Temperature Sensors (and Related Applications). In *Wireless Sensor Networks*; Jaydip, S., Mingqiang, Y., Fenglei, N., Hao, W., Eds.; IntechOpen: Rijeka, Croatia, 2023; Chapter 6.
23. Wei, Y.; Li, Z.; Liu, X.; Dai, K.; Zheng, G.; Liu, C.; Chen, J.; Shen, C. Temperature-resistivity characteristics of a segregated conductive CB/PP/UHMWPE composite. *Colloid Polym. Sci.* **2014**, *292*, 2891–2898. [CrossRef]
24. Li, X.; Guo, Q.; Yang, F.; Sun, X.; Li, W.; Yao, Z. Electrical Properties of LLDPE/LLDPE-g-PS Blends with Carboxylic Acid Functional Groups for Cable Insulation Applications. *ACS Appl. Polym. Mater.* **2020**, *2*, 3450–3457. [CrossRef]
25. Zhang, X.; Zheng, S.; Zheng, X.; Liu, Z.; Yang, W.; Yang, M. Distinct positive temperature coefficient effect of polymer–carbon fiber composites evaluated in terms of polymer absorption on fiber surface. *Phys. Chem. Chem. Phys.* **2016**, *18*, 8081–8087. [CrossRef] [PubMed]
26. Ohe, K.; Naito, Y. A New Resistor Having An Anomalously Large Positive Temperature Coefficient. *Jpn. J. Appl. Phys.* **1971**, *10*, 99. [CrossRef]
27. Zhang, C.; Ma, C.-A.; Wang, P.; Sumita, M. Temperature dependence of electrical resistivity for carbon black filled ultra-high molecular weight polyethylene composites prepared by hot compaction. *Carbon* **2005**, *43*, 2544–2553. [CrossRef]
28. Nakamura, S.; Tomimura, T.; Sawa, G. Dielectric properties of carbon black-polyethylene composites below the percolation threshold. In Proceedings of the 1999 Annual Report Conference on Electrical Insulation and Dielectric Phenomena (Cat. No.99CH36319), Austin, TX, USA, 17–20 October 1999; Volume 291, pp. 293–296.
29. Hu, X.; Shi, Y.; Wang, Y.; Liu, L.-Z.; Ren, Y.; Wang, Y. Crystallization, structure, morphology, and properties of linear low-density polyethylene blends made with different comonomers. *Polym. Eng. Sci.* **2021**, *61*, 2406–2415. [CrossRef]
30. Setnescu, R.; Lungulescu, M.; Bara, A.; Caramitu, A.; Mitrea, S.; Marinescu, V.; Culicov, O. Thermo-Oxidative Behavior of Carbon Black Composites for Self-Regulating Heaters. *Adv. Eng. Forum* **2019**, *34*, 66–80. [CrossRef]
31. Setnescu, R.; Lungulescu, E.-M.; Marinescu, V.E. Polymer Composites with Self-Regulating Temperature Behavior: Properties and Characterization. *Materials* **2023**, *16*, 157. [CrossRef]
32. Durmuş, A.; Woo, M.; Kaşgöz, A.; Macosko, C.W.; Tsapatsis, M. Intercalated linear low density polyethylene (LLDPE)/clay nanocomposites prepared with oxidized polyethylene as a new type compatibilizer: Structural, mechanical and barrier properties. *Eur. Polym. J.* **2007**, *43*, 3737–3749. [CrossRef]
33. Ilie, S.; Setnescu, R.; Lungulescu, E.M.; Marinescu, V.; Ilie, D.; Setnescu, T.; Mares, G. Investigations of a mechanically failed cable insulation used in indoor conditions. *Polym. Test.* **2011**, *30*, 173–182. [CrossRef]
34. Stancu, C.; Notingher, P.V.; Panaitescu, D.M.; Marinescu, V. Electrical Properties of Polyethylene Composites with Low Content of Neodymium. *Polym.-Plast. Technol. Eng.* **2015**, *54*, 1135–1143. [CrossRef]
35. Kita, Y.; Kasai, Y.; Hashimoto, S.; Iiyama, K.; Takamiya, S. Application of Brightness of Scanning Electron Microscope Images to Measuring Thickness of Nanometer-Thin SiO2 Layers on Si Substrates. *Jpn. J. Appl. Phys.* **2001**, *40*, 5861. [CrossRef]
36. Müller, M.T.; Dreyße, J.; Häußler, L.; Krause, B.; Pötschke, P. Influence of talc with different particle sizes in melt-mixed LLDPE/MWCNT composites. *J. Polym. Sci. Part B Polym. Phys.* **2013**, *51*, 1680–1691. [CrossRef]
37. Kundu, P.P.; Biswas, J.; Kim, H.; Choe, S. Influence of film preparation procedures on the crystallinity, morphology and mechanical properties of LLDPE films. *Eur. Polym. J.* **2003**, *39*, 1585–1593. [CrossRef]
38. Liu, M.; Horrocks, A.R. Effect of Carbon Black on UV stability of LLDPE films under artificial weathering conditions. *Polym. Degrad. Stab.* **2002**, *75*, 485–499. [CrossRef]
39. Lobo, H.; Bonilla, J.V. *Handbook of Plastics Analysis*; CRC Press: Boca Raton, FL, USA, 2003; Volume 68.
40. Jung, M.R.; Horgen, F.D.; Orski, S.V.; Rodriguez C, V.; Beers, K.L.; Balazs, G.H.; Jones, T.T.; Work, T.M.; Brignac, K.C.; Royer, S.-J.; et al. Validation of ATR FT-IR to identify polymers of plastic marine debris, including those ingested by marine organisms. *Mar. Pollut. Bull.* **2018**, *127*, 704–716. [CrossRef] [PubMed]
41. Yang, S.Y.; Bai, B.C.; Kim, Y.R. Effective Surface Structure Changes and Characteristics of Activated Carbon with the Simple Introduction of Oxygen Functional Groups by Using Radiation Energy. *Surfaces* **2024**, *7*, 12–25. [CrossRef]
42. Zappielo, C.D.; Nanicuacua, D.M.; dos Santos, W.N.; da Silva, D.L.; Dall'Antônia, L.H.; Oliveira, F.M.d.; Clausen, D.N.; Tarley, C.R. Solid phase extraction to on-line preconcentrate trace cadmium using chemically modified nano-carbon black with 3-mercaptopropyltrimethoxysilane. *J. Braz. Chem. Soc.* **2016**, *27*, 1715–1726. [CrossRef]
43. Veca, L.M.; Nastase, F.; Banciu, C.; Popescu, M.; Romanitan, C.; Lungulescu, M.; Popa, R. Synthesis of macroporous ZnO-graphene hybrid monoliths with potential for functional electrodes. *Diam. Relat. Mater.* **2018**, *87*, 70–77. [CrossRef]
44. Hristea, G.; Iordoc, M.; Lungulescu, E.-M.; Bejenari, I.; Volf, I. A sustainable bio-based char as emerging electrode material for energy storage applications. *Sci. Rep.* **2024**, *14*, 1095. [CrossRef]
45. Setnescu, R.; Jipa, S.; Setnescu, T.; Kappel, W.; Kobayashi, S.; Osawa, Z. IR and X-ray characterization of the ferromagnetic phase of pyrolysed polyacrylonitrile. *Carbon* **1999**, *37*, 1–6. [CrossRef]
46. Domingo-García, M.; Garzón, F.L.; Pérez-Mendoza, M. On the characterization of chemical surface groups of carbon materials. *J. Colloid Interface Sci.* **2002**, *248*, 116–122. [CrossRef]

47. Thomassin, J.-M.; Jérôme, C.; Pardoen, T.; Bailly, C.; Huynen, I.; Detrembleur, C. Polymer/carbon based composites as electromagnetic interference (EMI) shielding materials. *Mater. Sci. Eng. R Rep.* **2013**, *74*, 211–232. [CrossRef]
48. Ren, M.; Liu, X.; Jia, X.; Luo, C.; Zhang, L.; Alamo, R.G. Memory of crystallization in the melt of commercial linear low density polyethylenes processed in an open twin-screw extruder. *Thermochim. Acta* **2023**, *720*, 179423. [CrossRef]

Disclaimer/Publisher's Note: The statements, opinions and data contained in all publications are solely those of the individual author(s) and contributor(s) and not of MDPI and/or the editor(s). MDPI and/or the editor(s) disclaim responsibility for any injury to people or property resulting from any ideas, methods, instructions or products referred to in the content.

Article

Preparation and Characterization of Composites Based on ABS Modified with Polysiloxane Derivatives

Bogna Sztorch [1,*], Roksana Konieczna [1,2], Daria Pakuła [1,2], Miłosz Frydrych [1,2], Bogdan Marciniec [1,2] and Robert E. Przekop [1]

1. Centre for Advanced Technologies, Adam Mickiewicz University Poznan, Uniwersytetu Poznańskiego 10, 61-614 Poznan, Poland; rokkon@amu.edu.pl (R.K.); darpak@amu.edu.pl (D.P.); frydrych@amu.edu.pl (M.F.); bogdan.marciniec@amu.edu.pl (B.M.); rprzekop@amu.edu.pl (R.E.P.)
2. Faculty of Chemistry, Adam Mickiewicz University in Poznań, Uniwersytetu Poznańskiego 8, 61-614 Poznan, Poland
* Correspondence: bogna.sztorch@amu.edu.pl

Citation: Sztorch, B.; Konieczna, R.; Pakuła, D.; Frydrych, M.; Marciniec, B.; Przekop, R.E. Preparation and Characterization of Composites Based on ABS Modified with Polysiloxane Derivatives. *Materials* **2024**, *17*, 561. https://doi.org/10.3390/ma17030561

Academic Editors: Eduard-Marius Lungulescu, Radu Setnescu, Cristina Stancu and Andrea Sorrentino

Received: 11 August 2023
Revised: 21 November 2023
Accepted: 16 January 2024
Published: 24 January 2024

Copyright: © 2024 by the authors. Licensee MDPI, Basel, Switzerland. This article is an open access article distributed under the terms and conditions of the Creative Commons Attribution (CC BY) license (https://creativecommons.org/licenses/by/4.0/).

Abstract: In this study, organosilicon compounds were used as modifiers of filaments constituting building materials for 3D printing technology. Polymethylhydrosiloxane underwent a hydrosilylation reaction with styrene, octadecene, and vinyltrimethoxysilane to produce new di- or tri-functional derivatives with varying ratios of olefins. These compounds were then mixed with silica and incorporated into the ABS matrix using standard processing methods. The resulting systems exhibited changes in their physicochemical and mechanical characteristics. Several of the obtained composites (e.g., modified with VT:6STYR) had an increase in the contact angle of over 20° resulting in a hydrophobic surface. The addition of modifiers also prevented a decrease in rheological parameters regardless of the amount of filler added. In addition, comprehensive tests of the thermal decomposition of the obtained composites were performed and an attempt was made to precisely characterize the decomposition of ABS using FT-IR and optical microscopy, which allowed us to determine the impact of individual groups on the thermal stability of the system.

Keywords: ABS; FDM; polysiloxane; silica; 3D printing

1. Introduction

Additive technologies, commonly known as 3D printing, are an intensively developing field that involves the use of a three-dimensional virtual model to create a real object layer-by-layer. During printing, materials such as polymers, ceramics, or metal can be bonded together permanently when exposed to a desired temperature or a laser beam. The rapid development of additive technologies began in 2009 when most of the patents for 3D printing devices expired, which made it possible to use the technology on a wider scale [1].

Three-dimensional printing currently offers a wide spectrum of applications. It is used in the construction [2,3], automotive [4,5], medical [6–8], decoration, machinery, dental, and textile industries. Among the most important advantages of additive technologies is the ability to produce models with complex geometry without the need for multiple manufacturing tools. Disadvantages include limited dimensional accuracy, the problem of printing skewed surfaces, and the frequent need for additional surface treatments. Compared with conventional methods (i.e., milling and turning), additive techniques are a relatively young and rapidly growing discipline. Therefore, their disadvantages are gradually being eliminated through the introduction of design improvements and the development of a new range of materials suitable for 3D printing [9].

FDM (*Fused Deposition Modeling*) is one of the most commonly used incremental techniques. It creates three-dimensional objects by adding successive layers of molten material. The printout is based on the generated digital 3D object and provides us with the ability to manipulate the geometry of an object. The possibility of obtaining models with a complex geometry combined with a shorter design time and lower production costs make this technique very popular [10,11].

Thermoplastics are attractive materials for use in FDM as is the older technology of injection molding. Polymers such as acrylonitrile butadiene styrene (ABS), polylactide (PLA), and polyamide are among the best ones available [12]. ABS has been used in the electronics sector (e.g., for use in the bodies of electrical devices), sports, the automotive [4,5,13], household appliances (e.g., for use in the bodies of TV sets and radio receivers), and toys [14–16]. This material is characterized by its hardness and good mechanical strength, which have also been observed at low temperatures. In addition, ABS does not conduct electricity, has high impact resistance, and is resistant to high temperatures [17]. It owes its chemical resistance and thermal stability to the presence of acrylonitrile units in the polymer chain. Aromatic groups affects the stiffness and processability of the material. The butadiene phase, on the other hand, is responsible for improved impact strength and hardness [18]. One of the main disadvantages of using this material is its low adhesion to the surface during printing, and deformations occurring during its shape formation are another disadvantage [19]. Numerous publications have aimed to mitigate these effects through material modifications through the addition of different fillers while simultaneously enhancing mechanical or processing characteristics [20–24]. In our work, we presented the preparation of ABS composites with a modified silica nanofiller. There are well-known applications of nanofillers in the literature; Kim, I.-J. et al. [25] in their work described ABS nanocomposites with silica nanoparticles obtained through emulsion polymerization techniques and compress molding in their research. The composites produced through this method exhibited a noteworthy boost in impact strength, with an approximate enhancement of 30%.

In work [26] the influence of nanoparticles, i.e., montmorillonite, $CaCO_3$, silica, and multiwalled carbon nanotubes, on the mechanical properties of ABS composites produced using the FDM technique was examined. It was found that the addition of fillers improved the mechanical strength and thermal stability of the samples. Moreover, some of the obtained composites were characterized by increased bending strength and reduced mechanical anisotropy.

In the work of Bai Huang et al. [27], silica-based modifiers were also produced to modify ABS for 3D printing using FDM technology. They optimized the efficiency of 3D printing acrylonitrile–butadiene–styrene composites through cellulose nanocrystal/silica nanohybrids (CSNs), resulting in composites with increased adhesion between layers. CSNs obtained through the TEOS sol–gel method have uniformly distributed nanosilica on their surfaces, and nanohybrids demonstrated both an excellent efficiency and well-reinforcement effect in FDM. In our previous work, we thoroughly discussed the impact of functionalized organosilicon additives in the context of other thermoplastics such as polyethylene (PE) and polylactide acid (PLA) [28,29].

This paper presents the surface modification of silica using organosilicon compounds (polysiloxane derivatives) and then introduces the filler into the polymer matrix. The obtained composites were tested for processing properties (through the mass flow index) and physicochemical properties (through surface analysis and mechanical properties). Microscopic images (through an optical microscope) were taken to determine the morphology and dispersion of the filler in the matrix. Differential scanning calorimetry (DSC) and thermogravimetric analysis (TG) were also performed to determine thermal properties. The investigation conducted here facilitates the assessment of the impact of modified nanosilica on the resultant composites and also enables the evaluation of the potential applications of the material obtained.

2. Materials and Methods

2.1. Materials

ABS type TR558 was purchased from LG Chem (Seul, South Korea), Fumed silica filler AEROSIL® 200 (Aero) with a specific surface area of 200 m^2/g was purchased from Evonik (Essen, Germany)

The chemicals were purchased from the following sources:

Polymethylhydrosiloxane, trimethylsilyl-terminated, 15–25 cSt from Gelest (Morrisville, PA, USA); styrene (STYR), octadecene (OD), toluene, chloroform-d, Karstedt's catalyst xylene solution from Merck KGaA (Darmstadt, Germany); vinyltrimethoxysilane (VT) from BRB; and P_2O_5 from Avantor Performance Materials Poland S.A. (Gliwice, Poland) Toluene was degassed and dried by distilling it from P_2O_5 under an argon atmosphere.

2.2. Analyses

Fourier transform-infrared (FT-IR) spectra were recorded on a Nicolet iS 50 Fourier transform spectrophotometer (Thermo Fischer Scientific, Waltham, MA, USA) equipped with a diamond ATR unit with a resolution of 0.09 cm^{-1}.

^{1}H, ^{13}C, and ^{29}Si nuclear magnetic resonance (NMR) spectra were recorded at 25 °C on Bruker Ascend 400 and Ultra Shield 300 spectrometers using CDCl$_3$ as a solvent. Chemical shifts were reported in ppm concerning the residual solvent (CHCl$_3$) peaks for ^{1}H and ^{13}C.

The melt flow rate (MFR) was measured using the Instron CEAST MF20 melt flow tester according to the standard [30] at 220 °C for the load of 5 kg, and the time of cutting off the polymer stream was 30 s.

Water contact angle (WCA) analysis were performed through the sessile drop technique at room temperature and atmospheric pressure using a Krüss DSA100 goniometer. Three independent measurements were taken for each sample, each with a 5 μL water drop, and the obtained results were averaged.

Light microscopy images of the surface and fractures of the composites were taken using a KEYENCE VHX-7000 digital microscope (Keyence International, Mechelen, Belgium, NV/SA) with a 100–1000 VH-Z100T zoom lens. All images were recorded using a VHX 7020 camera.

Tensile and flexural strength tests were performed using the universal testing machine INSTRON 5969 with a maximum measuring capability of 50 kN. Seven samples were selected from each system, placed in the testing machine, and subjected to tensile and flexural tests. For each modifier, seven values of stress, modulus of elasticity, and elongation were obtained, which were then averaged. The traverse speed for the tensile strength measurements was set at 2 mm/min.

A Charpy impact test (with no notch) was performed on an Instron Ceast 9050 impact machine according to the [31] standard. For all the series, 6 measurements were performed for each material.

Hardness of the composite samples was tested through the Shore method using a durometer from Bareiss Prüfgerätebau GmbH (Oberdischingen, Germany).

Thermogravimetry (TGA) was performed using a NETZSCH 209 F1 Libra gravimetric analyzer (Selb, Germany). Samples of 9 ± 0.5 mg were cut from each granulate and placed in Al$_2$O$_3$ crucibles. Measurements were conducted under nitrogen and air (flow of 20 mL/min) within various temperature ranges, i.e., from 30 °C to 390 °C, from 30 °C to 400 °C, from 30 °C to 455 °C, or from 30 °C to 500 °C and a 10 °C/min heating rate.

Differential scanning calorimetry (DSC) was performed using a NETZSCH204 F1 Phoenix calorimeter. Samples of 6 ± 0.2 mg were placed in an aluminum crucible with a punctured lid. The measurements were performed under nitrogen within a temperature range of −20 °C to 310 °C and at a 10 °C/min heating rate.

2.3. The Procedure for Synthesis of Polysiloxane Derivatives

In a typical procedure, a 500 mL three-neck round bottom flask was charged with 30 g of polymethylhydrosiloxane, 250 mL of toluene, and calculated amounts of olefins (Table 1).

The reaction mixture was set at 70 °C and, before reaching boiling point, Karstedt's catalyst (10^{-5} eq Pt/mol SiH) solution was added, which resulted in a quick increase in temperature and the system starting to reflux. The reaction mixture was kept at reflux and samples were taken for FT-IR control until full Si–H group disappearance was observed. Then, the solvent was evaporated to dryness under a vacuum to obtain a pure analytical sample.

Table 1. Amounts of olefins used in the reactions.

Code	Amount of VT/g	Amount of STYR/g	Amount of OD/g
VT:6STYR	11.0	46.5	-
VT:4STYR:2OD	11.0	31.0	37.6
VT:3STYR:3OD	11.0	23.2	56.4
VT:2STYR:4OD	11.0	15.5	75.1
VT:6OD	11.0	-	112.6

2.4. The Procedure of Mixing Nanosilicate with a Modifier

A total of 30 g of Aerosil200 and 30 g of an organosilicon modifier were weighed in a ceramic vessel with alumina balls. In the next step, the system was placed on rollers and mixing was continued for 24 h. The ABS copolymer, after pre-drying in an oven for 2 h at 70 °C, was loaded into a V-mixer, and then a pre-weighed silica modifier mixture (5% Aerosil 200, 5% modifier, 90% ABS) was added. Each of the systems was mixed in a mixer for 20 min. Microscopic images for all Aerosil + modifier mixtures looked analogous and are presented in Figure 1.

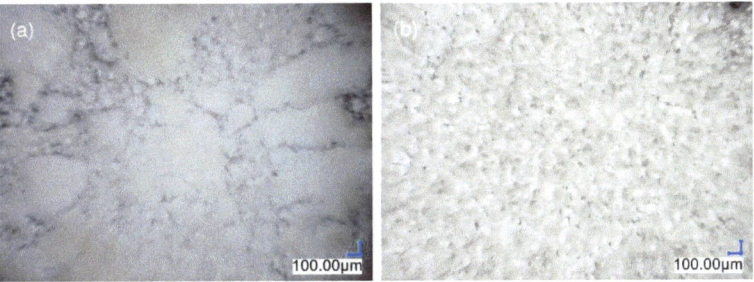

Figure 1. Optical microscopic images of (**a**) Aerosil and (**b**) Aerosil with modifier.

2.5. Injection Molding

The samples were homogenized through the injection molding process. An Engel E-victory 170/80 tie-bar-less injection molding machine produced the test samples. The parameters of the injection process are presented in Table 2. Standardized type 1A fittings were obtained according to the [32] standard (Figure 2) for further testing.

Table 2. The parameters of the injection process.

Properties	Parameters
Maximum dispensing time	15.0 s
Dispensing volume	31.00 cm^3
Holding time	7 s
Cooling time	35.00 s
Holding pressure	500.0–1100.0 bar
Mold temperature	70 °C
Dosing efficiency	0.71 cm^3/s

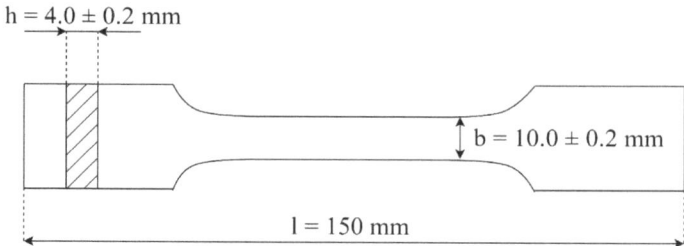

Figure 2. A 1A fitting with dimensions as follows: total length, l = 150 mm; thickness, h = 4 ± 0.2 mm; width of the measuring part, b = 10.0 ± 0.2 mm [33].

2.6. Preparation of Filament

The injection samples were milled. The obtained granulate was homogenized during the extrusion of the stream with the HAAKE Rheomex OS.

The filament obtained in the previous stage was ground and extruded into a filament for 3D printing. Appropriate amounts of granules containing 5% modifier and ABS were weighed on a laboratory balance, and each system was diluted into six concentrations as follows: 0.1%, 0.25%, 0.5%, 1%, 1.5%, and 2.5%. The filament was extruded on a Filabot EX6 single-screw extruder with four heating zones (Table 3), an L/D 24 screw, and a nozzle with a diameter of 1.75 mm.

Table 3. The parameters of the filament extrusion.

Properties	Parameters
Zone 1 temperature	215 °C
Zone 2 temperature	240 °C
Zone 3 temperature	230 °C
Filling zone temperature	90 °C
Voltage	20–25 V
Current	1–1.5 A

2.7. 3D Printing (FDM)

An extruded filament was used for FDM printing. The samples were printed on a Flashforge Guider IIs with a closed working chamber, a heated bed, and an extruder heating up to a maximum temperature of 300 °C. Parameters of printing are collected together and shown in Table 4. Bars for impact and bending tests were printed (Figure 3) as were paddles for stretching tests (Figure 4).

Table 4. The parameters used for 3D printing.

Properties	Parameters
Nozzle diameter	0.4 mm
Extruder temperature	225 °C
Bed temperature	105 °C
Layer height	0.2 mm
Bottom and top layer style	linear
Fill style	linear
Infill density	100%
Printing speed	60 mm/s

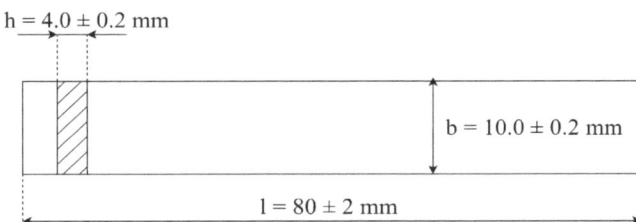

Figure 3. Dimensions of samples used for the bending test as follows: total length, l = 80 ± 2 mm; thickness, h = 4.0 ± 0.2 mm; width of the measuring part b = 10.0 ± 0.2 mm [33].

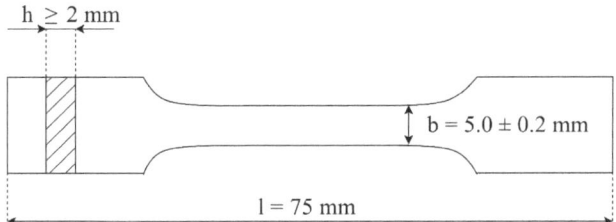

Figure 4. Dimensions of samples used for stretching test: total length l = 75 mm, thickness h ≥ 2 mm, the width of the measuring part, b = 5.0 ± 0.2 mm [33].

3. Results and Discussion

3.1. Chemical Characterization of Modifiers

The hydrosilylation process was carried out until the disappearance of the signal from the Si–H group (the strong band from stretching vibrations at 2140 cm^{-1} and the bending vibrations at the wavelength of 888 cm^{-1}), which indicates the complete conversion of polysiloxane. All compounds were obtained at a high yield (>94%) (Figures 5–9). To validate compound structure, purity, and conversion, NMR analysis (^{1}H, ^{13}C, ^{29}Si) was performed. The following signals were assigned:

VT:6STYR

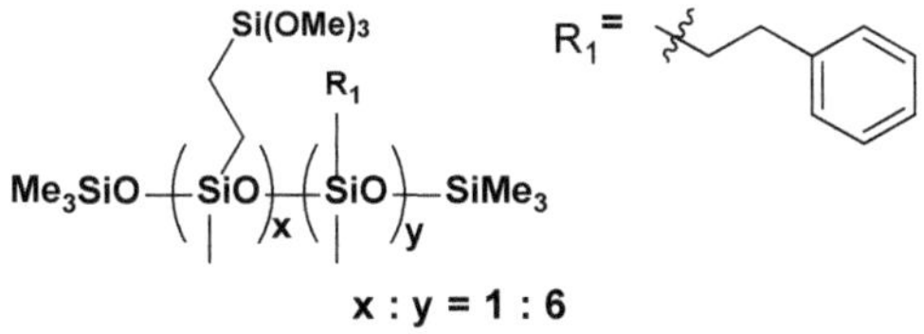

Figure 5. The formula of VT:6STYR.

^{1}**H NMR** (400 MHz, CDCl$_3$): σ(ppm) = 7.44–7.16 (m, Ph), 6.78, 6.75, 6.74, 6.71, 5.80, 5.75, 5.28, 5.25, 3.55 (s, -Si(OCH$_3$)$_3$), 2.71 (m, Ph-CH$_2$CH$_2$-Si-), 2.62, 2.38 (solvent), 2.18, 1.52, 1.41, 1.15, 0.94 (m, Ph-CH$_2$-CH$_2$-Si), 0.86, 0.64 (m, -SiCH$_2$CH$_2$Si-), 0.57, 0.15, 0.07 (-Si(CH$_3$)$_3$, -SiCH$_3$)

^{13}**C NMR** (101 MHz, CDCl$_3$): σ(ppm) = 144.61, 144.33, 144.15, 137.98, 137.69, 137.01, 129.17 (solvent), 128.63–124.88 (m, Ph), 113.90, 50.63, 30.86, 30.72, 29.27, 29.15, 29.02, 21.59, 19.73, 19.52 (-CH$_2$-Ph), 15.76, 15.01 (-Si-CH$_2$), 14.80, 8.50, 2.05, −0.17, −1.97 (m, -Si(CH$_3$)$_3$, -SiCH$_3$)

^{29}**Si NMR** (79.5 MHz, CDCl$_3$): σ(ppm) = −22.68–(−23.01) (SiMe, SiMe$_3$), −41.96 (SiOMe$_3$)

VT:4STYR:2OD

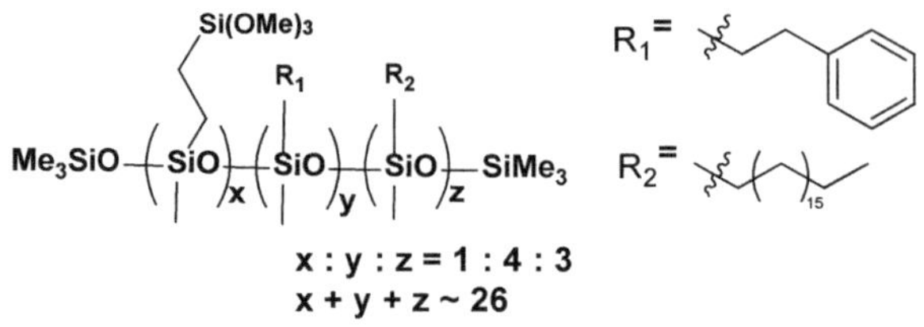

Figure 6. The formula of VT:4STYR:2OD.

^{1}**H NMR** (400 MHz, CDCl$_3$): σ(ppm) = 7.19, 7.13 (m, Ph), 3.53 (s, -Si(OCH$_3$)$_3$), 2.67 (m, Ph-CH$_2$CH$_2$-Si-), 2.36 (solvent), 1.55–1.26 (Si-CH$_2$(CH$_2$)$_{16}$CH$_3$), 0.88 (t, -CH$_2$-CH$_3$), 0.52 (-SiCH$_2$(CH$_2$)$_{16}$CH$_3$), 0.08 (-Si(CH$_3$)$_3$, -SiCH$_3$)
^{13}**C NMR** (101 MHz, CDCl$_3$): σ(ppm) = 129.18 (solvent), 128.37–127.81 (m, Ph), 32.09, 29.89, 29.53 (-CH$_2$-), 22.85 (-CH$_3$), 14.27 (-Si-CH$_2$), 0.12 (-Si(CH$_3$)$_3$, -SiCH$_3$)
^{29}**Si NMR** (79.5 MHz, CDCl$_3$): σ(ppm) = −22.08–(−23.53) (SiMe, SiMe$_3$), −41.92 (SiOMe$_3$)

VT:3STYR:3OD

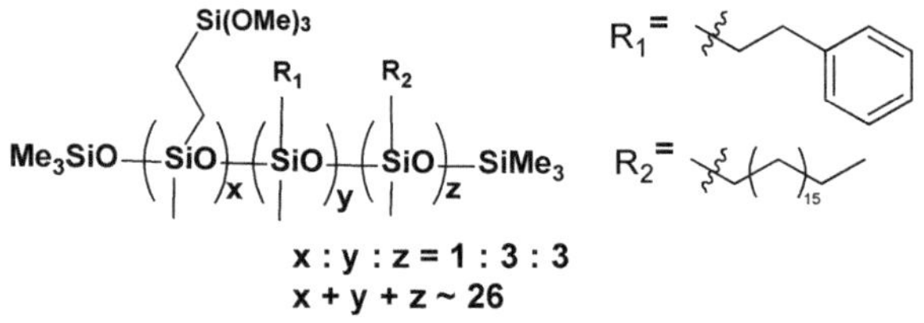

Figure 7. The formula of VT:3STYR:3OD.

^{1}**H NMR** (400 MHz, CDCl$_3$): σ(ppm) = 7.14 (m, Ph), 3.53 (s, -Si(OCH$_3$)$_3$), 2.66 (m, Ph-CH$_2$CH$_2$-Si-), 2.36 (solvent), 1.54–1.25 (Si-CH$_2$(CH$_2$)$_{16}$CH$_3$), 0.88 (t, -CH$_2$-CH$_3$), 0.51 (-SiCH$_2$(CH$_2$)$_{16}$CH$_3$), 0.08–0.07 (-Si(CH$_3$)$_3$, -SiCH$_3$)
^{13}**C NMR** (101 MHz, CDCl$_3$): σ(ppm) = 129.18 (solvent), 128.37–127.81 (m, Ph), 32.09, 29.90, 29.53 (-CH$_2$-), 22.85 (-CH$_3$), 14.27 (-Si-CH$_2$)
^{29}**Si NMR** (79.5 MHz, CDCl$_3$): σ(ppm) = −22.53–(−24.08) (SiMe, SiMe$_3$), −41.78 (SiOMe$_3$)

VT:2STYR:4OD

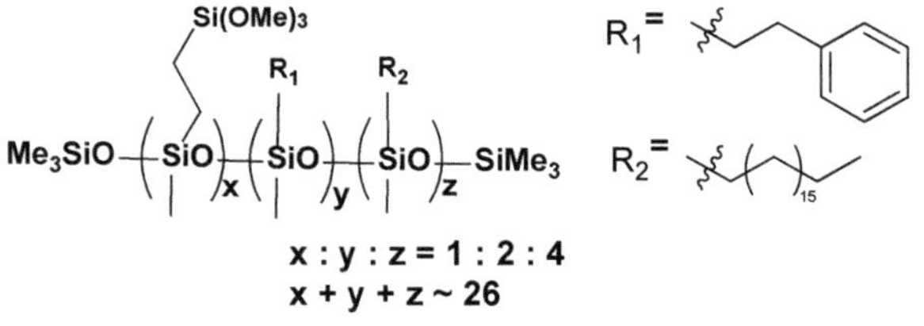

Figure 8. The formula of VT:2STYR:4OD.

^{1}**H NMR** (400 MHz, CDCl$_3$): σ(ppm) = 7.18 (solvent), 3.55 (s, -Si(OCH$_3$)$_3$), 2.36 (solvent), 1.54, 1.25 (m, Si-CH$_2$(CH$_2$)$_{16}$CH$_3$), 0.88–0.86 (t, -CH$_2$-CH$_3$), 0.51 (m, -SiCH$_2$CH$_2$Si-), 0.07 (m, -Si(CH$_3$)$_3$, -SiCH$_3$)
^{13}**C NMR** (101 MHz, CDCl$_3$): σ(ppm) = 128.37–127.81 (m, Ph), 32.10, 29.91, 29.54 (-CH$_2$-), 22.85 (-CH$_3$), 14.27 (-Si-CH$_2$)
^{29}**Si NMR** (79.5 MHz, CDCl$_3$): σ(ppm) = −22.49–(−23.94) (SiMe, SiMe$_3$), −41.59 (SiOMe$_3$)

VT:6OD

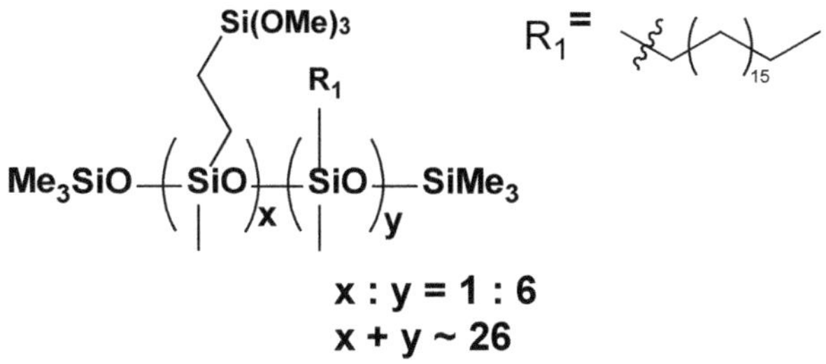

Figure 9. Formula of VT:6OD.

^{1}H NMR (400 MHz, CDCl$_3$): σ(ppm) = 7.18 (solvent) 3.55 (s, -Si(OCH$_3$)$_3$), 2.36 (solvent), 1.55, 1.26 (m, Si-CH$_2$(CH2)$_{16}$CH$_3$), 0.88 (t, -CH$_2$-CH$_3$), 0.50 (m, -SiCH$_2$CH$_2$Si-), 0.04 (m, -Si(CH$_3$)$_3$, -SiCH$_3$)

^{13}C NMR (101 MHz, CDCl$_3$): σ(ppm) = 129.18 (solvent), 32.11, 29.93, 29.86, 29.56 (-CH$_2$-), 22.86 (-CH$_3$), 14.27 (-Si-CH$_2$)

^{29}Si NMR (79.5 MHz, CDCl$_3$): σ(ppm) = −22.06–(−23.01) (SiMe, SiMe$_3$), −41.78 (SiOMe$_3$)

3.2. Thermal Analysis (TGA and DSC)

Thermogravimetric analysis (TGA) and differential scanning calorimetry (DSC) were performed for both reference and ABS with unmodified SiO$_2$ as well as ABS/SiO$_2$/organosilicon modifier composites.

3.2.1. Thermogravimetric Analysis (TGA)

The determined parameters, including temperatures of 1% and 5% mass loss, the temperature of the start of degradation, and the temperature of the maximum rate of mass loss, are summarized in Tables 5 and 6. The process of the thermal decomposition of the samples was carried out in air and nitrogen atmospheres (Figure 10).

Table 5. Thermogravimetric analysis results in an air atmosphere.

Code	1% of Weight Loss/°C	5% of Weight Loss/°C	Onset Temperature/°C	Temperature at the Maximum Rate of Mass Loss/°C
neat ABS	235.7	346.6	363.7	392.6
ABS + 0.1%Aero	205.5	342.8	364.0	396.3
ABS + 0.1% VT:6STYR/Aero	191.9	341.2	365.6	397.1
ABS + 0.1% VT:4STYR:2OD/Aero	217.2	342.1	363.9	395.0
ABS + 0.1% VT:3STYR:3OD/Aero	207.8	341.0	366.9	399.7
ABS + 0.1% VT:2STYR:4OD/Aero	187.3	340.7	365.3	397.1
ABS + 0.1% VT:6OD/Aero	187.1	340.6	364.5	396.3

The study of thermal stability in an air atmosphere shows that the temperature of the onset of degradation is highest for the system that has the same molar ratio of styrene and octadecyl substituents (ABS + VT:3STYR:3OD/Aero); an excess within any of these groups causes a decrease in this temperature.

Table 6. Thermogravimetric analysis results in a nitrogen atmosphere.

Code	1% of Weight Loss/°C	5% of Weight Loss/°C	Onset Temperature/°C	Temperature at the Maximum Rate of Mass Loss/°C
neat ABS	212.9	350.2	361.9	384.3
ABS + 0.1%Aero	200.6	349.7	367.9	393.2
ABS + 0.1% VT:6STYR/Aero	209.6	351.3	365.1	389.7
ABS + 0.1% VT:4STYR:2OD/Aero	196.3	350.1	365.4	390.0
ABS + 0.1% VT:3STYR:3OD/Aero	191.3	349.1	370.3	400.1
ABS + 0.1% VT:2STYR:4OD/Aero	200.2	352.5	366.7	391.5
ABS + 0.1% VT:6OD/Aero	205.8	353.8	368.5	393.2

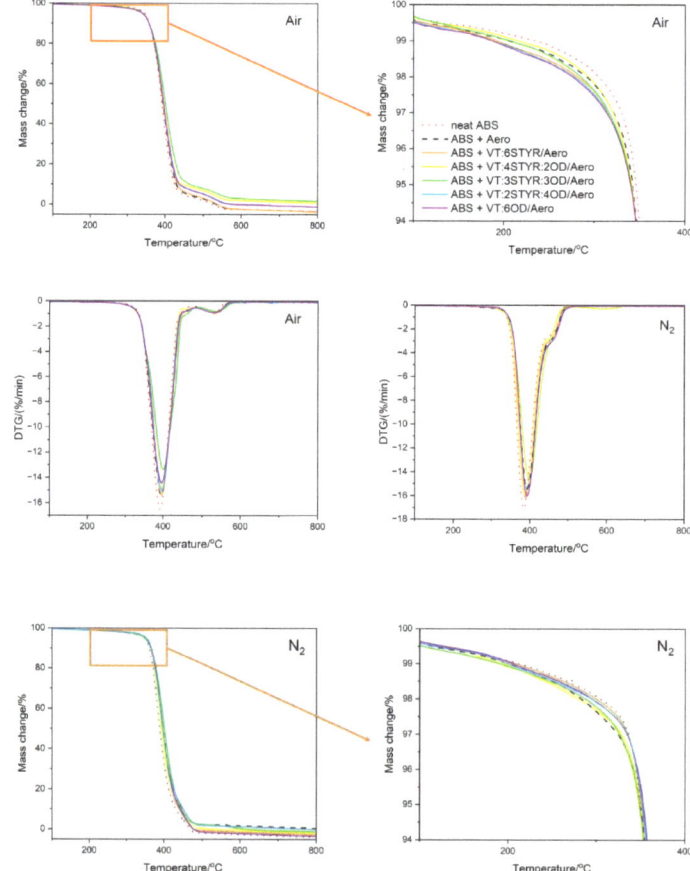

Figure 10. TGA and DTG curves of the ABS/SiO$_2$/modifier composite in air and nitrogen atmospheres.

The thermal stability study in an inert gas atmosphere proved that the degradation onset temperature is utmost for the system that has the same molar ratio of styrene and octadecyl substituents (ABS + VT:3STYR:3OD/Aero); an excess of any of these groups causes a decrease in onset temperature. The temperature of the onset of degradation in the nitrogen atmosphere is the lowest for the reference sample of neat ABS, and the modification further increases the thermal stability by 8.4 °C. The highest temperature at the maximum rate of mass loss occurs for the ABS + VT:3STYR:3OD/Aero composite. The difference between ABS + VT:3STYR:3OD/Aero and neat ABS is 15.8 °C. A loss of 1% of mass occurs latest (at the highest temperature) for a neat ABS sample. On the other hand, the temperature at 5% mass loss oscillates around 350 °C for both the reference samples and the modified ABS.

The temperature at the maximum weight loss is higher for modified ABS (in nitrogen and air) due to the higher Si—O bond energy (488 kJ/mol) in silica and the modifier than that of the C—C bonds (348 kJ/mol) and the C—H bonds (415 kJ/mol) in ABS copolymer.

3.2.2. Differential Scanning Calorimetry (DSC)

DSC analysis was performed to determine the effect of additives on phase transition temperatures such as the glass transition temperature (T_g) of the ABS samples. Figure 11 shows the thermograms of pure ABS and ABS with the addition of 0.1% by weight of Aerosil 200 as well as with the addition of 0.1% of both silica and the organosilicon modifier.

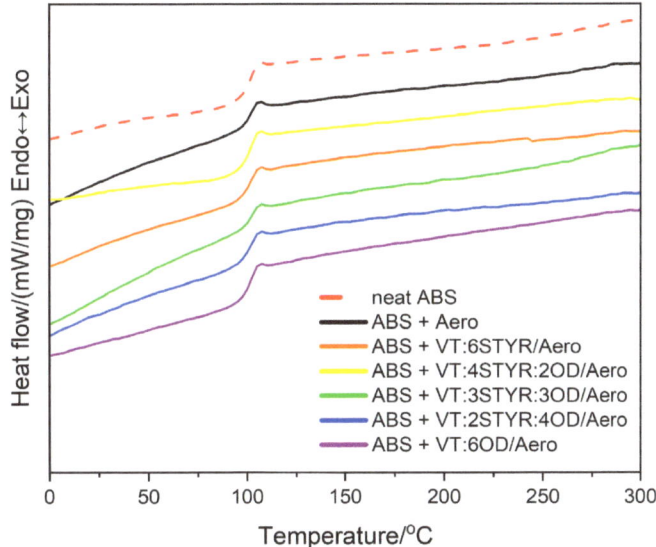

Figure 11. Results of the DSC analysis of neat ABS and its composites.

The transition at a temperature of about 106 °C, which is visible in the thermographs, corresponds to the glass transition temperature (T_g) that belongs to the second-order transformations. It is interpreted as a transition from a glassy state to a liquid state and is associated with the transition from the glassy phase to the rubber phase in this case. The addition of nanosilica and organosilicon modifiers does not significantly affect the change of the glass transition temperature of pure ABS (105.8 °C). The highest glass transition temperature is observed for ABS + 0.1% VT:2STYR:4OD/Aero (106.2 °C). The elevated T_g value following the introduction of the filler into the matrix is associated with the constrained thermal mobility of acrylonitrile butadiene styrene (ABS) chain segments. This phenomenon significantly influences the polymer's processability by inducing a rise in system viscosity (see Section 3.4) [26]. ABS is an amorphous polymer, so no visible melting point is observed for the thermogram.

3.3. Testing the Composition of the Sample during Temperature Decomposition

Thermogravimetric tests were carried out both in nitrogen and in the air for the ABS + VT:4STYR:2OD/Aero samples, which were carried out within various temperature ranges, i.e., from 30 °C to 390 °C, from 30 °C to 400 °C, from 30 °C to 455 °C, and from 30 °C to 500 °C (Figures 12 and 13). The percentage of residual mass is summarized in Table 7.

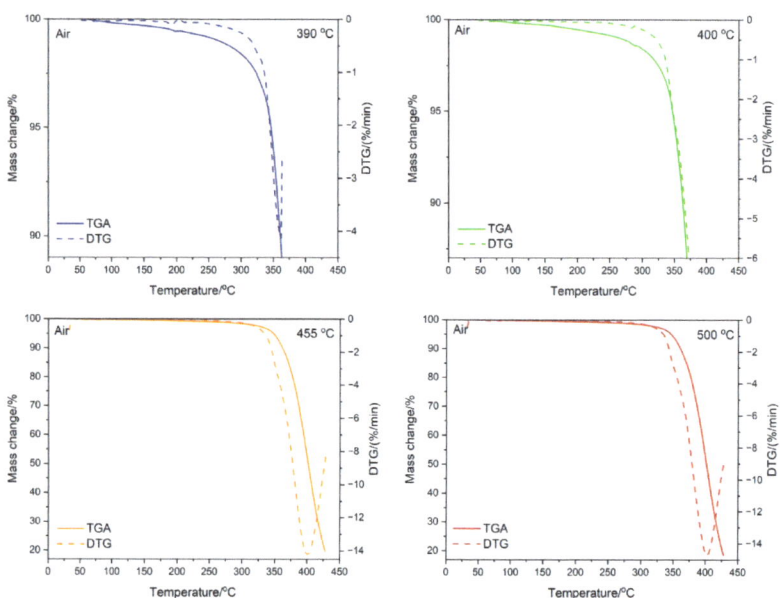

Figure 12. Results of thermogravimetric analysis for the ABS/VT:4STYR:2OD/Aero sample in an air atmosphere.

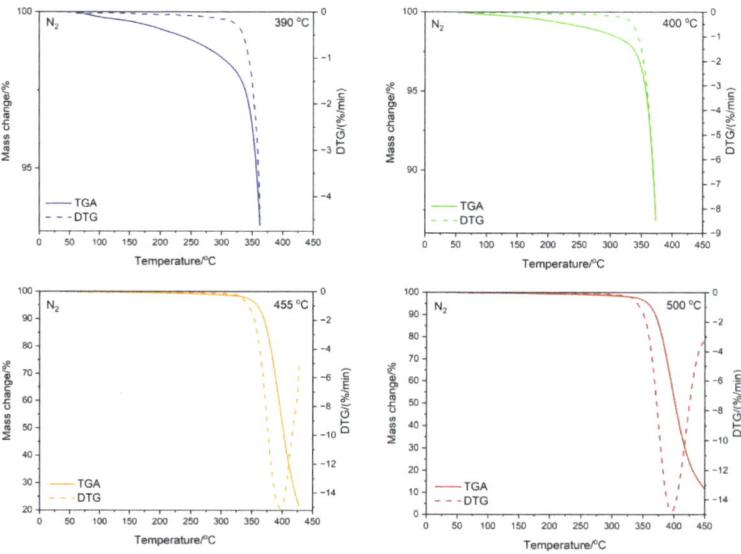

Figure 13. Results of thermogravimetric analysis for the ABS/VT:4STYR:2OD/Aero sample in a nitrogen atmosphere.

Table 7. Results of thermogravimetric analysis of the ABS + VT:4STYR:2OD/Aero sample at different temperatures.

Conditions	Residual Mass/°C	
	N_2	Air
390 °C	92.75	88.86
400 °C	86.40	84.35
455 °C	21.27	18.60
500 °C	5.14	17.46

The images of the samples after the TGA test were made using an optical microscope at 100 and 200 times magnification, and the results are shown in Figures 13 and 14. Transmission measurements were carried out for samples subjected to thermogravimetric analysis within various temperature ranges. Sample weights alongside KBr were ground in an agate mortar, and then pellets were made using a hydraulic press. The spectra of the samples, together with their microscopic images, are shown in Figure 14 (TGA in the air) and Figure 15 (TGA in nitrogen).

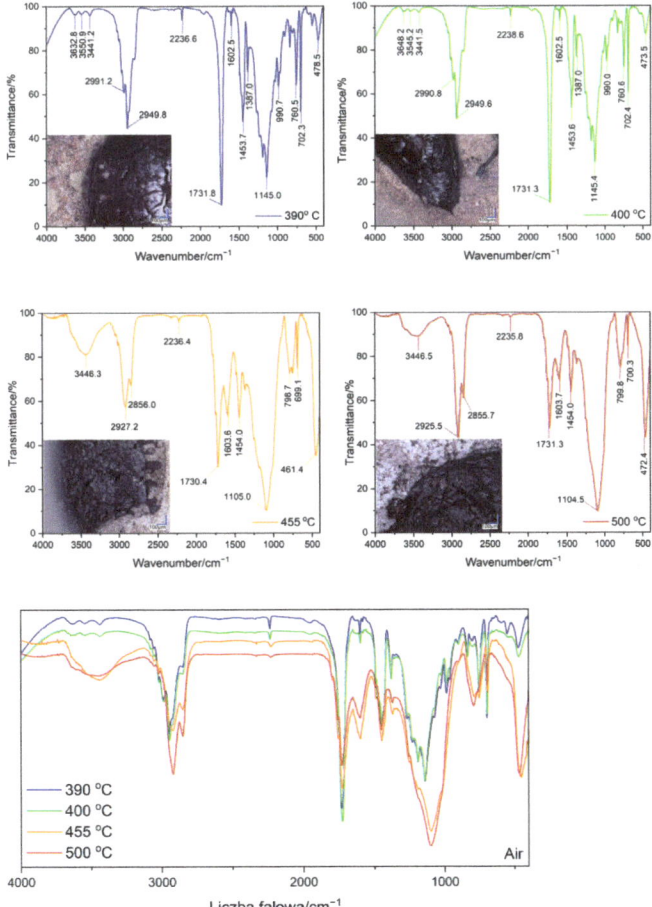

Figure 14. Results of FT-IR analysis with optical microscopic images of samples after TGA measurements in air atmosphere.

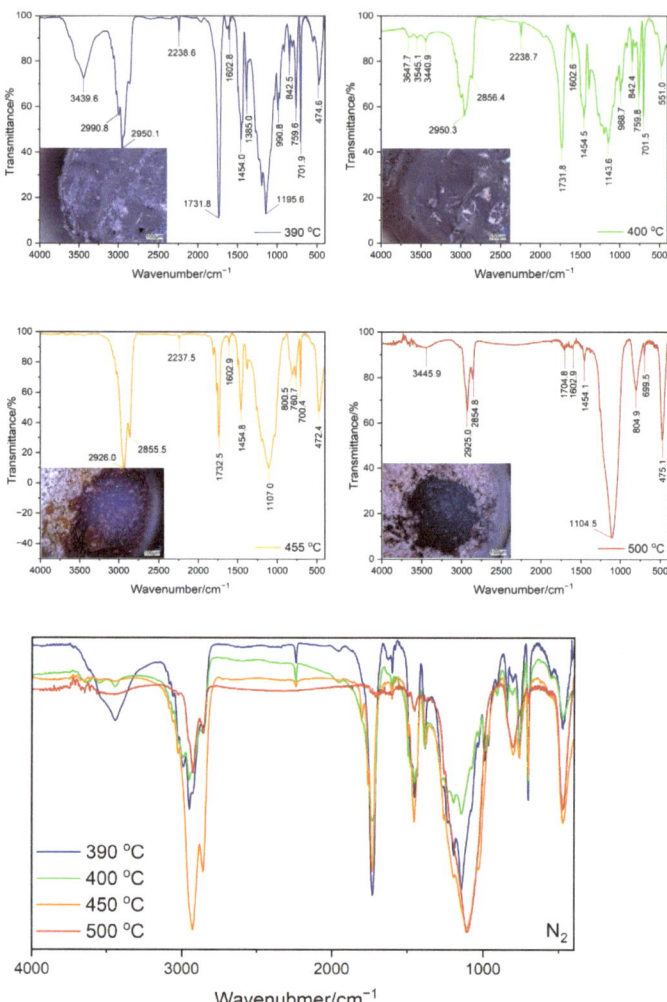

Figure 15. Results of FT-IR analysis with optical microscopic images of samples after TGA testing in a nitrogen atmosphere.

The broad peak near the value of 3440 cm^{-1} corresponds to the O–H stretching vibrations found in silica. As the degradation temperature increases, Si-O-Si stretching (about 1100 cm^{-1} and 800 cm^{-1}) and the bending (460/470 cm^{-1}) vibrations of nanosilica increase in intensity. This means that the mass percentage of Aerosil increases with the loss of volatile decomposition observed among products of the sample at the given temperatures. As the temperature increases, signals from vibrations in the aromatic ring, mainly that from C-H stretching vibrations near 3025 cm^{-1} originating in styrene, disappear in both ABS and the modifier. For a peak near 2235 cm^{-1} coming from the stretching vibrations of the carbon–nitrogen triple bond present in the polymer chain, complete disappearance is observed at 500 °C in a nitrogen atmosphere. Peak 1730 cm^{-1} corresponds to the stretching vibrations of the C–O double bond present in the decomposition products.

3.4. Rheology

Neat ABS's melt flow rate (MFR) at 220 °C is 8.664 g/10 min. The addition of small amounts of modifiers does not significantly change the MFR value (Figure 16). The highest value of the MFR index is characterized by samples containing VT:3Styr:3OD/Aero with a concentration of 0.1%. The styryl groups are responsible for interactions with the ABS matrix through weak π-stacking interactions, while the octadecyl group causes a better plasticization effect of the system; therefore, they slightly improve the melt flow ratio. The addition of modifiers also prevents a decrease in the rheological parameters regardless of the amount of filler (Aero) added, which is important from the perspective of plastic processing.

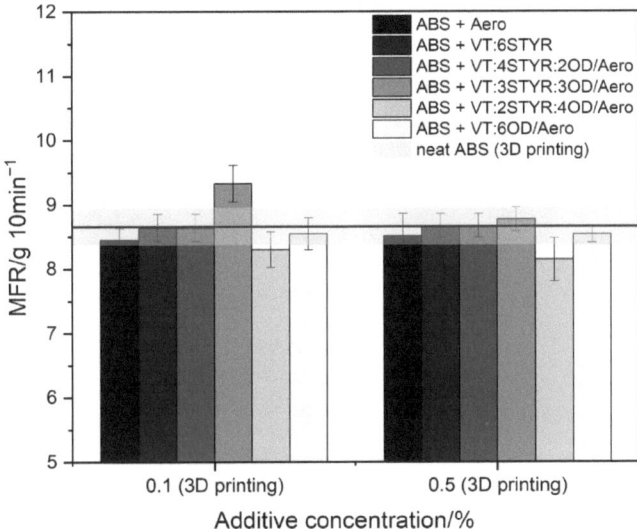

Figure 16. Result of MFR measurements.

3.5. Microscopy

3.5.1. Optical Microscopy

Figures 17 and 18 illustrate the structure of the injected and printed samples, respectively. In the case of samples obtained through injection molding, thanks to their transparency inside the polymer matrix, it is possible to observe the agglomerated particles of nanosilica in the photos.

The microscopic image of the reference sample (Figure 17a) shows nanosilica agglomerates present in the composite after the first homogenization step. Imaging was performed to pre-evaluate the dispersion of additives in the matrix prior to the extruder blending process. The addition of the modifier changed the degree of silica dispersion in the polymer matrix. The largest silica agglomerates are shown in image (Figure 17c), which presents a system containing the VT:4STYR:OD modifier. As the content of the OD groups in the modifier increases in subsequent samples (Figure 17d–f), there is improved dispersion of the additive in the polymer matrix.

Figure 18 shows the fractures of FDM-printed samples after being subjected to Charpy impact testing. At a higher concentration, more protrusions are visible between and within the layers. Neat ABS fractures exhibit a highly compact structure, with a significant interlayer contact area and minimal free spaces (Figure 18a).

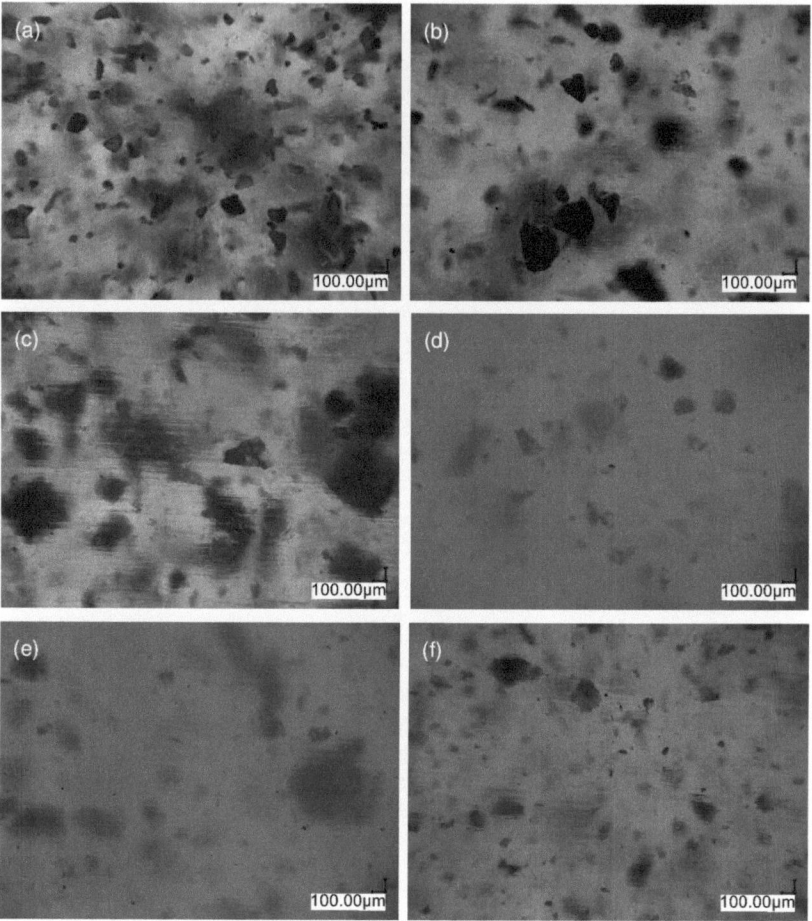

Figure 17. Optical microscopic images of injected ABS samples containing 5% by mass of (**a**) Aero, (**b**) VT:6STYR/Aero, (**c**) VT:4STYR:2OD/Aero, (**d**) VT:3STYR:3OD/Aero, (**e**) VT:2STYR:4OD/Aero, and (**f**) VT:6OD/Aero.

The fracture of samples with a concentration of 0.1% additives has a more compact structure and more homogeneity than samples with a concentration of 0.5%, which is caused by a lower concentration of nanosilica. The inclusion of modified silica lessens the adhesive force between the layers of the substance. The ABS + VT:4STYR:2OD/Aero system (Figure 18f,g) features a less compact structure and numerous voids for both concentrations. It was discovered that the additive did not disperse well in the polymer matrix, as evidenced by the lack of homogeneity after the initial step of mixing (Figure 18c).

With the increase in the content of the OD groups in the modifier, cohesion between the layers is higher, and smaller free spaces are observed due to the additional OD groups having a plasticizing effect (Figure 18h–m). ABS + VT:6OD/Aero samples have the most solid structures, a relatively large contact area between the material layers, and small free spaces (Figure 18l,m).

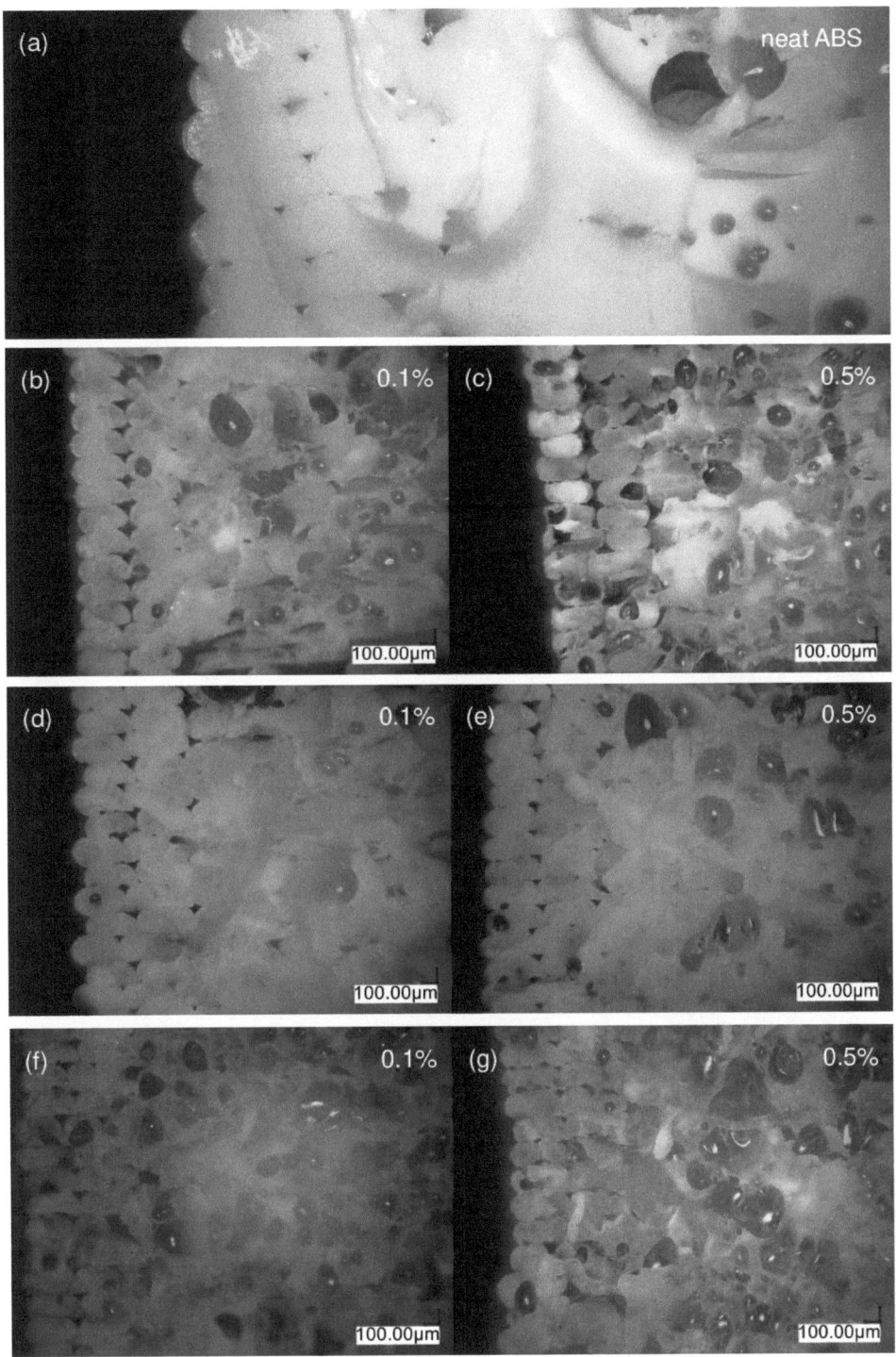

Figure 18. *Cont.*

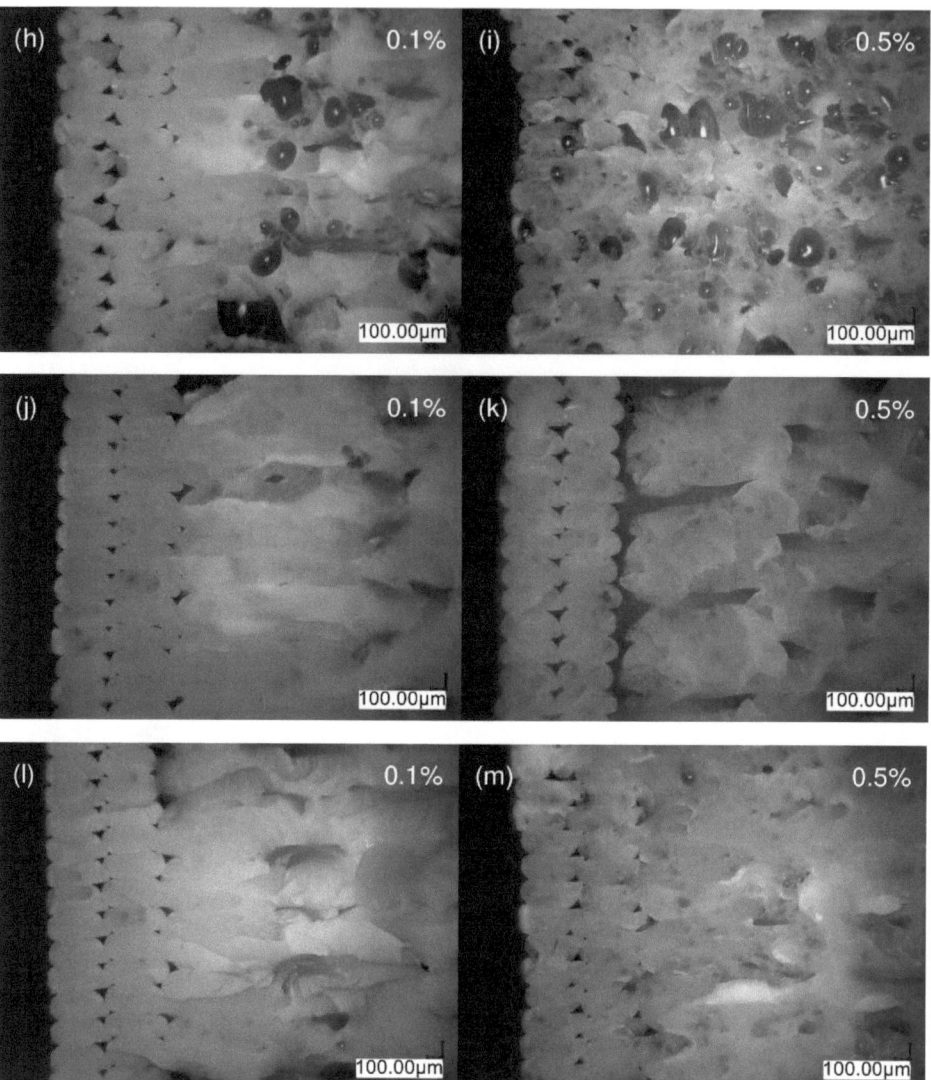

Figure 18. Optical microscopic images of the cross-sections of printed ABS samples after impact testing of (**a**) neat ABS and two selected concentrations as follows: (**b**,**c**) ABS/Aero, (**d**,**e**) VT:6STYR/Aero, (**f**,**g**) VT:4STYR:2OD/Aero, (**h**,**i**) VT:3STYR:3OD/Aero, (**j**,**k**) VT:2STYR:4OD/Aero, (**l**,**m**) VT:6OD/Aero.

3.5.2. SEM-EDS

SEM-EDS microscopy was used to examine dispersion of filler in the matrix, and identify any agglomerates present. Figure 19. shows the mapping of silicon atoms in composites. Mapping was performed both for ABS with the addition of only nanosilica and for systems containing the following modifiers: VT:6STYR/Aero, VT:4STYR:2OD/Aero, VT:3STYR:3OD/Aero, VT:2STYR:4OD/Aero, and VT:6OD/Aero. Microscopic images were taken from the section of the injection-molded bar. All the systems are characterized by the presence of larger agglomerates, both for the reference sample without the addition of the modifier and for the modified samples. The SEM-EDS photos validate the observations made through optical microscopy. Specifically, these observations highlight that a higher content of alkyl groups, coupled with a lower content of styryl groups in the organosilicon modifiers, results in better filler dispersion.

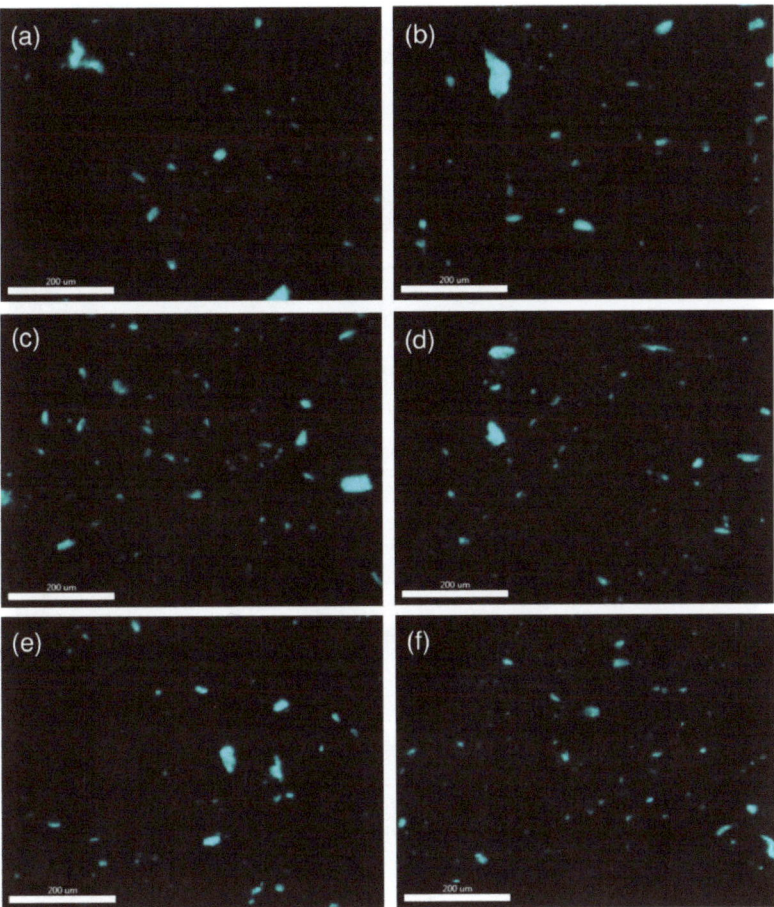

Figure 19. SEM with Si mapping (EDS) for samples of (**a**) ABS + 2.5% Aero, (**b**) ABS + 2.5% (VT:6STYR/Aero), (**c**) ABS + 2.5% (VT:4STYR:2OD/Aero), (**d**) ABS + 2.5% (VT:3STYR:3OD/Aero), (**e**) ABS + 2.5% (VT:2STYR:4OD/Aero), and (**f**) ABS + 2.5% (VT:6OD/Aero).

3.6. Contact Angle Measurements

Measurements of the water contact angle (WCA) were performed for the modifier/Aero/ABS composites obtained through the FDM method (Figure 20). Additionally, measurements were also carried out for the reference samples, i.e., those of neat ABS and ABS with unmodified silica. The contact angle of neat ABS was 70.9°, indicating the hydrophilic nature of the copolymer surface. The addition of silica changed the microstructure (increased roughness) of the composite, which resulted in achieving higher values of the contact angle. The contact angle analysis allows for the evaluation of the potential functional characteristics of novel materials with regard to their hydrophilic–hydrophobic properties.

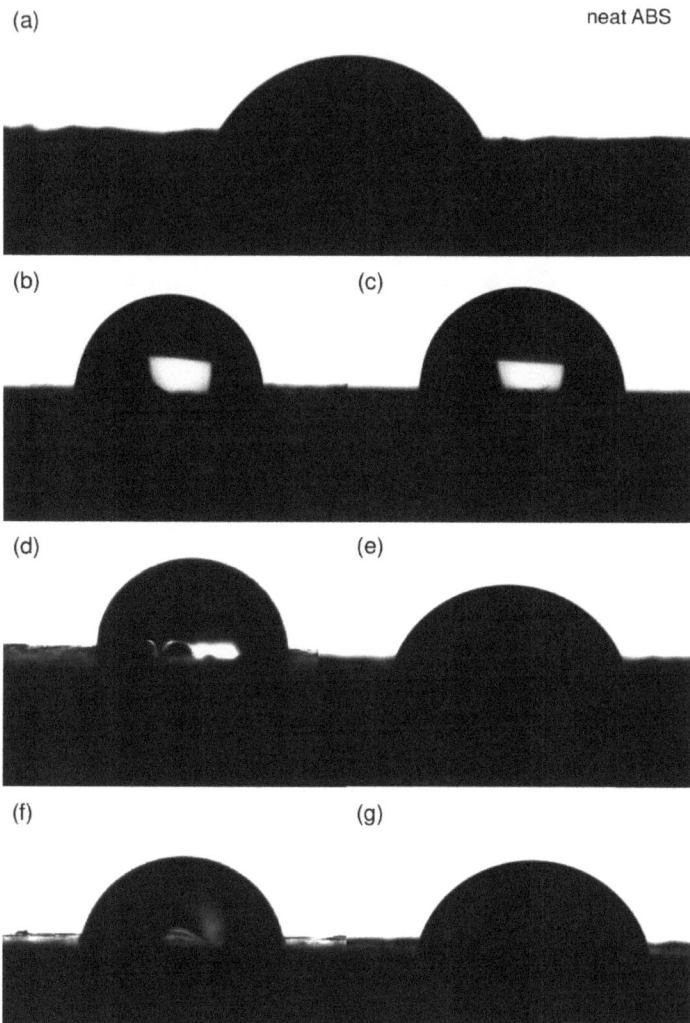

Figure 20. The water contact angle for samples of (**a**) neat ABS, (**b**) ABS + 2.5% Aero, (**c**) ABS + 2.5% (VT:6STYR/Aero), (**d**) ABS + 2.5% (VT:4STYR:2OD/Aero), (**e**) ABS + 2.5% (VT:3STYR:3OD/Aero), (**f**) ABS + 2.5% (VT:2STYR:4OD/Aero), and (**g**) ABS + 2.5% (VT:6OD/Aero).

For ABS samples modified by VT:6STYR/Aero, a significant increase in the contact angle is observed. For 2.5% of the modifier concentration, the contact angle reached a value above 90°, which proves the hydrophobic properties of the material. High contact angle values were also achieved for VT:4STYR:2OD (92.9° for 2.5% concentration). VT:6STYR and VT:4STYR:2OD modifiers have phenylethylene groups in their structure, which are compatible with the polymer matrix. Additionally, OD consists of long alkyl chains that are responsible for hydrophobic properties.

Adding other organosilicon modifiers to ABS + Aero systems results in forming a hydrophilic surface for most systems (Table 8).

Table 8. Water contact angle.

Code	Contact Angle/°						
		Concentration of Additives/%					
	-	0.1	0.25	0.5	1	1.5	2.5
neat ABS	70.9 ± 3.2	-	-	-	-	-	-
ABS + Aero	-	86.2 ± 1.3	89.6 ± 1.4	93.9 ± 2.0	89.8 ± 3.5	81.2 ± 0.9	91.7 ± 2.4
ABS + VT:6STYR/Aero	-	76.5 ± 0.5	76.6 ± 0.6	74.9 ± 2.0	79.2 ± 0.5	86.4 ± 3.1	92.9 ± 2.0
ABS + VT:4STYR:2OD/Aero	-	90.5 ± 0.5	86.8 ± 1.2	87.4 ± 3.7	84.8 ± 3.9	85.0 ± 0.5	92.9 ± 1.9
ABS + VT:3STYR:3OD/Aero	-	86.6 ± 2.5	84.1 ± 2.1	88.6 ± 2.6	82.1 ± 0.8	73.0 ± 1.6	71.7 ± 1.2
ABS + VT:2STYR:4OD/Aero	-	85.3 ± 0.8	85.5 ± 3.2	83.2 ± 1.2	82.8 ± 1.4	72.1 ± 2.6	79.1 ± 3.3
ABS + VT:6OD/Aero	-	74.9 ± 0.5	77.4 ± 1.8	76.6 ± 1.1	79.5 ± 2.6	77.2 ± 0.4	79.1 ± 2.6

Based on the analysis of microscopic images and contact angle values, a correlation can be observed. Neat ABS exhibits hydrophilic properties and has a compact and uniform internal structure (as shown in Figure 18). This results in a relatively flat surface when printed. However, the addition of Aerosil causes the surface to become more irregular with the formation of cracks and large protrusions. The degree of irregularity is higher at 0.5% compared with 0.1%, resulting in a more hydrophobic surface with a larger water contact angle. ABS + VT:6OD/Aero displays the most homogeneous cross-section and a hydrophilic surface. For the other systems, the impact of the modification on surface properties is minimal, which is likely due to the formation of modified silica agglomerates.

3.7. Mechanical Properties

A summary of the results of mechanical tests carried out on modified samples obtained through both 3D printing and traditional injection molding will be discussed.

3.7.1. Tensile Strength

Printed samples with a concentration of between 0.1% and 0.5% additives as well as injected samples with a concentration of 5% additives were subjected to a tensile strength test.

The tensile strength values for neat ABS were 36.4 ± 1.1 MPa for samples obtained through FDM printing and 42.0 ± 0.3 MPa for samples obtained through injection molding (Figure 21). The higher value of tensile strength of a neat ABS sample obtained by injection molding than that by FDM method results from the specificity of a given process and the related more compact structure of the injected samples. Similar results were obtained by Dawoud et al. in their earlier research [34]. The tensile stress values of injected samples are highest for Aero/ABS and VT:6Styr/Aero/ABS; the addition of other modifiers reduces the stress value.

The tensile stress values of 3D printed samples are highest for systems containing organosilicon modifiers. In addition, at a lower modifier concentration of 0.1%, these values exceed the values of neat ABS samples obtained through injection. Even a small amount of the additive improves the mechanical properties of composites, which are beneficial in economic terms.

The highest tensile strength values among printed composites are observed among systems with a concentration of 0.1% as follows: VT:6STYR/Aero, VT:3STYR:3OD/Aero, and VT:6OD/Aero. Due to the compatibility of the phenylethylene groups of the modifier (VT:6STYR/Aero) with the ABS chain, the strength of the material is higher. The increase in the tensile strength parameters of the composite containing the largest number of the OD groups is due to the increased plasticizing effect that resulted in a greater level of homogeneity within the internal structure of the sample. The VT:3STYR:3OD/Aero composite has an indirect influence on both groups and, hence, also has high strength values.

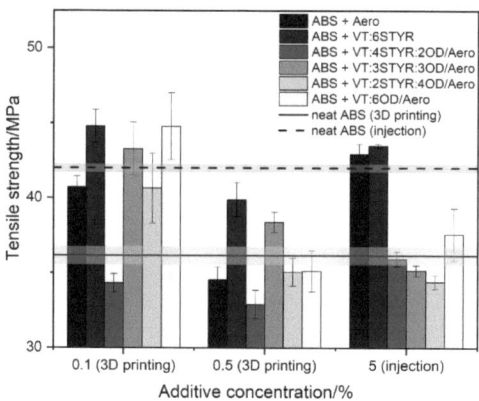

Figure 21. Tensile strength of ABS/modifier/Aero in 3D printing and injection molding.

At higher concentrations, due to a higher proportion of silica, as shown in microscopic images of cross-sections, it causes greater heterogeneity in the structure of composites, which results in a decrease in their tensile strength.

3.7.2. Elongation

According to Figure 22, elongation at the maximum load for neat ABS samples is 3.39% and 4.21% for 3D printed and injection molded samples, respectively. The injection molded samples with silica show lower elongation values due to increased brittleness and silica agglomeration in the material matrix. Silica is added to plastics in order to increase the strength and hardness of materials because it acts as a filler that increases resistance to abrasion and mechanical damage. Values of tensile strength (See Section 3.7.1) and hardness (See Section 3.7.4) are higher for composites compared with neat ABS, which is associated with lower flexibility. However, the effect of silica in printed samples is limited, and the modifier has a greater influence on elongation values that are around the reference sample value. In samples VT:2STYR:4OD and VT:6OD with a concentration of 0.5%, the elongation value is higher due to the plasticizing effect of the OD groups, resulting in increased flexibility.

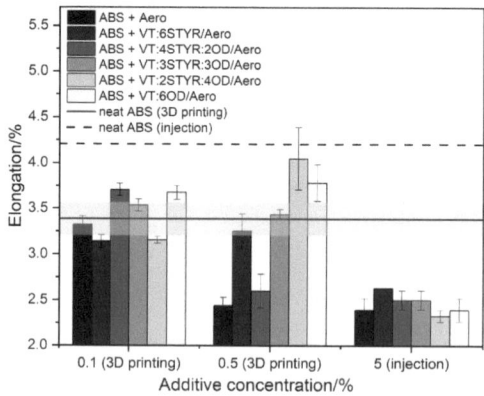

Figure 22. Elongation at maximum load for ABS/modifier/Aero in 3D printing and injection molding.

3.7.3. Young's Modulus

The addition of modifiers or filler has no significant impact on Young's modulus. The modulus of linear deformation (Figure 23) reaches the highest values for samples obtained through injection molding. The addition of modifiers slightly reduces the module value. In the case of samples obtained through the FDM technique, a slight increase in the modulus value is observed at the lowest concentration of 0.1%.

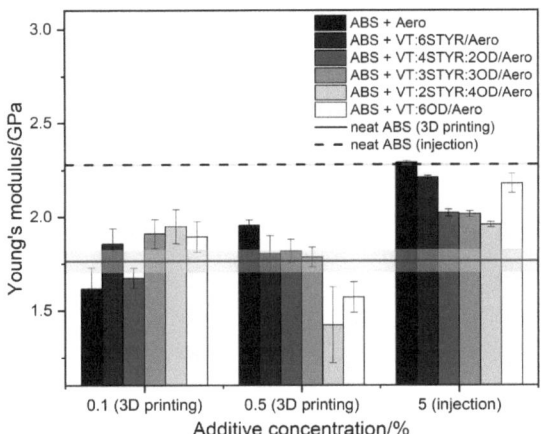

Figure 23. Tensile modulus of ABS samples in 3D printing and injection molding.

3.7.4. Impact Test and Hardness

The impact strength of 3D printed neat ABS is 27.1 ± 1.3 [kJ/m^2]. However, during testing, most systems showed a lower level of fracture toughness compared with the neat ABS reference sample (Figure 24). This can be attributed to the lower homogeneity and cohesion of the printed samples containing nanosilica, as seen in the cross-sectional images (Figure 18).

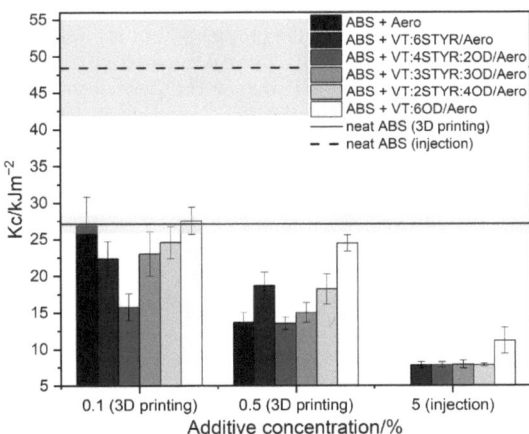

Figure 24. Impact strength of ABS samples in 3D printing and injection molding.

The presence of silica as a filler increases the hardness and stiffness of plastics, resulting in a loss of flexibility and lower impact strength. However, the addition of modifiers containing 6OD and 2STYR:4OD improves the impact strength of samples with

Aerosil. The lowest concentration of 0.1% VT:6OD/Aero even shows values above the neat ABS range.

The impact strength of neat ABS obtained through injection molding is 48.4 ± 6.6 [kJ/m^2]. The addition of nanosilica to ABS samples results in a deterioration in fracture toughness. This is due to the formation of agglomerates in systems above 3 wt%, as shown in the optical microscopic images (Figure 17).

Shore hardness tests were conducted to confirm the reinforcing effect of silica. The Shore D hardness (Table 9) of printed neat ABS is 59. The obtained composites have higher hardness compared with neat ABS, as expected. Silica has a high hardness, which reinforces the plastic to which it is added. The ABS + VT:3STYR:3OD/Aero system has the highest values.

Table 9. Hardness in Shore D scale.

Code	Shore Hardness						
	Concentration of Additives/%						
	-	0.1	0.25	0.5	1	1.5	2.5
neat ABS	59	-	-	-	-	-	-
ABS + Aero	-	71	66	71	71	66	63
ABS + VT:6STYR/Aero	-	68	66	72	73	66	68
ABS + VT:4STYR:2OD/Aero	-	72	74	79	74	73	74
ABS + VT:3STYR:3OD/Aero	-	75	77	79	78	77	79
ABS + VT:2STYR:4OD/Aero	-	76	69	75	76	72	78
ABS + VT:6OD/Aero	-	78	68	76	77	78	77

4. Conclusions

The obtained results confirm the effect of the addition of modified silica with organosilicon compounds on the improvement of the mechanical properties of printed composites based on the ABS matrix. This is important in the case of FDM technology because 3D printed objects usually have much lower levels of resistance than injected ones do, which are related to different specificities used in the operation of both methods.

To the best of our knowledge, an endeavor has been undertaken to delineate the thermal decomposition process of ABS composites in this study for the first time. The mechanism of the thermal decomposition of the tested systems was determined using additional techniques (FT-IR, optical microscope). The presented results constitute an important contribution to research on the thermal stability of ABS.

Microscopic examination (digital microscope and SEM-EDS) has proven the positive effect of the addition of octadecyl groups on the increased dispersion of silica in composites. The best results were obtained with the lowest concentration of additives (0.1%). The properties of the composites can be notably influenced by the addition of a minimal quantity of additives, thus presenting economic advantages.

The compatibility of the introduced phenylethylene groups and the plasticizing effect of the OD groups affected both the tensile strength and the surface character of the composites. Silica in the composite increases the contact angle of the surface, making it hydrophobic due to structural irregularities. The modifier altered the contact angle value by smoothing the composite surface, thereby making it hydrophilic.

The obtained results indicate the potential benefits of using silica that is functionalized with organosilicon compounds as a modifier in 3D printing, introducing significant changes in both the surface, strength, and thermal properties of ABS composites. These findings may be used in further work on improving 3D printing technology and modifying composite materials.

Author Contributions: Conceptualization, B.S. and R.E.P.; methodology, B.S.; software, R.K., D.P., M.F. and B.S.; validation, B.S.; formal analysis, R.K., B.S. and D.P.; investigation, R.K., M.F., D.P. and B.S.; resources, B.S.; data curation, B.S.; writing—original draft preparation, R.K. and B.S.; writing—review and editing, B.S., R.E.P. and B.M.; supervision, B.S.; project administration, B.S.; funding acquisition, B.S. All authors have read and agreed to the published version of the manuscript.

Funding: Scientific research in the field of designing organosilicon modifiers of thermoplastic properties for the incremental FDM technique financed from the sources of the National Center for Research and Development under the LIDER X project (LIDER/01/0001/L-10/18/NCBR/2019).

Institutional Review Board Statement: Not applicable.

Informed Consent Statement: Not applicable.

Data Availability Statement: Data are contained within the article.

Conflicts of Interest: The authors declare no conflicts of interest.

References

1. Matias, E.; Rao, B. 3D Printing: On Its Historical Evolution and the Implications for Business. In Proceedings of the Portland International Conference on Management of Engineering and Technology (PICMET), Portland, OR, USA, 2–6 August 2015. [CrossRef]
2. Hossain, M.A.; Zhumabekova, A.; Paul, S.C.; Kim, J.R. A Review of 3D Printing in Construction and Its Impact on the Labor Market. *Sustainability* **2020**, *12*, 8492. [CrossRef]
3. Holt, C.; Edwards, L.; Keyte, L.; Moghaddam, F.; Townsend, B. Construction 3D Printing. In *3D Concrete Printing Technology*; Butterworth-Heinemann: Oxford, UK, 2019; pp. 349–370. [CrossRef]
4. Palinkas, I.; Pekez, J.; Desnica, E.; Rajic, A.; Nedelcu, D. Analysis and Optimization of UAV Frame Design for Manufacturing from Thermoplastic Materials on FDM 3D Printer. *Mater. Plast.* **2022**, *58*, 238–249. [CrossRef]
5. Chambon, P.; Curran, S.; Huff, S.; Love, L.; Post, B.; Wagner, R.; Jackson, R.; Green, J. Development of a Range-Extended Electric Vehicle Powertrain for an Integrated Energy Systems Research Printed Utility Vehicle. *Appl. Energy* **2017**, *191*, 99–110. [CrossRef]
6. Pugliese, R.; Beltrami, B.; Regondi, S.; Lunetta, C. Polymeric Biomaterials for 3D Printing in Medicine: An Overview. *Ann. 3D Print. Med.* **2021**, *2*, 100011. [CrossRef]
7. Liaw, C.-Y.; Guvendiren, M. Current and Emerging Applications of 3D Printing in Medicine. *Biofabrication* **2017**, *9*, 024102. [CrossRef] [PubMed]
8. Fang, J.-J.; Lin, C.-L.; Tsai, J.-Y.; Lin, R.-M. Clinical Assessment of Customized 3D-Printed Wrist Orthoses. *Appl. Sci.* **2022**, *12*, 11538. [CrossRef]
9. Vakharia, V.S.; Singh, M.; Salem, A.; Halbig, M.C.; Salem, J.A. Effect of Reinforcements and 3-D Printing Parameters on the Microstructure and Mechanical Properties of Acrylonitrile Butadiene Styrene (ABS) Polymer Composites. *Polymers* **2022**, *14*, 2105. [CrossRef] [PubMed]
10. Li, Z.; Chen, G.; Lyu, H.; Ko, F. Experimental Investigation of Compression Properties of Composites with Printed Braiding Structure. *Materials* **2018**, *11*, 1767. [CrossRef]
11. Mohamed, O.A.; Masood, S.H.; Bhowmik, J.L. Optimization of Fused Deposition Modeling Process Parameters: A Review of Current Research and Future Prospects. *Adv. Manuf.* **2015**, *3*, 42–53. [CrossRef]
12. Valino, A.D.; Dizon, J.R.C.; Espera, A.H.; Chen, Q.; Messman, J.; Advincula, R.C. Advances in 3D Printing of Thermoplastic Polymer Composites and Nanocomposites. *Prog. Polym. Sci.* **2019**, *98*, 101162. [CrossRef]
13. Yadav, D.K.; Srivastava, R.; Dev, S. Design & Fabrication of ABS Part by FDM for Automobile Application. *Mater. Today Proc.* **2020**, *26*, 2089–2093. [CrossRef]
14. Aung, T.K.; Churei, H.; Tanabe, G.; Kinjo, R.; Togawa, K.; Li, C.; Tsuchida, Y.; Tun, P.S.; Hlaing, S.; Takahashi, H.; et al. Air Permeability, Shock Absorption Ability, and Flexural Strength of 3D-Printed Perforated ABS Polymer Sheets with 3D-Knitted Fabric Cushioning for Sports Face Guard Applications. *Polymers* **2021**, *13*, 1879. [CrossRef] [PubMed]
15. Galatas, A.; Hassanin, H.; Zweiri, Y.; Seneviratne, L. Additive Manufactured Sandwich Composite/ABS Parts for Unmanned Aerial Vehicle Applications. *Polymers* **2018**, *10*, 1262. [CrossRef]
16. Reggio, D.; Saviello, D.; Lazzari, M.; Iacopino, D. Characterization of Contemporary and Historical Acrylonitrile Butadiene Styrene (ABS)-Based Objects: Pilot Study for Handheld Raman Analysis in Collections. *Spectrochim. Acta Part A Mol. Biomol. Spectrosc.* **2020**, *242*, 118733. [CrossRef]
17. Whelan, A. *Polymer Technology Dictionary*; Springer Science & Business Media: Berlin/Heidelberg, Germany, 2012.
18. Kumar, V.; RamKumar, J.; Aravindan, S.; Malhotra, S.K.; Vijai, K.; Shukla, M. Fabrication and Characterization of ABS Nano Composite Reinforced by Nano Sized Alumina Particulates. *Int. J. Plast. Technol.* **2009**, *13*, 133–149. [CrossRef]

19. Rosli, A.A.; Shuib, R.K.; Ishak, K.M.K.; Hamid, Z.A.A.; Abdullah, M.K.; Rusli, A. Influence of Bed Temperature on Warpage, Shrinkage and Density of Various Acrylonitrile Butadiene Styrene (ABS) Parts from Fused Deposition Modelling (FDM). In Proceedings of the 3rd International Postgraduate Conference on Materials, Minerals & Polymer (Mamip), Penang, Malaysia, 31 October–1 November 2019; AIP Publishing: Long Island, NY, USA, 2020. [CrossRef]
20. Alhallak, L.M.; Tirkes, S.; Tayfun, U. Mechanical, Thermal, Melt-Flow and Morphological Characterizations of Bentonite-Filled ABS Copolymer. *Rapid Prototyp. J.* **2020**, *26*, 1305–1312. [CrossRef]
21. Andrzejewski, J.; Misra, M. Development of Hybrid Composites Reinforced with Biocarbon/Carbon Fiber System. The Comparative Study for PC, ABS and PC/ABS Based Materials. *Compos. B Eng.* **2020**, *200*, 108319. [CrossRef]
22. Vidakis, N.; Petousis, M.; Maniadi, A.; Koudoumas, E.; Kenanakis, G.; Romanitan, C.; Tutunaru, O.; Suchea, M.; Kechagias, J. The Mechanical and Physical Properties of 3D-Printed Materials Composed of ABS-ZnO Nanocomposites and ABS-ZnO Microcomposites. *Micromachines* **2020**, *11*, 615. [CrossRef]
23. Vidakis, N.; Moutsopoulou, A.; Petousis, M.; Michailidis, N.; Charou, C.; Papadakis, V.; Mountakis, N.; Dimitriou, E.; Argyros, A. Rheology and Thermomechanical Evaluation of Additively Manufactured Acrylonitrile Butadiene Styrene (ABS) with Optimized Tungsten Carbide (WC) Nano-Ceramic Content. *Ceram. Int.* **2023**, *49*, 34742–34756. [CrossRef]
24. Vidakis, N.; Petousis, M.; Velidakis, E.; Maniadi, A. Mechanical Properties of 3D-Printed ABS with Combinations of Two Fillers: Graphene Nanoplatelets, TiO$_2$, ATO Nanocomposites, and Zinc Oxide Micro (ZnOm). In *Recent Advances in Manufacturing Processes and Systems: Select Proceedings of RAM 2021*; Springer Nature: Singapore, 2022; pp. 635–645.
25. Kim, I.-J.; Kwon, O.-J.; Park, J.B.; Joo, H. Synthesis and Characterization of ABS/Silica Hybrid Nanocomposites. *Curr. Appl. Phys.* **2006**, *6*, e43–e47. [CrossRef]
26. Meng, S.; He, H.; Jia, Y.; Yu, P.; Huang, B.; Chen, J. Effect of Nanoparticles on the Mechanical Properties of Acrylonitrile-Butadiene-Styrene Specimens Fabricated by Fused Deposition Modeling. *J. Appl. Polym. Sci.* **2017**, *134*, 44470. [CrossRef]
27. Huang, B.; He, H.; Meng, S.; Jia, Y. Optimizing 3D Printing Performance of Acrylonitrile-Butadiene-Styrene Composites with Cellulose Nanocrystals/Silica Nanohybrids. *Polym. Int.* **2019**, *68*, 1351–1360. [CrossRef]
28. Brząkalski, D.; Przekop, R.E.; Sztorch, B.; Jakubowska, P.; Jałbrzykowski, M.; Marciniec, B. Silsesquioxane Derivatives as Functional Additives for Preparation of Polyethylene-Based Composites: A Case of Trisilanol Melt-Condensation. *Polymers* **2020**, *12*, 2269. [CrossRef] [PubMed]
29. Brząkalski, D.; Sztorch, B.; Frydrych, M.; Pakuła, D.; Dydek, K.; Kozera, R.; Boczkowska, A.; Marciniec, B.; Przekop, R.E. Limonene Derivative of Spherosilicate as a Polylactide Modifier for Applications in 3D Printing Technology. *Molecules* **2020**, *25*, 5882. [CrossRef]
30. ISO 1133; Plastics—Determination of the Melt Mass-Flow Rate (MFR) and Melt Volume-Flow Rate (MVR) of Thermoplastics. International Organization for Standardization: Geneva, Switzerland, 2005.
31. ISO 179; Plastics—Determination of Charpy Impact Properties. International Organization for Standardization: Geneva, Switzerland, 2010.
32. ISO 527; Plastics—Determination of Tensile Properties. International Organization for Standardization: Geneva, Switzerland, 2019.
33. Sztorch, B.; Brząkalski, D.; Głowacka, J.M.; Pakuła, D.J.; Frydrych, M.; Przekop, R.E. Trimming flow, plasticity, and mechanical proper-ties by cubic silsesquioxane chemistry. *Sci. Rep.* **2023**, *13*, 14156. [CrossRef]
34. Dawoud, M.; Taha, I.; Ebeid, S.J. Mechanical Behaviour of ABS: An Experimental Study Using FDM and Injection Moulding Techniques. *J. Manuf. Process.* **2016**, *21*, 39–45. [CrossRef]

Disclaimer/Publisher's Note: The statements, opinions and data contained in all publications are solely those of the individual author(s) and contributor(s) and not of MDPI and/or the editor(s). MDPI and/or the editor(s) disclaim responsibility for any injury to people or property resulting from any ideas, methods, instructions or products referred to in the content.

Article

A Comparative Study between Blended Polymers and Copolymers as Emitting Layers for Single-Layer White Organic Light-Emitting Diodes

Despoina Tselekidou [1,*], Kyparisis Papadopoulos [1], Vasileios Foris [1], Vasileios Kyriazopoulos [2], Konstantinos C. Andrikopoulos [3], Aikaterini K. Andreopoulou [3], Joannis K. Kallitsis [3], Argiris Laskarakis [1], Stergios Logothetidis [1,2] and Maria Gioti [1,*]

1. Nanotechnology Lab LTFN, Department of Physics, Aristotle University of Thessaloniki, GR-54124 Thessaloniki, Greece; kypapado@physics.auth.gr (K.P.); alask@physics.auth.gr (A.L.); logot@auth.gr (S.L.)
2. Organic Electronic Technologies P.C. (OET), 20th KM Thessaloniki—Tagarades, GR-57001 Thermi, Greece; v.kyriazopoulos@oe-technologies.com
3. Department of Chemistry, University of Patras, Caratheodory 1, University Campus, GR-26504 Patras, Greece; k.andrikopoulos@ac.upatras.gr (K.C.A.); andreopo@upatras.gr (A.K.A.); j.kallitsis@upatras.gr (J.K.K.)
* Correspondence: detselek@physics.auth.gr (D.T.); mgiot@physics.auth.gr (M.G.)

Citation: Tselekidou, D.; Papadopoulos, K.; Foris, V.; Kyriazopoulos, V.; Andrikopoulos, K.C.; Andreopoulou, A.K.; Kallitsis, J.K.; Laskarakis, A.; Logothetidis, S.; Gioti, M. A Comparative Study between Blended Polymers and Copolymers as Emitting Layers for Single-Layer White Organic Light-Emitting Diodes. *Materials* 2024, 17, 76. https://doi.org/10.3390/ma17010076

Academic Editors: Eduard-Marius Lungulescu, Radu Setnescu and Cristina Stancu

Received: 1 December 2023
Revised: 18 December 2023
Accepted: 19 December 2023
Published: 23 December 2023

Copyright: © 2023 by the authors. Licensee MDPI, Basel, Switzerland. This article is an open access article distributed under the terms and conditions of the Creative Commons Attribution (CC BY) license (https://creativecommons.org/licenses/by/4.0/).

Abstract: Extensive research has been dedicated to the solution-processable white organic light-emitting diodes (WOLEDs), which can potentially influence future solid-state lighting and full-color flat-panel displays. The proposed strategy based on WOLEDs involves blending two or more emitting polymers or copolymerizing two or more emitting chromophores with different doping concentrations to produce white light emission from a single layer. Toward this direction, the development of blends was conducted using commercial blue poly(9,9-di-n-octylfluorenyl2,7-diyl) (PFO), green poly(9,9-dioctylfluorenealt-benzothiadiazole) (F8BT), and red spiro-copolymer (SPR) light-emitting materials, whereas the synthesized copolymers were based on different chromophores, namely distyryllanthracene, distyrylcarbazole, and distyrylbenzothiadiazole, as yellow, blue, and orange–red emitters, respectively. A comparative study between the two approaches was carried out to examine the main challenge for these doping systems, which is ensuring the proper balance of emissions from all the units to span the entire visible range. The emission characteristics of fabricated WOLEDs will be explored in terms of controlling the emission from each emitter, which depends on two possible mechanisms: energy transfer and carrier trapping. The aim of this work is to achieve pure white emission through the color mixing from different emitters based on different doping concentrations, as well as color stability during the device operation. According to these aspects, the WOLED devices based on the copolymers of two chromophores exhibit the most encouraging results regarding white color emission coordinates (0.28, 0.31) with a CRI value of 82.

Keywords: WOLEDs; single-layer; blend polymers; copolymers; energy transfer

1. Introduction

White organic light-emitting diodes (WOLEDs), produced through a solution deposition process, offer a formidable opportunity for low-cost, environmentally friendly, and flexible light sources capable of meeting the spectral, functional, and budgetary needs of emerging application fields. These fields include next-generation full-color displays and solid-state lighting [1–7]. Until now, the reported high-efficiency WOLEDs with multilayer device structures have been mostly fabricated through vacuum evaporation [8]. Although such complex device structures can achieve high efficiency, the complicated process and high economic cost are not suitable for further industrial application. In this context, a device processing procedure based on solutions, such as spin-coating, slot die coating, etc., provides numerous advantages compared to a vacuum deposition process. Aside from the low cost and mass production, solution-processed OLEDs hold an edge over other techniques, since they can also be designed in many different shapes and designs, opening an enormous array of design possibilities [9].

The most straightforward way to obtain white light is to integrate luminophores of different wavelengths in the same device or even in a single emitting layer [10–12]. Especially, the widely adopted methods for fabricating white OLEDs are either stacking multiple emitting layers or mixing different colors in one emitting layer (EML). Among these strategies, multilayer polymer LED fabrication is more laborious, as their fabrication is more difficult and there is a risk that the solvent may destroy the underlayers. On the other hand, the alternative strategy based on single-emissive layer WOLEDs has shown to be a more reliable option in scientific and industrial communities because of the low fabrication costs, simple structure, and easy solution processing [4,13,14].

More specifically, the output spectral range of an OLED must span the entire visible range (400–800 nm) in order to be considered a preferred and optimal white light source. Therefore, a combination of two or three emissive materials is usually required for efficient and easy color tuning to achieve white EL emission, because only one emissive material cannot cover a sufficient spectral range. According to this demand, the white light emission could be generated by mixing three primary colors (blue, green, and red) or complementary colors (red or yellow and blue) to cover the entire visible region. There have been several proposed luminescent material systems, which have been devised resulting in the exciton regulation to emit white light in a balanced manner. Regarding this issue, two of the proposed strategies involve either the blend of two or more light-emitting polymers [15,16] or the use of multicolor-emitting copolymers bearing different chromophores [4,14]. It is worth noting that for producing environmentally friendly and practical WOLEDs, fluorescent materials are superior to phosphorescent materials, as the former do not contain any heavy metals [17]. In addition, compared to small molecules, the emitting polymers are deemed more suitable for low-cost and large-area wet methods [18].

More specifically, in the case of blended materials, some well-studied binary blends, based on blue and red emitters, are the combination of poly [9,9-dioctylfluorenyl-2,7-diyl] (PFO) with poly(2-methoxy-5(2-ethylhexyl)-1,4-phenylenevinylene) (MEH-PPV) [19] or poly [2-methoxy-5-(3,7-dimethyl-octyloxy)-1,4-phenylenevinylene] capped with dimethylphenyl (MDMO-PPV-DMP) [20]. More specifically, Prakash et al. suggested, as an emissive layer, the binary blended PFO-MEH-PPV for WOLED and exhibiting Commission Internationale de l'Enclairage (CIE) coordinates of (0.30, 0.38) and luminance value reaching 1234 cd/m^2 at 8.5 V [19]. The ternary blended materials are based on PFO, MEH-PPV, and incorporation into a green emitter, such as poly 9,9-dioctylfluorene-altbenzothiadiazole (F8BT) [21] or 2-butyl-6-(butylamino)benzo(de)isoquinoline-1,3-dione (F7GA) [22]. Al-Asbahi studied the latter case PFO-F7GA-MEH-PPV and obtained CIE coordinates of (0.31, 0.24) and the luminance value of 2295 cd/m^2 [22]. For the synthesized materials, Kim et al. present, as an emissive layer, the copolymer, which consists of polyfluorene (PFO) and polytriarylamine (PTAA) with a benzothiadiazole (BZ) moiety and exhibiting CIE coordinates of (0.34, 0.41) and a luminance value of 320 cd/m^2 [23]. However, for all studied cases, there is still a need to optimize the emission and operating characteristics of WOLEDs.

Both the scenarios of blended and copolymers as emitting layers mentioned above are founded on a host–guest system. An ideal host–guest system should emit a continuous spectrum of different colors spanning the entire visible range. The efficient control and tuning of emission colors are crucial. The diffusion of excitons is a vital factor in the operation of OLEDs, particularly WOLEDs, and their control is influenced by the doping concentration. More specifically, for the efficient WOLED, excitons need to be distributed to various emitters. It is well established that the white emission can be realized through the Förster resonance energy transfer (FRET) from a wide-band-gap host molecule to a low-band-gap dopant molecule. The requirements for efficient energy transfer are as follows: (i) the donor should possess a high photoluminescence quantum yield (PLQY), (ii) the acceptor species should be located in close proximity to the donor species, and most importantly, (iii) it should satisfy the fundamental criterion of excellent spectral overlap in between the emission of the donor and absorption of the acceptor [22,24,25]. Part of the energy is transferred to the guest emitter, allowing for color mixing with its self-emission. However, the efficient energy transfer from the host to the dopant material often requires precise control of the guest doping concentration, thus increasing the challenge of device preparation. Hence, the key challenge lies in discovering a host emission and dopants most fitting while considering the doping concentration, which is an equally critical aspect to attain emission in the whole visible range. Accomplishing pure white light emission that offers commendable CRI values and CIE coordinates proximate to the ideal white point of (0.33, 0.33) remains the ultimate objective.

With the aim of advancing solution-processable WOLEDs toward their full potential, herein, we investigate light-emitting blended polymers or multicolor copolymers, bearing different chromophores, as a route to achieve the emission in the whole visible range for lighting applications. In particular, commercial blue poly(9,9-di-n-octylfluorenyl2,7-diyl) (PFO), green poly(9,9-dioctylfluorenealt-benzothiadiazole) (F8BT), and red spiro-copolymer (SPR) light-emitting materials are used to develop blends, and distyrylanthracene, distyryl-carbazole, and distyrylbenzothiadiazole chromophores as yellow, blue, and orange–red emitters, respectively, are used to synthesize novel copolymers. Since the host and dopant molecules emit at different visible-spectrum wavelengths, adjusting the concentration ratio could facilitate white light emission. The proper tuning of this ratio holds significant potential for generating white light. The optical and photophysical properties of the emitting films are thoroughly investigated through Spectroscopic Ellipsometry and Photoluminescence. Following this, these materials are applied as emitting layers in solution-processable single-layer OLEDS, via the spin coating method, and the device architecture is the simplest and most commonly used. We assess the WOLEDs produced based on their electroluminescence properties, including their luminance, CIE coordinates, CRI, and CCT values. This study aims to compare various emitting systems with different doping concentrations to investigate the photo- and electroluminescence characteristics. Subsequently, the most suitable emitting system will be evaluated based on its most promising results toward the attainment of pure white light emission.

2. Materials and Methods

2.1. Ink Formulation

Poly-3,4-ethylene dioxythiophene:poly-styrene sulfonate (PEDOT:PSS, purchased from Clevios Heraus, Hanau, Germany) Al 4083, mixed with ethanol in the ratio of 2:1, was prepared and used as the hole transport layer (HTL) in produced OLEDs. For the blend emitting polymers, the blue emitting poly(9,9-di-n-octylfluorenyl-2,7-diyl) PFO (Mw = 114,050) and the green emitting poly(9,9-dioctylfluorene-alt-benzothiadiazole) F8BT (Mw = 237,460) were purchased from Ossila (Sheffield, UK), and the red emitting spiro-copolymer-001 SPR (Mw = 720,000) was purchased from Sigma-Aldrich Chemie GmbH, Taufkirchen, Germany. All materials were dissolved in a toluene solvent. SPR and PFO solutions with different weight ratios, stirred under heating at 50 °C for 24 h, were used to prepare the homogeneous binary blends. In the same manner, SPR, F8BT, and PFO solutions were used to prepare the ternary blends. Analytically, the ratios used are listed in Table 1, where the codes of studied samples and the blend type are included. For all solutions, the final concentration was fixed at 1% w/w.

Table 1. Blend weight ratios and code-names.

Blend Type	Blend Code					
	Binary#1	Binary#2	Binary#3	Ternary#1	Ternary#2	Ternary#3
PFO:SPR	95:5	97.5:2.5	99:1			
PFO:F8BT:SPR				97:2.5:0.5	98:1.5:0.5	99:0.5:0.5

In the case of the synthesized copolymers Cz:BTZ with ratios 90:10, 99:1, and 99.5:0.5, the molecular weights of the copolymers were estimated at Mn = 49,800, 45,100, and 34,900, respectively. More details on the synthetic processes are given in refs. [4,14,26]. The molecular weights for terpolymers with ratios 98.5:1.0:0.5 and 99:0.5:0.5 were estimated at 9540 and 8800. The chemical structures of copolymers and terpolymers are shown in Scehmes 1 and 2, respectively. The copolymers were dissolved in N, N-dimethylformamide, DMF, with a resulting concentration of 1% wt, and the terpolymers were dissolved in o-DCB:CHCl$_3$ (9:1) with a resulting concentration of 1% wt [4,26]. The ratios for the copolymers and terpolymers are summarized in Table 2, where the codes of studied samples and the copolymer type are included.

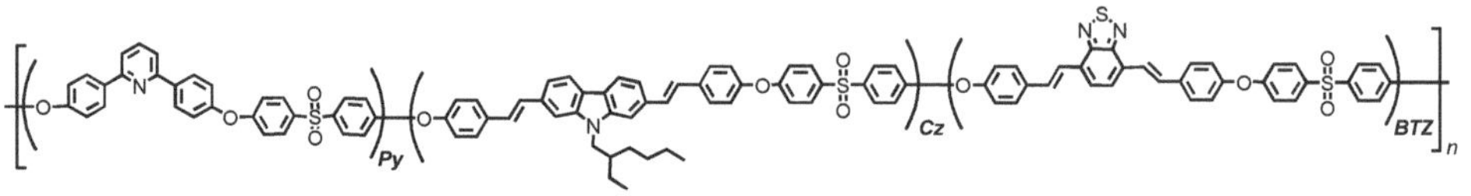

Cz:BTZ
Copolymer#1: 90:10
Copolymer#2: 99:1
Copolymer#3: 99.5:0.5

Scheme 1. Chemical structure of copolymers bearing two emissive chromophores.

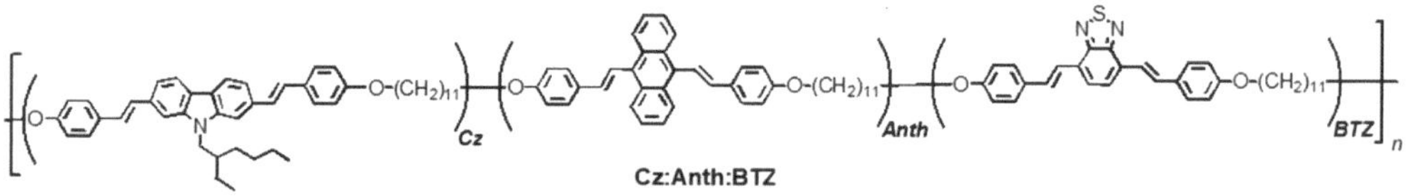

Cz:Anth:BTZ
Terpolymer#1: 98.5:1:0.5
Terpolymer#2: 99:0.5:0.5

Scheme 2. Chemical structure of terpolymers bearing three emissive chromophores.

Table 2. The co-monomer ratios of the copolymers and code-names.

Copolymer Type	Copolymer Code				
	Copolymer#1	Copolymer#2	Copolymer#3	Terpolymer#1	Terpolymer#2
Cz:BTZ	90:10	99:1	99.5:0.5		
Cz:Anth:BTZ				98.5:1.0:0.5	99:0.5:0.5

2.2. OLED Fabrication

The solution-processed WOLED structure is shown in Scheme 3. Initially, the glass substrates (provided by Ossila, Sheffield, UK) precoated with indium tin oxide (ITO) layers were cleaned in an ultrasonic bath by sequentially immersing them in deionized water, acetone, and ethanol for 10 min each. The substrates were then dried with ultrapure nitrogen gas and subsequently underwent oxygen plasma treatment at 40 W for 3 min. Afterward, the PEDOT:PSS ink was filtered (using PTFE filters with a 0.45 μm pore size) and spin-coated on the substrates at 4500 rpm for 60 s. The samples were then annealed on a hot plate glass/ITO/PEDOT:PSS at 120 °C for 5 min. The photoactive layer solution was then spin-coated at 2000 rpm for 25 s onto the samples without further annealing. Finally, by using a shadow mask, a 6 nm-thick Ca layer and a 125 nm-thick Ag layer were deposited via vacuum thermal evaporation (VTE) to form the bilayer cathode.

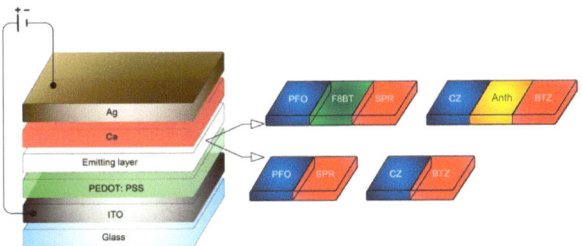

Scheme 3. The architecture of fabricated OLED devices with either blends or copolymers forming the photoactive layers.

2.3. Thin Film and Device Characterization

The emissive thin films based on blended emitting polymers or the copolymers were characterized in terms of their optical, electronic, and photophysical properties using Spectroscopic Ellipsometry (SE) and Photoluminescence (PL). In addition, in order to examine the possible energy transfer mechanism between the different units of both blends and copolymers, absorbance measurements were conducted for dopants and PL for the hosts. Then, the emissive layers were inserted in the produced OLEDs and were investigated through Electroluminescence (EL) Spectroscopy.

The SE measurements were conducted using a phase-modulated ellipsometer (Horiba Jobin Yvon, UVISEL, Europe Research Center—Palaiseau, France) from the near IR to far UV spectral region 1.5–6.5 eV with a step of 20 meV at a 70° angle of incidence. The SE experimental data were fitted to model-generated data using the Levenberg–Marquardt algorithm taking into consideration all the fitting parameters of the applied model.

The absorbance measurements were conducted using the set-up of Theta Metrisis (model FR UV/VIS) (Theta Metrisis S.A., Athens, Greece). Combining deuterium and halogen light sources, all measurements were performed in the 300–700 nm spectral range.

The PL and EL measurements of the active layers and the final OLED devices, respectively, were measured using the Hamamatsu Absolute PL Quantum Yield measurement system (C9920-02) and the External Quantum Efficiency (EQE) system (C9920-12) (Joko-cho, Higashi-ku, Hamamatsu City, Japan) which measures the luminance and light distribution of the devices. The PL spectra are recorded upon the excitation wavelength of 380 nm.

3. Results and Discussion
3.1. Spectroscopic Ellipsometry

The Spectroscopic Ellipsometry technique is a well-established characterization tool for determining optical constants. It is a non-destructive method that is free from the limitations associated with other methods that require physical contact with the film. In this study, the optical properties of the emissive materials were examined using the Vis-fUV SE method in the spectral region 1.5–6.5 eV. The Tauc-Lorentz model was used to derive the optical properties of each emitting phase, which was used and incorporated into either the copolymers or blended materials [27–29]. In addition, we can also obtain valuable information through the SE analysis, such as the thickness values of emissive films.

After the fitting process, the absorption coefficient versus wavelength was calculated for each emissive polymer contained in the blended two-phase and three-phase systems and for each monomer–chromophore incorporated in both copolymers and terpolymers, as shown in Figure 1a,b, respectively. Thus, Figure 1a,b depict the resulting spectra of the absorption coefficient used for the calculation of the spectral overlap integral $J(\lambda)$, which are presented as follows.

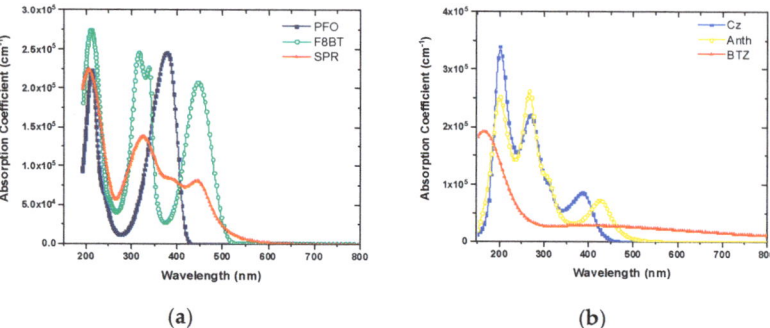

Figure 1. The absorption coefficient of each phase, which was incorporated in the (**a**) two and three blended systems and (**b**) copolymers and terpolymers.

3.2. Förster Resonance Energy Transfer (FRET)

Control in the Förster resonance energy transfer mechanism between emitters is regarded as the most important factor for achieving white light emission. It is well known that in the host–guest system, the absorbed energy in fluorescent organic molecules excites electrons from the ground state to a higher excited state, but a part of the energy can be released to lower excited states throughout the internal conversion of the electron before relaxation and light emission. The spectral overlap between the host's PL emission and the absorbance of the dopants in all studied emitting films is investigated further to find out if the energy transfer mechanism can possibly take place between the host and the dopants.

Figure 2a displays the absorption and PL emission profiles of the dopants, F8BT and SPR, accompanied by the PL emission of PFO. It is obvious that the absorption spectra of the F8BT and SPR overlap strongly with the emission spectrum of the PFO, whereas a lower spectral overlap is observed in the case of SPR absorption and F8BT PL. The degree of overlap between the absorption and PL emission peaks of the emissive materials is used to predict the energy transfer efficiency. It is possible to quantify the amount of energy transferred between the hosts and dopants by calculating the relative area of the overlap $\mathcal{E}^{overlap}$ between the residual absorption of each dopant within the wavelength range of each host's respective PL emission in UV-Vis absorption spectra and the total area of the PL emission of the hosts $\mathcal{E}^{emission}$ [29]. Finally, the calculated percentage is given using the following equation:

$$\mathcal{E} = \frac{\mathcal{E}^{overlap}}{\mathcal{E}^{emission}} \times 100\% \tag{1}$$

Figure 2b–d show the spectral overlap between the PL emission of PFO and absorption of F8BT, the PL emission of PFO and absorption of SPR, and the PL emission of F8BT and absorption of SPR, respectively. As shown in Figure 2b, the emission spectrum of PFO overlaps very well with the absorption band of F8BT in the region between 400 and 525 nm, and this is confirmed by the calculation of a 62% percentage. In the case of PFO-SPR, a smaller spectral overlap within the range of 400–600 nm is observed, and the percentage is calculated as 33%. Therefore, the energy-transfer mechanism is feasible, and this fact is significant for both approaches, either two- or three-phase blend polymers. Finally, the frustrated overlap of F8BT-SPR is observed, and it is much lower than that of PFO-SPR and PFO-F8BT, as indicated by the 6% calculated percentage.

For the synthesized emitting copolymers, we follow the same methodology in order to study if it is possible for the energy transfer between the host and dopants to occur. Figure 3a depicts the absorption and PL spectra of Anthracene (Anth) and Benzothiadiazole (BTZ) together with the PL emission of Carbazole (Cz). It is evident that spectral overlaps are observed between the Cz and Anth and BTZ but also between Anth and BTZ.

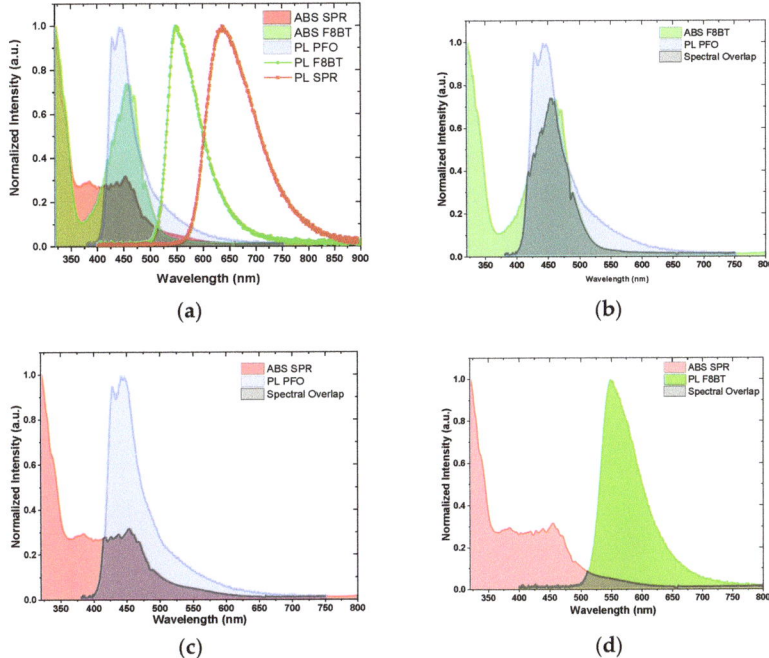

Figure 2. The overall correlation between (**a**) the absorbance of the SPR and F8BT units with the PL emissions of PFO, F8BT, and SPR. The examined spectral overlap between (**b**) PFO and F8BT, (**c**) PFO and SPR, and (**d**) F8BT and SPR.

To be clearer, Figure 3b–d show the spectral overlaps between the PL emission of Cz and absorbance of Anth, the PL emission of Cz and absorbance of BTZ, and the PL emission of Anth and absorbance of BTZ, respectively. The overlap between Cz and Anth covers the spectra range 400–550 nm, and the degree is calculated at 24%, indicating sufficient energy transfer from the Cz to the Anthracene. The spectral overlap between Cz and BTZ is significantly broader at the spectral range 400–750 nm. The calculated percentage for this case is equal to 58%, and this confirms that it is possible for more efficient energy transfer to take place. At this point, it is noteworthy that this huge spectral overlap manifests as a large fraction of energy being transferred from the Cz to the BTZ, and this fact is favorable for both cases of the copolymers based on two and three chromophores. The last combination of Anth–BTZ proves the lower spectral overlap within the range 470–750 nm, and this is also verified from the respective percentage value, which is calculated at 19%. According to these results, we assume that the sequential energy transfer happens from the Cz to Anthracene and BTZ, but also that there is a possibility that the energy transfer occurs from the Anthracene to BTZ. All the after mentioned results concerning the calculation of $\mathcal{E}$ values for all the possible combinations of host and dopants are written in Table 3.

A second approach is to estimate the overlap degree between the absorption spectrum of the acceptor with the emission spectrum of the donor, by calculating the overlap integral between them. More specifically, the overlap integral, J, is the integral of the spectral overlap between the donor emission and the acceptor absorption, which is given by the following:

$$J(\lambda) \sim \int_0^\infty \varepsilon_A(\lambda) F_D(\lambda) \lambda^4 d\lambda \tag{2}$$

where $F_D(\lambda)$ is the donor emission spectrum normalized to its area, and $\varepsilon_A(\lambda)$ is the molar extinction coefficient of the acceptor, measured in $M^{-1} \cdot cm^{-1}$ [20,25,30]. The calculated values $J(\lambda)$ are listed in Table 3. According to the results, the Förster type of energy transfer between the PFO monomers and each monomer of SPR and F8BT, as well as between SPR and F8BT, is confirmed. The calculated values $J(\lambda)$ of copolymers also validate the Förster energy transfer between the Cz unit and each unit of Anth and BTZ, as well as between Anth and BTZ.

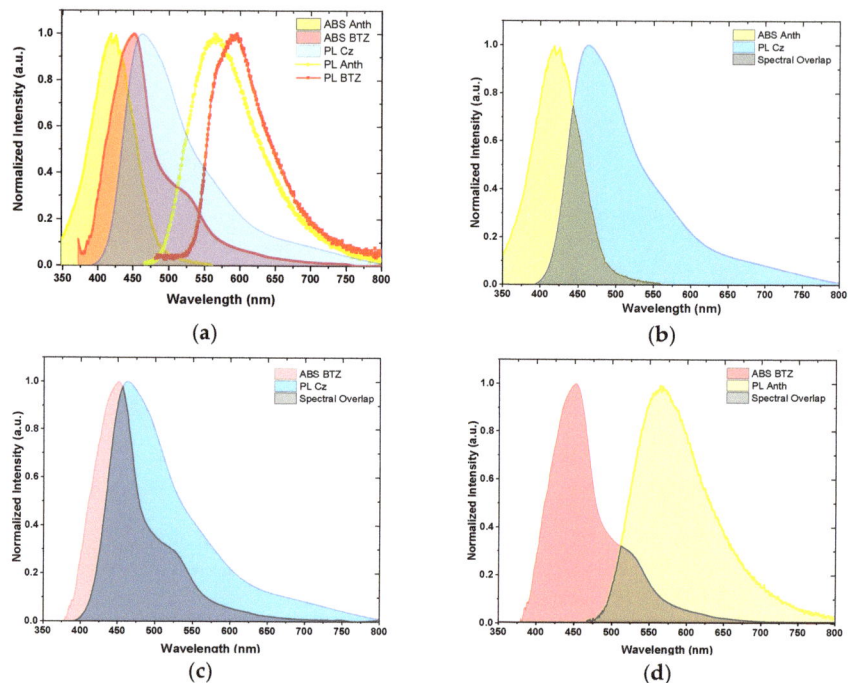

Figure 3. The overall correlation between (**a**) the absorbance of the Anth and SPR units with the PL emissions of Cz, Anth, and BTZ. The examined spectral overlap between (**b**) Cz and Anth, (**c**) Cz and BTZ, and (**d**) Anth and BTZ.

Table 3. Calculated values of the parameter $\mathcal{E}$ and overlap integral $J(\lambda)$.

Material	$\mathcal{E}$ (%)	$J(\lambda)$ ($M^{-1} \cdot cm^{-1} \cdot nm^4$)
PFO-F8BT	62	2.164×10^{10}
PFO-SPR	33	9.451×10^{9}
F8BT-SPR	6	6.176×10^{8}
Cz-Anth	24	2.001×10^{10}
Cz-BTZ	58	1.426×10^{12}
Anth-BTZ	19	2.017×10^{12}

3.3. Photoluminescence

Figure 4 shows the emission spectra of the produced (a) binary and (b) ternary blends with different doping concentrations, respectively. In all cases, the characteristic emission of the PFO (400–500 nm) exhibits a strong influence on the doping concentration. In particular, in binary blends (PFO-SPR), a gradual decrease in the PFO emission intensity with the addition of SPR is obtained, and at the same time, the intensity of the SPR unit (550–800 nm) is strengthened. The quenching mechanism is due to the partial energy transfer from PFO to SPR. On the other hand, in ternary blends (PFO-F8BT-SPR), a significant quenching of the PFO fluorescence intensity by increasing the total doping concentration of F8BT-SPR is observed (Figure 4b). This indicates an almost complete energy transfer from PFO to dopants.

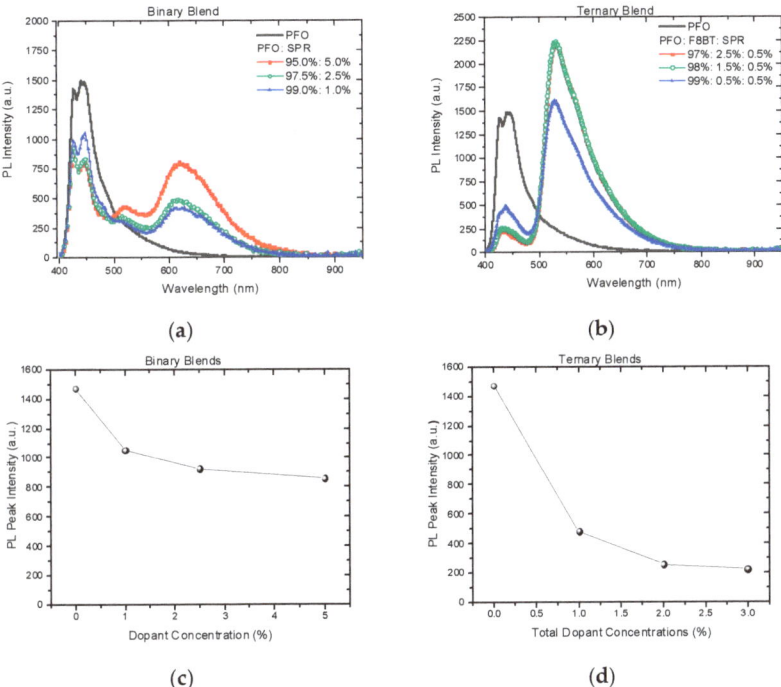

Figure 4. Emission spectra of PFO in the absence and the presence of (**a**) SPR and (**b**) F8BT-SPR and PFO PL peak intensity versus dopant concentration for the blended systems (**c**) PFO-SPR and (**d**) PFO-F8BT-SPR.

Figure 4 also shows the PFO PL peak intensity versus dopant concentration for the respective examined (c) binary and (d) ternary systems. In the case of a binary system, the PL intensity peak decreases gradually as the dopant concentration increases. On the other hand, in the ternary system, the PL peak intensity is reduced significantly, when the total doping concentration of F8BT-SPR is increased. This fact may be attributed to the energy transfer mechanism that quenches the fluorescence intensity of the PFO and reduces its excited-state lifetime while increasing the emission intensity of the acceptors F8BT-SPR [20,25,30,31].

The same analysis was conducted for copolymer and terpolymer emissive layers. Figure 5 shows the emission spectra of the produced (a) copolymers and (b) terpolymers in different concentrations, respectively. Firstly, the general result is that the maximum emission intensities of the copolymers (Cz-BTZ) decrease with a decreasing acceptor concentration. The possible reason for this phenomenon is radiative migration due to self-absorption. When the concentration of BTZ reaches 10% w/v, the emission of Cz is almost completely quenched and the emission of BTZ is enhanced and is the dominant one. The damping of Cz is attributed to the complete energy transfer from Cz to the BTZ unit. Similarly, in the case of terpolymers (Cz-Anth-BTZ), the emission intensity of Cz decreases in the presence of Anth-BTZ. Increasing the total concentration of Anth-BTZ results in the decrease in the PL intensity of Cz and in the enhancement of the combined PL intensity of Anth-BTZ.

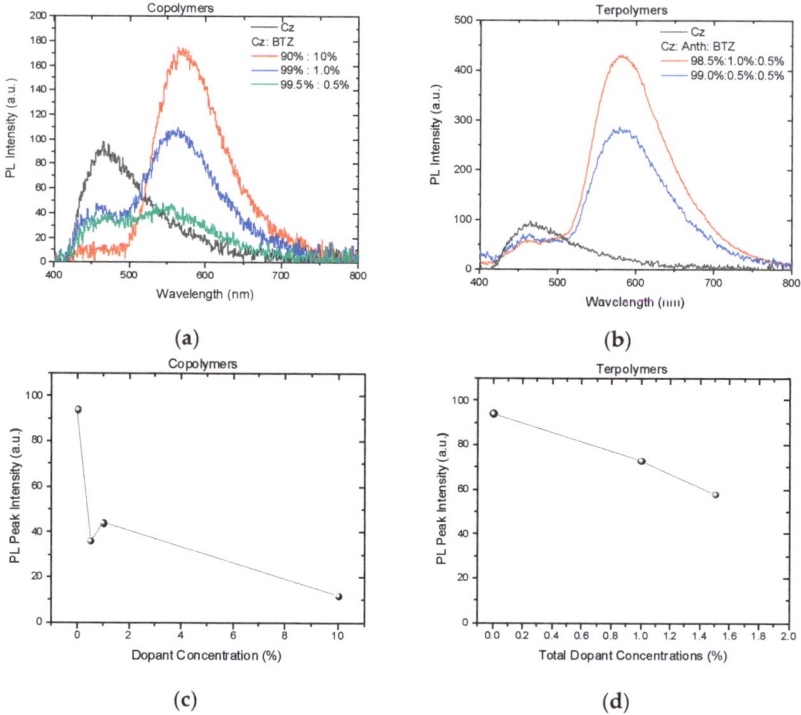

Figure 5. Emission spectra of Cz in the absence and the presence of (**a**) BTZ, and (**b**) Anth-BTZ and Cz PL peak intensity versus the dopant concentration for the copolymers (**c**) Cz-BTZ and (**d**) Cz-Anth-BTZ.

Further analysis, regarding the correlation of the Cz maximum PL intensity values with the dopant concentration, is illustrated in Figure 5 for the (c) copolymers and (d) terpolymers. In both cases, it is obvious that the intensity of the PL peak originating from the Cz emission decreases significantly with an increasing doping concentration. Thus, relying on these results, it is confirmed that the reduction in the PL intensity peak, which is assigned to the host unit, occurs due to the efficient energy transfer between the host and the dopants.

The potential use of the studied films in achieving broad emission in the visible spectral region is determined by the controllable FRET mechanisms between the host and dopants. The emission tuning based on two different emitters, the blue and red color, enables the definition of white light emission. Figure 6a shows the normalized PL emission spectra of the binary blended polymers in comparison with the PL emission of copolymers. It is obvious that the emission profile of both binary blended polymers and copolymers covers the whole visible range from 400 to 800 nm. We further analyzed the PL spectra using the deconvolution method using the required number of Gauss oscillators. The extracted results for λ^{max} and full width half maximum (FWHM) of each oscillator are summarized in Table 4. The respective thicknesses of the studied films, derived by SE analysis, are also included in Table 4.

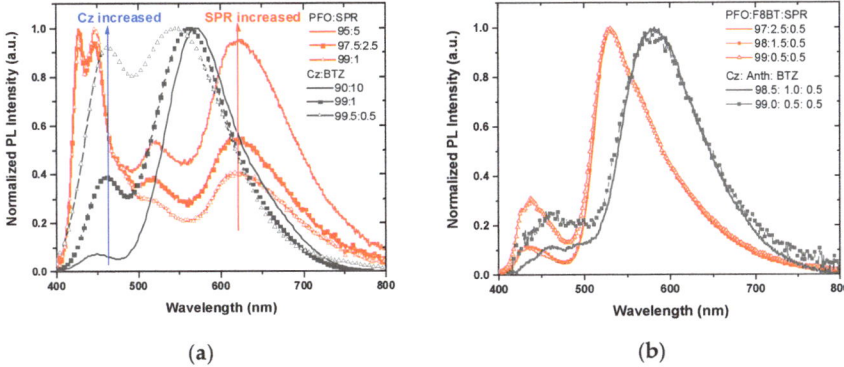

Figure 6. Comparison of normalized PL spectra between the (**a**) binary PFO-SPR with Cz-BTZ copolymers and (**b**) ternary PFO-F8BT-SPR with terpolymers Cz-Anth-BTZ.

Table 4. The obtained thickness values, based on the SE analysis, parameters of the deconvolution analysis of PL spectra, and the values of PL CIE coordinates.

Material	Ratio (%)	Thickness (nm)	λ_{PL}^{max} (nm)	FWHM (nm)	CIE Coordinates	
					x	y
PFO-SPR	95:5	48.3	424, 444, 510, 615, 651	12, 30, 70, 59, 119	0.3621	0.2877
	97.5:2.5	42.8	424, 444, 510, 615, 651	11, 28, 81, 49, 118	0.3156	0.2473
	99:1	40.4	424, 444, 510, 615, 651	12, 25, 74, 22, 131	0.2762	0.209
Cz-BTZ	90:10	21.0	445, 567, 634	35, 68, 113	0.4097	0.4378
	99:1	32.0	454, 560, 638	41, 84, 168	0.3473	0.3946
	99.5:0.5	9.9	459, 537, 616	51, 83, 89	0.3282	0.3826
PFO-F8BT-SPR	97:2.5:0.5	42.6	438, 525, 555, 602	30, 29, 54, 111	0.3737	0.5471
	98:1.5:0.5	43.5	439, 524, 554, 602	32, 29, 55, 112	0.3688	0.54
	99:0.5:0.5	40.4	439, 523, 551, 586	33, 28, 54, 129	0.3422	0.4832
Cz-Anth-BTZ	98.5:1.0:0.5	53.7	467, 570, 619	43, 66, 114	0.4607	0.5081
	99:0.5:0.5	60.5	467, 574, 622	44, 66, 110	0.4607	0.4888

According to our results, in the binary blended polymers PFO-SPR, the main emission peak attributed to the PFO unit is observed, and a weaker emission peak can be also detected, which is assigned to SPR. Here, it is notable that the characteristic emission intensity (λ^{max} = 615 nm) of SPR was gradually enhanced by increasing the SPR concentration. The highest doping concentration of SPR led to two characteristic emission peaks covering the entire visible range, but in the region between 490 and 570 nm, the intensity peak remains low. The PL peak (λ^{max} = 510 nm) can be attributed to the PFO defects, as it is well established that polyfluorene-based materials often exhibit a long-range emission around 2.2–2.3 eV [32]. On the contrary, in the copolymers Cz-BTZ, the dominant PL peak is located at approximately 560 nm, which is associated with the red chromophore BTZ. This means that in copolymers Cz-BTZ, the energy transfer mechanism plays a significant role in the PL emission profile, as the lower intensity peak comes from the blue chromophore. It is also remarkable to refer that in the case of the copolymer with the lowest concentration of BTZ, two main peaks are distinguished to span the entire visible range. The emission intensities of the blue and red chromophores are evenly balanced. This suggests that the energy-transfer mechanism between the two chromophores is optimized, which efficiently contributes to the mixing of blue and orange–red emission. Therefore, it is evident that the PL spectra can be tuned based on the ratios both in the binary blends and copolymers.

Figure 6b displays the normalized PL spectra of the ternary blended polymers (PFO-F8BT-SPR) and the terpolymers bearing three chromophores (Cz-Anth-BTZ), to facilitate a comparison. One can see that the PL emission spectra based on three emitters present a dominant peak, which originates from the dopant's units. This fact can be interpreted based on the assumption that the final intensity of the emission spectrum is a sum of the total energy transfer from the host to the dopants. More specifically, for the ternary blended polymers, the PL intensity of PFO is lower compared to the other two emitters F8BT and SPR due to the efficient energy transfer. In the case of the highest concentration of PFO in ternary blends, the contribution from the PFO emission was much more evident. However, the PFO intensity is weak, so there is no balanced tuning between the emitters. For the other approach based on terpolymers, the PL profiles reveal that the main peak appearing in all cases is attributed to the dopant emission. The emission intensity of Cz is attenuated when the doping concentration is increased. Thus, the quenching of the Cz emission can be supported by the efficient energy transfer mechanism from Cz to Anthracene and BTZ.

Note that in the case of both the binary blend and copolymer, the dopant concentration significantly affected the peak intensities in the PL spectra, which were attributed to the two emitters (blue-red). More specifically, in the case of the binary, the different doping concentrations influence the PL intensity peak of the red emitter, whereas, in copolymers, the different doping concentrations influence the PL intensity peak of the blue emitter. Therefore, efficient energy transfer from the wide-gap homopolymer to the narrow band-gap unit may happen due to a strong interchain interaction. This difference between the two emitting systems may be assigned to the rate of energy transfer. This rate is highly dependent on many factors, such as the spectral overlap, the relative orientation of the transition dipoles, and, most importantly, the distance between the donor and acceptor molecules. Regarding the three emitters, the different doping concentrations did not effectively contribute to the tuning of the emission of each emitter to span the entire visible range. Thus, adjusting the doping concentration in two emitters leads to easier tuning and results in a balanced two-color emission compared to three emitters.

As a means of elucidating which combination of emitting systems will generate white light, CIE (Commission Internationale de l'Éclairage) coordinates were calculated through the respective PL emission spectra. The CIE color system is the most widely used colorimetric standard so far. The CIE-1931 coordinates describe how the human eye perceives the color emission from any light source using Cartesian coordinates. The CIE specifies that the coordinates for the point of white light are (0.33, 0.33) [33,34].

Figure 7 shows the CIE diagram of coordinates corresponding to the PL spectra. As can be seen, by increasing concentrations of the red-emitting polymer SPR (denoted by arrows) in the binary blends, the CIE coordinates tend to approach the white region. On the other hand, the opposite occurs in the case of copolymers Cz-BTZ, in which by reducing the concentration of the red dopant, BTZ (the arrow denotes the increase of BTZ), the CIE coordinates approach the ideal white light emission. Furthermore, concerning the CIE coordinates of both ternary blends and terpolymers, they are quite far from the white light emission. Here, it is also observed that in both cases, the reduced doping concentration shifts the coordinates toward the white region.

3.4. Electroluminescence

The blended and synthesized polymers were applied as emitting layers in single-layer OLEDs, having the device structure of ITO/PEDOT:PSS/emitting layer/Ca/Ag. The electro-optical properties and the performance of the fabricated devices were investigated. Figure 8 shows the comparison of the respective normalized EL spectra between the (a) binary blended system and copolymers and (b) ternary blended system and terpolymers. Concerning the analysis of EL spectra using the deconvolution method, the calculated values of λ^{max} and FWHM are summarized in Table 5. It is obvious that for all studied cases, the EL emission spectra covered the spectral region from 400 to 800 nm.

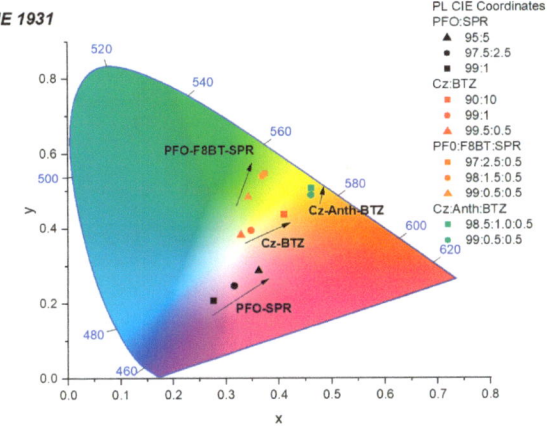

Figure 7. CIE diagram of PL emissions of the studied materials. The arrows denote the increase in the percentage of the dopant.

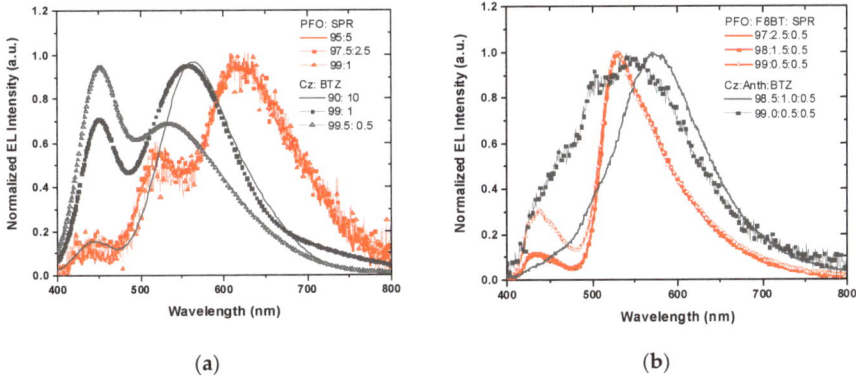

Figure 8. Comparison of normalized EL spectra between (a) binary PFO-SPR with Cz-BTZ copolymers and (b) ternary PFO-F8BT-SPR with terpolymers Cz-Anth-BTZ.

For a better evaluation of the EL characteristics, Figure 9 shows, comparatively, the EL (lines) and PL (lines with symbol) spectra of emitting layers based on the (a) binary, (b) copolymers, (c) ternary, and (d) terpolymers, respectively. In all studied emitting layers, a blue shift of the EL spectra is observed, when comparing to that of PL. If FRET was the only mechanism controlling the emission of WOLED devices, the relative intensities of host and dopant emissions would be equivalent in the corresponding PL and EL spectra. However, this is not valid in all cases examined in our studies, and the relative intensities are substantially different between the PL and EL spectra. In the case of PL emission, it has been proven that the energy-transfer process plays a significant role. Particularly, for all cases, energy transfer can possibly occur, as found based on the evaluation of the spectral overlap between the PL emission of each host and the absorbance of each dopant. On the other hand, it is well established that the EL emission depends not only on the energy transfer mechanism but also on the charge trapping.

Table 5. The obtained parameters of deconvolution analysis of EL spectra and electrical-operational characteristics of the produced OLED.

Material	Ratio (%)	λ_{EL}^{max} (nm)	FWHM (nm)	EL CIE Coordinares (x, y)	Luminance (cd/m^2)	CRI	CCT (K)	* V_{on} (V)
PFO-SPR	95:5	438,517,613,642	36,47,61,142	(0.47, 0.44)	1033	94	2515	4.1
	97.5:2.5	442, 517,612,651	43,44,72,145	(0.46, 0.42)	1038	94	2663	6.5
	99:1	441,519,613,654	43,41,73,150	(0.48, 0.42)	561	95	2620	4.5
Cz-BTZ	90:10	443,555,613	60,64,104	(0.43, 0.50)	37	60	4063	7.1
	99:1	448,551,614	41,91,138	(0.33, 0.38)	91	70	5498	6.1
	99.5:0.5	447, 520,580	17,40,75	(0.28, 0.31)	10	82	8000	5.1
PFO-F8BT-SPR	97:2.5:0.5	421,440,526,572	14,16,47,94	(0.34, 0.56)	7474	44	5264	4
	98:1.5:0.5	423,441,525,573	13,17,47,98	(0.34, 0.56)	9838	49	5697	3.1
	99:0.5:0.5	424,448,522,562	16,27,56,168	(0.31, 0.47)	3321	74	7698	5.1
Cz-Anth-BTZ	98.5:1.0:0.5	481,569,614	60,83,142	(0.44, 0.48)	107	60	2787	7
	99:0.5:0.5	477,561,607	56,115,132	(0.34, 0.42)	134	74	2974	9

* V_{on}: Turn-on Voltage at 1 cd/m^2

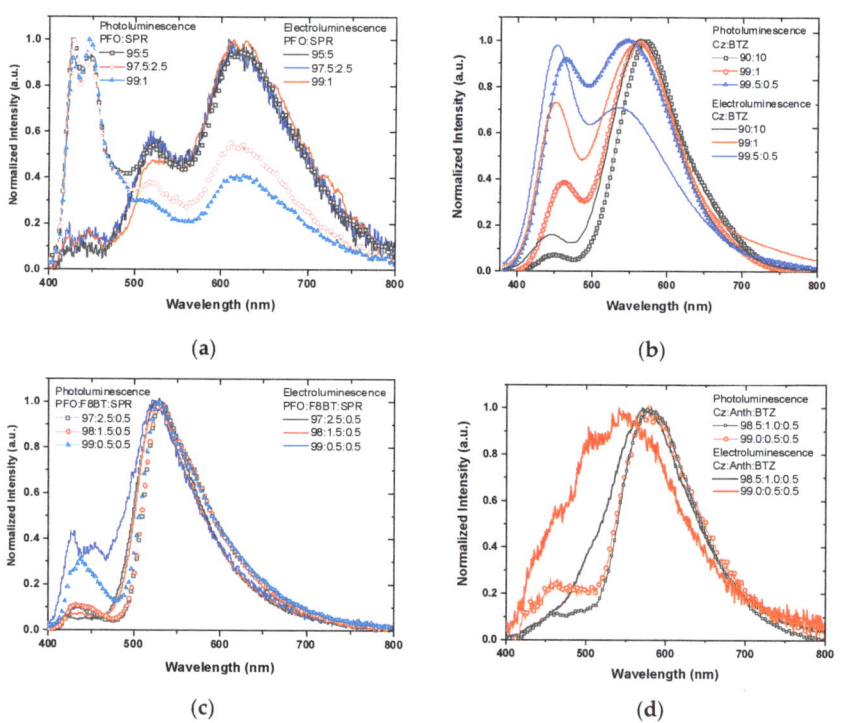

Figure 9. Normalized EL spectra (lines) of OLEDs based on the emitting layers (**a**) binary, (**b**) copolymer, (**c**) ternary, and (**d**) terpolymer, respectively; for a better evaluation of the EL characteristics, the corresponding PL spectra are also incorporated (lines with symbols).

As depicted in Figure 9a, in the case of binary PFO-SPR, a comparison between the PL and EL profiles clearly shows substantial differences. The EL emission profiles of all binary blends present a dominant peak at approximately 610 nm, which is attributed to the

SPR moiety, and this peak is independent of the doping concentration. Furthermore, the emission of PFO is completely quenched compared to the PL emission profile. Therefore, we can assume that the EL emission originated from other exciton species than those responsible for the PL. This fact can account for the charge trapping in the SPR moiety, apart from energy transfer from the PFO to SPR. In Figure 9b, one can see that significant differences between the PL and EL spectra are also observed in the studied copolymers Cz-BTZ. More specifically, in all copolymers, the EL emission spectra exhibit the main peak at 560 nm, which is associated with the BTZ moiety. It is also important to mention that the relative intensity of Cz is reinforced strongly compared to the PL emission when the doping concentration of BTZ decreased. These EL spectra reveal that the dual emission is a result of the combination of the energy-transfer mechanism and the charge trapping in the Cz unit. It was found that when the concentration of BTZ is generally low (e.g., <1%) apparent dual emissions are displayed at 450 and 560 nm, respectively, which are strong enough to render color balance between blue and red–orange emission to achieve white light.

In all ternary blends (Figure 9c), the tendency of the EL emission band presents similarities with the PL. However, upon decreasing the doping concentrations of both F8BT and SPR, the relative intensity of the PFO peak is more pronounced in the EL spectrum compared to the PL spectrum, and this means that the charge trapping into the PFO unit is more efficient. Finally, the EL emission of terpolymers (Figure 9d) is broader and featureless compared to that of PL. The increase in the Cz concentration enhances the broadening, as well as the blue-shifting, of the emission.

In conclusion, the EL emission of the host is decreased, mainly for the cases of binary and ternary blends, which may be due to the cascade mechanism for the charge injection that favors the exciton formation in the emitter with the lower band gap. On the contrary, in the EL emission profile of copolymers and the ternary blend with the ratio 99:0.5:0.5, the intensity of the host is enhanced, and here, the possible reason is the partial energy transfer from the host to the dopants due to the charge trapping in the host. Finally, in the terpolymers, a featureless and broad EL emission is derived covering the visible spectral range.

Another important feature to evaluate is the color characteristics, including CIE color coordinates (x, y), of the fabricated OLEDs. To better explore this aspect, Figure 10 depicts the CIE coordinates of the studied materials derived from the EL spectra. The CIE coordinates of binary blends is measured representatively at (0.46, 0.42), which are not close to the white region due to the stronger emission of SPR in relation to PFO. On the other hand, the CIE coordinates of copolymers tend to be the values of the ideal white emission (0.33, 0.33). By adjusting the ratio between the two units, Cz and BTZ, and with a dopant ratio <1%, a balanced blue and red emission achieved, and the CIE coordinates are located very close to the pure white-light point with values (0.28, 0.31). These values indicate that this emitting system can be used to generate white-light emission that is well-suited for lighting applications.

As can be seen, the CIE coordinates of ternary blends approach the white region, by increasing the relative ratio of PFO. The perceived emission color of the mixed emitters can be adjusted from green to nearly white, and the obtained CIE coordinates are calculated (0.31, 0.47). Concerning the CIE coordinates of terpolymers, it has been observed that elevating the concentration of Cz leads to the white region being approached, and the obtained coordinates values are equal to (0.34, 0.42).

3.5. Performance of OLED Devices

Figure 11 shows the logarithmic plot of (a) Current density–Voltage (J–V) characteristic curves and (b) Luminance–Voltage of the fabricated WOLEDs in order to provide additional insights for the operational-electrical characteristics of the devices. The blended materials exhibit higher Luminance values compared to the novel emitting copolymers and terpolymers. Generally, further investigation is needed to enhance the functionalization of produced WOLED devices in order to attain higher efficiency. At this point, it is also remarkable to see that the materials based on three emitters present better luminance performance than the two.

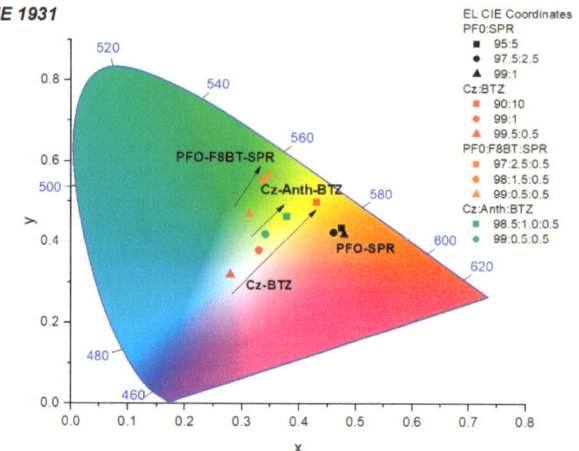

Figure 10. CIE diagram of coordinates derived from EL emissions of the fabricated WOLEDs. The arrows denote the increase in the percentage of the dopant.

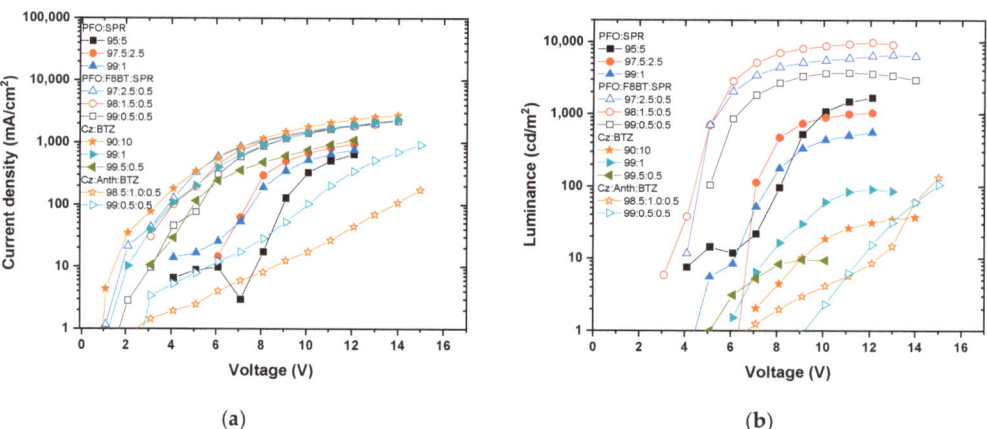

Figure 11. The logarithmic plot of characteristic curves of (**a**) Current density–Voltage (J–V) and (**b**) Luminance–Voltage of the fabricated OLED devices.

For lighting applications, the color quality and stability must also be considered towards the performance of OLEDs. Detailed characteristics of produced devices, such as CIE color coordinates (x, y), color rendering indexes (CRI), and correlated color temperatures (CCT) [35–37], are listed in Table 5. Among all the studied cases, the most encouraging results for the CRI values are obtained from the binary blends and copolymers.

The effect of increasing the applied voltage on the shift in the color of the emitted light from WOLED is also examined. This is a crucial factor since the increment in voltage usually results in an increase in emitted light intensity, which may end up causing the unbalanced mixing of emitters. Figure 12 shows the evolution of the EL CIE coordinates only for the WOLED devices, which showed the closest coordinates to the ideal white point, for each examined case. Specifically, the coordinates were measured during device operation and recorded after the turn-on voltage. One can see that the CIE coordinates of copolymer Cz-BTZ demonstrate superior color stability with no variations since the voltage is increased. These materials are promising candidates for lighting applications, as the CIE coordinates are very close to the ideal white point and the CRI value reaches 82. It is a great achievement for WOLED technology, as it is well known that the CRI value of binary-emitter WOLEDs can be boosted from the 63 to 80 threshold [38].

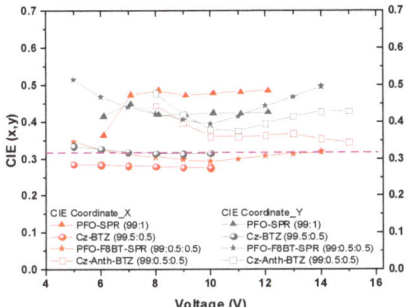

Figure 12. The evolution of EL Chromaticity Coordinates for the WOLED devices, which showed the closest coordinates to the ideal white point at an increasing applied voltage (The dashed line represents the CIE Coordinates of the ideal white point (0.33, 0.33).

4. Conclusions

In conclusion, we have presented two distinct approaches that rely on light-emitting blended polymers or copolymers to realize pure white light emission. Two potential options were explored for blended systems. The first approach was based on mixing two emitters, a red, SPR, with a blue one, PFO. The second approach involved blending three emitters, red SPR with green F8BT and blue PFO. In all cases, PFO was dominated in the percentage and acted as a host. In the case of novel copolymers, the blue Cz derivative was used as the host, which was incorporated with the red chromophore BTZ, and for the terpolymers, the yellow Anth chromophore was added to them. In all doping systems, different ratios of the constituents were examined to tune the emissions and realize white light emission.

At first, for all emitting systems, it was verified that the energy transfer was feasible between the host and the dopants, highlighting the possibility of the management of the emission of each constituent to attain the coverage of the entire visible range. Following this, the emitting polymers were applied as emitting layers in single-layer WOLEDs. Comparing PL and EL emission revealed that charge trapping had a significant impact on EL emission, in addition to the energy transfer mechanism. Mainly, in the case of the Cz-BTZ copolymer, incomplete energy transfer from the blue Cz backbone to the orange-red BTZ moiety resulted in two significant emissions with equal intensity when the doping concentration of BTZ is less than 1%. Specifically, controlling the device's emission color through a balance of emission between the two units, Cz and BTZ, was achieved, thereby leading to a broad white light emission with satisfying CIE coordinates of (0.28,0.31). Moreover, the color coordinates of the resulting white-light emission remained extremely stable over a wide range of increasing driving voltages. Also noteworthy is that the CRI value obtained for the WOLED was 80, which makes it highly promising for lighting applications. Thus, compared to doping based on blended polymers, chemical synthesis was more efficient, which was conducive to achieving white light emission, as the emission tuning of chromophores was easier by controlling the doping concentration. Overall, the solution-processable WOLED based on copolymers offers an innovative pathway for high-quality emission.

Author Contributions: Conceptualization, J.K.K., S.L. and M.G.; Methodology, D.T., K.P., V.F., V.K., K.C.A. and M.G.; Validation, D.T., K.P., V.F., V.K., A.K.A., A.L. and M.G.; Formal analysis, D.T., K.P., V.F., V.K., A.K.A. and M.G.; Investigation, D.T., K.P., K.C.A., A.K.A., J.K.K., A.L. and M.G.; Resources, J.K.K., S.L. and M.G.; Data curation, D.T., K.P., V.F., V.K. and K.C.A.; Writing—original draft, D.T.; Writing—review & editing, D.T., K.C.A., A.K.A. and M.G.; Visualization, D.T., V.K., A.K.A., J.K.K. and M.G.; Supervision, M.G.; Project administration, M.G.; Funding acquisition, J.K.K., S.L. and M.G. All authors have read and agreed to the published version of the manuscript.

Funding: This research has been co-financed by the European Regional Development Fund of the European Union and Greek national funds through the Operational Program Competitiveness, Entrepreneurship and Innovation, under the call RESEARCH—CREATE—INNOVATE (project code:T1EDK-01039).

Institutional Review Board Statement: Not applicable.

Informed Consent Statement: Not applicable.

Data Availability Statement: Data are contained within the article.

Conflicts of Interest: Author Vasileios Kyriazopoulos was employed by the company Organic Electronic Technologies P.C. (OET). The remaining authors declare that the research was conducted in the absence of any commercial or financial relationships that could be construed as a potential conflict of interest.

References

1. Luo, D.; Chen, Q.; Liu, B.; Qiu, Y. Emergence of Flexible White Organic Light-Emitting Diodes. *Polymers* **2019**, *11*, 384. [CrossRef] [PubMed]
2. Das, D.; Gopikrishna, P.; Barman, D.; Yathirajula, R.B.; Iyer, P.K. White light emitting diode based on purely organic fluorescent to modern thermally activated delayed fluorescence (TADF) and perovskite materials. *Nano Converg.* **2019**, *6*, 31. [CrossRef] [PubMed]
3. Wang, Q.; Bai, J.; Zhao, C.; Ali, M.U.; Miao, J.; Meng, H. Simplified dopant-free color-tunable organic light-emitting diodes. *Appl. Phys. Lett.* **2021**, *118*, 253301. [CrossRef]
4. Tselekidou, D.; Papadopoulos, K.; Kyriazopoulos, V.; Andrikopoulos, K.C.; Andreopoulou, A.K.; Kallitsis, J.K.; Laskarakis, A.; Logothetidis, S.; Gioti, M. Photophysical and Electro-Optical Properties of Copolymers Bearing Blue and Red Chromophores for Single-Layer White OLEDs. *Nanomaterials* **2021**, *11*, 2629. [CrossRef] [PubMed]
5. Bozkus, V.; Aksoy, E.; Varlikli, C. Perylene Based Solution Processed Single Layer WOLED with Adjustable CCT and CRI. *Electronics* **2021**, *10*, 725. [CrossRef]
6. Guo, F.; Karl, A.; Xue, Q.-F.; Tam, K.C.; Forberich, K.; Brabec, C.J. The fabrication of color-tunable organic light-emitting diode displays via solution processing. *Light Sci. Appl.* **2017**, *6*, e17094. [CrossRef] [PubMed]
7. Zhang, T.; Sun, J.; Liao, X.; Hou, M.; Chen, W.; Li, J.; Wang, H.; Li, L. Poly(9,9-dioctylfluorene) based hyperbranched copolymers with three balanced emission colors for solution-processable hybrid white polymer light-emitting devices. *Dye Pigment.* **2017**, *139*, 611–618. [CrossRef]
8. Cho, W.; Kim, Y.; Lee, C.; Park, J.; Gal, Y.S.; Lee, J.W.; Jin, S.H. Single emissive layer white phosphorescent organic light-emitting diodes based on solution-processed iridium complexes. *Dye Pigment.* **2014**, *108*, 115–120. [CrossRef]
9. Witkowska, E.; Glowacki, I.; Ke, T.-H.; Malinowski, P.; Heremans, P. Efficient OLEDs Based on Slot-Die-Coated Multicomponent Emissive Layer. *Polymers* **2022**, *14*, 3363. [CrossRef]
10. Xiang, H.; Wang, R.; Chen, J.; Li, F.; Zeng, H. Research progress of full electroluminescent white light-emitting diodes based on a single emissive layer. *Light Sci. Appl.* **2021**, *10*, 206. [CrossRef]
11. Ying, L.; Ho, C.L.; Wu, H.; Cao, Y.; Wong, W.Y. White Polymer Light-Emitting Devices for Solid-State Lighting: Materials, Devices, and Recent Progress. *Adv. Mater.* **2014**, *26*, 2459–2473. [CrossRef] [PubMed]
12. Reineke, S.; Thomschke, M.; Lussem, B.; Leo, K. White organic light-emitting diodes: Status and perspective. *Rev. Mod. Phys.* **2013**, *85*, 1245–1293. [CrossRef]
13. Wang, H.; Xu, Y.; Tsuboi, T.; Xu, H.; Wua, Y.; Zhang, Z.; Miao, Y.; Hao, Y.; Liu, X.; Xu, B.; et al. Energy transfer in polyfluorene copolymer used for white-light organic light emitting device. *Org. Electron.* **2013**, *14*, 827–838. [CrossRef]
14. Gioti, M.; Kokkinos, D.; Stavrou, K.; Simitzi, K.; Andreopoulou, A.K.; Laskarakis, A.; Kallitsis, J.K.; Logothetidis, S. Fabrication and study of white-light OLEDs based on novel copolymers with blue, yellow, and red chromophores. *Phys. Status Solidi RRL* **2019**, *13*, 1800419–1800424. [CrossRef]
15. Xu, J.; Yu, L.; Sun, Z.; Li, T.; Chen, H.; Yang, W. Efficient, stable and high color rendering index white polymer light-emitting diodes by restraining the electron trapping. *Org. Electron.* **2020**, *84*, 105785–105792. [CrossRef]
16. Deus, J.F.; Faria, G.C.; Faria, R.M.; Iamazaki, E.T.; Atvars, T.D.Z.; Cirpan, A.; Akcelrud, L. White light emitting devices by doping polyfluorene with two red emitters. *J. Photochem. Photobiol. A Chem.* **2013**, *253*, 45–51. [CrossRef]

17. Chen, K.; Wei, Z. High-efficiency single emissive layer color-tunable all-fluorescent white organic light-emitting diodes. *Chem. Phys. Lett.* **2022**, *786*, 139145. [CrossRef]
18. Mei, D.; Yan, L.; Liu, X.; Zhao, L.; Wang, S.; Tian, H.; Ding, J.; Wang, L. De novo design of single white-emitting polymers based on one chromophore with multi-excited states. *Chem. Eng. J.* **2022**, *446*, 137004. [CrossRef]
19. Prakash, A.; Katiyar, M. White polymer light emitting diode using blend of fluorescent polymers. *Proc. SPIE* **2012**, *8549*, 854938. [CrossRef]
20. Al-Asbahi, B.A.; AlSalhi, M.S.; Fatehmulla, A.; Jumali, M.H.H.; Qaid, S.M.H.; Mujamammi, W.M.; Ghaithan, H.M. Controlling the Emission Spectrum of Binary Emitting Polymer Hybrids by a Systematic Doping Strategy via Förster Resonance Energy Transfer for White Emission. *Micromachines* **2021**, *12*, 1371. [CrossRef]
21. Gioti, M.; Tselekidou, D.; Foris, V.; Kyriazopoulos, V.; Papadopoulos, K.; Kassavetis, S.; Logothetidis, S. Influence of Dopant Concentration and Annealing on Binary and Ternary Polymer Blends for Active Materials in OLEDs. *Nanomaterials* **2022**, *12*, 4099. [CrossRef]
22. Al-Asbahi, B.A. Dual Förster resonance energy transfer in ternary PFO/MEH-PPV/F7GA hybrid thin films for white organic light-emitting diodes. *Dye Pigment.* **2021**, *186*, 109011. [CrossRef]
23. Kim, K.; Inagaki, Y.; Kanehashi, S.; Ogino, K. Incorporation of benzothiadiazole moiety at junction of polyfluorene–polytriarylamime block copolymer for effective color tuning in organic light emitting diode. *J. Appl. Polym. Sci.* **2017**, *134*, 45393. [CrossRef]
24. Joshi, N.K.; Polgar, A.M.; Steer, R.P.; Paige, M.F. White light generation using Förster resonance energy transfer between 3-hydroxyisoquinoline and Nile Red. *Photochem. Photobiol. Sci.* **2016**, *15*, 609–617. [CrossRef] [PubMed]
25. Al-Asbahi, B.A.; AlSalhi, M.S.; Jumali, M.H.H.; Fatehmulla, A.; Qaid, S.M.H.; Mujamammi, W.M.; Ghaithan, H.M. Conjugated Polymers-Based Ternary Hybrid toward Unique Photophysical Properties. *Molecules* **2022**, *27*, 7011. [CrossRef] [PubMed]
26. Gioti, M.; Papadopoulos, K.; Kyriazopoulos, V.; Andreopoulou, A.K.; Kallitsis, J.K.; Laskarakis, A.; Logothetidis, S. Optical and emission properties of terpolymer active materials for white OLEDs (WOLEDs). *Mater. Today Proc.* **2021**, *37*, 141101. [CrossRef]
27. Gioti, M.; Tselekidou, D.; Panagiotidis, L.; Kyriazopoulos, V.; Simitzi, K.; Andreopoulou, A.K.; Kalitsis, J.K.; Gravalidis, C.; Logothetidis, S. Optical characterization of organic light-emitting diodes with selective red emission. *Mater. Today Proc.* **2021**, *37*, A39–A45. [CrossRef]
28. Jellison, G.E.; Modine, F.A. Parameterization of the optical functions of amorphous materials in the interband region. *Appl. Phys. Lett.* **1996**, *69*, 371–373. [CrossRef]
29. Tselekidou, D.; Papadopoulos, K.; Zachariadis, A.; Kyriazopoulos, V.; Kassavetis, S.; Laskarakis, A.; Gioti, M. Solution-processable red phosphorescent OLEDs based on Ir(dmpq)2(acac) doped in small molecules as emitting layer. *Mater. Sci. Semicond. Process.* **2023**, *163*, 107546. [CrossRef]
30. Al-Asbahi, B.A. Energy transfer mechanism and optoelectronic properties of (PFO/TiO2)/Fluorol 7GA nanocomposite thin films. *Opt. Mater.* **2017**, *72*, 644–649. [CrossRef]
31. Kim, N.H.; Kim, Y.H.; Yoon, J.A.; Lee, S.Y.; Ryu, D.H.; Wood, R.; Moon, C.B.; Kim, W.Y. Color optimization of single emissive white OLEDs via energy transfer between RGB fluorescent dopants. *J. Lumin.* **2013**, *143*, 723–728. [CrossRef]
32. Kuik, M.; Wetzelaer, G.-J.A.H.; Laddé, J.G.; Nicolai, H.T.; Wildeman, J.; Sweelssen, J.; Blom, P.W.M. The Effect of Ketone Defects on the Charge Transport and Charge Recombination in Polyfluorenes. *Adv. Funct. Mater.* **2011**, *21*, 4502–4509. [CrossRef]
33. Zhang, L.; Li, X.L.; Luo, D.; Xiao, P.; Xiao, W.; Song, Y.; Ang, Q.; Liu, B. Strategies to Achieve High-Performance White Organic Light-Emitting Diodes. *Materials* **2017**, *10*, 1378. [CrossRef] [PubMed]
34. Ye, S.H.; Hu, T.Q.; Zhou, Z.; Yang, M.; Quan, M.H.; Mei, Q.B.; Zhai, B.C.; Jia, Z.H.; Laia, W.Y.; Huang, W. Solution processed singlee-mission layer white polymer light-emitting diodes with high color quality and high performance from a poly(N-vinyl)carbazole host. *Phys. Chem. Chem. Phys.* **2015**, *17*, 8860–8869. [CrossRef] [PubMed]
35. Meng, F.; Chen, D.; Xiong, W.; Tan, H.; Wang, Y.; Zhu, W.; Su, S.J. Tuning color-correlated temperature and color rendering index of phosphorescent white polymer light-emitting diodes: Towards healthy solid-state lighting. *Org. Electron.* **2016**, *34*, 18–22. [CrossRef]
36. Kang, J.; Cho, Y.; Jang, W. Long-Term Reliability Characteristics of OLED Panel and Luminaires for General Lighting Applications. *Appl. Sci.* **2021**, *11*, 74. [CrossRef]
37. Miao, Y.; Wang, K.; Zhao, B.; Gao, L.; Tao, P.; Liu, X.; Hao, Y.; Wang, H.; Xu, B.; Zhu, F. High-efficiency/CRI/color stability warm white organic light-emitting diodes by incorporating ultrathin phosphorescence layers in a blue fluorescence layer. *Nanophotonics* **2018**, *7*, 295–304. [CrossRef]
38. Tang, X.; Liu, X.-Y.; Jiang, Z.-Q.; Liao, L.-S. High-Quality White Organic Light-Emitting Diodes Composed of Binary Emitters with Color Rendering Index Exceeding 80 by Utilizing Color Remedy Strategy. *Adv. Funct. Mater.* **2019**, *29*, 1807541. [CrossRef]

Disclaimer/Publisher's Note: The statements, opinions and data contained in all publications are solely those of the individual author(s) and contributor(s) and not of MDPI and/or the editor(s). MDPI and/or the editor(s) disclaim responsibility for any injury to people or property resulting from any ideas, methods, instructions or products referred to in the content.

Article

Exploration of Methodologies for Developing Antimicrobial Fused Filament Fabrication Parts

Sotirios Pemas [1], Eleftheria Xanthopoulou [2], Zoi Terzopoulou [2,*], Georgios Konstantopoulos [3], Dimitrios N. Bikiaris [2], Christine Kottaridi [3], Dimitrios Tzovaras [1] and Eleftheria Maria Pechlivani [1,*]

1. Centre for Research and Technology Hellas, Information Technologies Institute, 6th km Charilaou-Thermi Road, 57001 Thessaloniki, Greece; sopemas@iti.gr (S.P.); dimitrios.tzovaras@iti.gr (D.T.)
2. Laboratory of Chemistry and Technology of Polymers and Colors, Department of Chemistry, Aristotle University of Thessaloniki, 54124 Thessaloniki, Greece; elefthxanthopoulou@gmail.com (E.X.); dbic@chem.auth.gr (D.N.B.)
3. Laboratory of General Microbiology, Department of Genetics, Development and Molecular Biology, School of Biology, Aristotle University of Thessaloniki, 54124 Thessaloniki, Greece; konstanto@bio.auth.gr (G.K.); ckottaridi@bio.auth.gr (C.K.)
* Correspondence: terzozoi@chem.auth.gr (Z.T.); riapechl@iti.gr (E.M.P.); Tel.: +30-2310-997818 (Z.T.); +30-2311-257751 (E.M.P.); Fax: +30-2310-474128 (E.M.P.)

Abstract: Composite 3D printing filaments integrating antimicrobial nanoparticles offer inherent microbial resistance, mitigating contamination and infections. Developing antimicrobial 3D-printed plastics is crucial for tailoring medical solutions, such as implants, and cutting costs when compared with metal options. Furthermore, hospital sustainability can be enhanced via on-demand 3D printing of medical tools. A PLA-based filament incorporating 5% TiO_2 nanoparticles and 2% Joncryl as a chain extender was formulated to offer antimicrobial properties. Comparative analysis encompassed PLA 2% Joncryl filament and a TiO_2 coating for 3D-printed specimens, evaluating mechanical and thermal properties, as well as wettability and antimicrobial characteristics. The antibacterial capability of the filaments was explored after 3D printing against Gram-positive Staphylococcus aureus (*S. aureus*, ATCC 25923), as well as Gram-negative Escherichia coli (*E. coli*, ATCC 25922), and the filaments with 5 wt.% embedded TiO_2 were found to reduce the viability of both bacteria. This research aims to provide the optimal approach for antimicrobial and medical 3D printing outcomes.

Keywords: additive manufacturing (AM); 3D printing; fused filament fabrication (FFF); filament; antimicrobial properties; *Escherichia coli*; *Staphylococcus aureus*; titanium dioxide (TiO_2); poly(lactic acid) (PLA); mechanical properties

1. Introduction

Fused Filament Fabrication (FFF) technology is one of the most prevalent techniques in additive manufacturing (AM) [1–3] that has been developed exponentially [4,5]. This can be attributed to its rapid prototyping capabilities and cost-effective nature [6]. Due to the versatility that it provides, it can be applied for building every type of geometry from a variety of materials. The majority of the materials utilized in additive manufacturing applications consist of polymer-based composite materials and polymer blends [7] with the most commonly used in 3D printing of composite materials being Poly(lactic acid) (PLA) and acrylonitrile butadiene styrene (ABS) [8,9].

Poly(lactic acid) (PLA), a biobased and compostable polyester, finds diverse applications in tissue engineering, drug delivery, and medical applications [10–12]. It has been proven that when polymers are combined with specific nanoparticles, such as TiO_2, they acquire antimicrobial properties [13,14]. Titanium dioxide (TiO_2) has the ability to tune polymer properties such as antimicrobial activity, UV resistance, opacity, gas barrier, and color stability [7,15].

Numerous efforts have been documented in the literature with the aim of creating antimicrobial filaments for Fused Filament Fabrication (FFF) technology. Vidakis et al. [7] published a study that used (TiO_2) nanoparticles as nanofillers in order to enhance the properties of polypropylene (PP). The findings demonstrated that the characteristics (mechanical properties) of the nanocomposites had improved and it was demonstrated that PP/TiO_2 could be a nanocomposite system for use in AM applications. González et al. [16] examined PLA filled with TiO_2 nanoparticles in various particle content concentrations in a bacterial culture of *E. coli* and found that TiO_2 nanoparticles decreased the amount of extracellular polymeric substance and reduced bacterial growth. Also, no significant differences were observed for higher contents than 1% TiO_2 nanoparticles.

However, while most studies suggest that 3D-printed parts can be produced using antimicrobial filaments containing additives like TiO_2 with antimicrobial properties, this study examines a coating methodology to determine if it offers comparable results to the development of antimicrobial filaments. Implementing this coating technique in 3D-printed PLA components offers an alternative approach to imparting antimicrobial properties, bypassing the need to develop antimicrobial composite filaments for subsequent part fabrication through 3D printing. Although coating methodology has been used in scaffold applications to improve cell attachment/proliferation [17], here it is proposed as a method to provide antibacterial properties to medical 3D-printed parts and components or daily devices used in agri-food sector [6,18].

This study introduces an innovative approach to producing antimicrobial 3D-printed components through FFF AM technology. For the experiments, PLA was selected as the matrix material due to its biobased nature, as opposed to petroleum-based alternatives [19], and its favorable mechanical properties [20]. Given that FFF technology employs feedstock materials in filament form, the integration of TiO_2 nanoparticles into the filament is pursued to achieve antibacterial properties. The inclusion of Joncryl in small quantities as a chain extender enhances PLA's printability and mechanical properties by elevating molecular weight, complex viscosity, and melt flow index [21–23].

This work compares two methodologies for developing antibacterial 3D-printed components. The first method involves developing a composite filament that combines PLA, TiO_2, and Joncryl, suitable for use in all FFF 3D printers. A reference point is established by developing an additional filament made of PLA and Joncryl alone. The second method entails a coating process (dispersion immersion method) applied to the final parts manufactured using the PLA and Joncryl filament. These filaments were developed to produce specimens as a proof of concept for the future production of DIY (do-it-yourself) customized or pre-existing antimicrobial parts, medical tools, and more. All three specimen categories (1. PLA/Joncryl filament, 2. PLA/Joncryl/TiO_2 filament, and 3. PLA/Joncryl filament with a coating process) underwent comprehensive analysis, including physiochemical characterization and mechanical and antibacterial property assessments.

This paper is organized as follows: Section 2 outlines the materials and methods employed in the development of the present study. Section 3 encompasses the results, focusing on the characterization of the developed filaments and the coating methodology and presents data concerning their antibacterial activity and mechanical properties. Finally, the study concludes in Section 4. Figure 1 presents the architectural diagram of the methodology steps that were followed. These steps include the development of the filaments and the necessary tests conducted to extract results for characterizing the properties of the 3D-printed specimens.

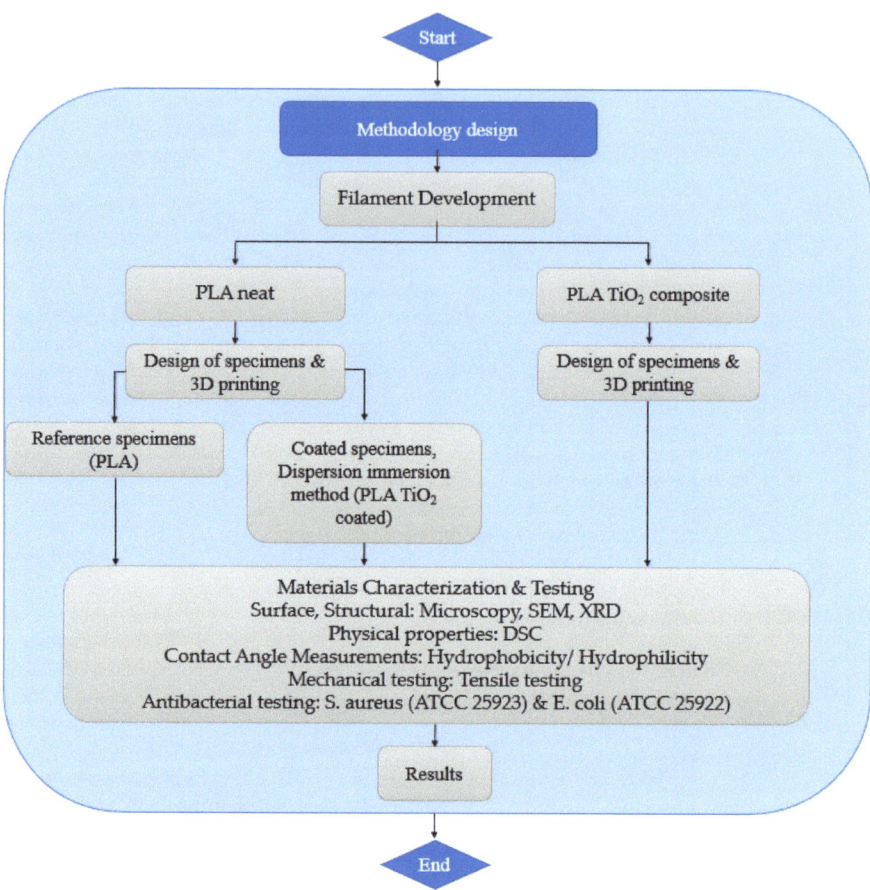

Figure 1. Architectural diagram.

2. Materials and Methods

2.1. Materials

In this study, the experimental materials used for the development of the filament included PLA in pellet form, Joncryl as a chain extender, and titanium dioxide (TiO_2) nanoparticles. The PLA pellets used were of PLA 4043D type, supplied by 3devo (Utrecht, The Netherlands). The chain extender Joncryl ADR® 4400 was supplied by BASF (Ludwigshafen, Germany). It possesses an epoxy equivalent weight of 485 g/mol and a weight-average molecular weight of 7100 g/mol. Aeroxide® TiO_2 P25 with a nanoparticle size of 21 nm and a specific surface area of 35–65 m^2/g was supplied from Sigma-Aldrich (Saint Louis, MO, USA).

2.2. Development of a PLA-Based TiO_2 Filament

In order to develop the two distinct filaments based on PLA pellets, the PLA pellets were vacuum-dried overnight at 40 °C. The dried PLA was subsequently mixed with Joncryl to formulate the PLA/Joncryl filament (named as PLA), and with both Joncryl and TiO_2 to fabricate the potentially antimicrobial PLA/Joncryl/TiO_2 filament (named as PLA TiO_2 comp). Table 1 provides the composition of each filament. In total, 2 wt.% of Joncryl was used, and it was proven by Grigora et al. [24] in a previous work that it gives the best physicochemical properties to the final filament.

Table 1. Summary of Fabricated Filaments.

Composite Filaments	Experimental Materials		
	PLA (wt.%)	TiO_2 (wt.%)	Joncryl (wt.%)
PLA/Joncryl	98%	-	2%
PLA/Joncryl/TiO_2	93%	5%	2%

The filaments were fabricated using the 3devo Composer Series 350/450 filament maker (Utrecht, The Netherlands). Parameters were configured for extrusion of a 1.75 mm diameter filament. However, due to variations, the filament thickness deviation was ±0.06, resulting in a final filament diameter ranging between 1.69 mm and 1.81 mm. The machine contains a mixing screw that aids in material passage through four heating zones. Upon melting, the material is extruded as filament through a nozzle.

The four heating zones of the extruder can be independently set to distinct temperature values. In this experiment, temperature ranged from 175 °C to 192 °C. Various combinations were tested to optimize filament extrusion. Ultimately, the optimal temperature combination was found to be Heater 1—180 °C, Heater 2—192 °C, Heater 3—187 °C, and Heater 4—175 °C. Heater 1 is closest to the nozzle, while Heater 4 is situated near the hopper. The extruder's screw rotational speed, driving material through the heating zones and extruder, was set at 4.1 rpm. For filament cooling, integrated fans were adjusted to 60%, facilitating timely solidification for spool collection during fabrication. Figure 2 illustrates the characteristic components of the 3devo filament maker.

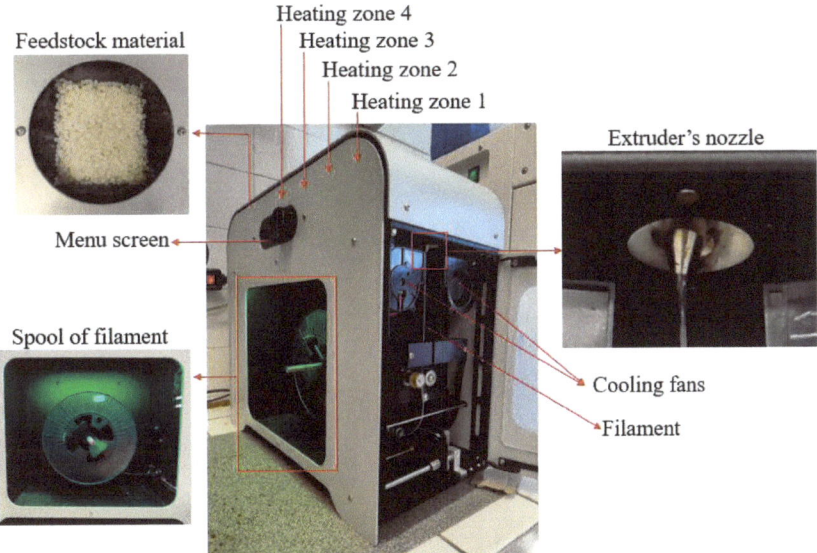

Figure 2. 3devo Filament Maker Overview.

2.3. Fabrication of 3D-Printed Specimens

After the successful development of both filaments, the FFF technology was used to fabricate the specimens. All specimens were designed using SOLIDWORKS® CAD Software (2022 SP2.0 Professional version) and manufactured utilizing an Original Prusa i3 MK3S+ 3D printer. Each part's 3D printing parameters were established using Prusa Slicer 2.5.0 software. For the PLA/Joncryl filament, the nozzle temperature was set at 220 °C, while for the PLA/Joncryl/TiO_2 filament, it was set at 240 °C. The bed temperature for both filaments was set at 60 °C. In all cases, a 0.4 mm nozzle and 0.2 mm layer height were employed. The specimens from both filaments were printed with a 100% fill density and concentric fill pattern for infill. Settings not explicitly mentioned in the FFF process were maintained at their default values as per the Prusa Slicer software™ (Version 2.6.1) utilized in this study. Table 2 presents the specimens, along with their dimensions, used in the present study to characterize the properties of the 3D-printed specimens fabricated with the developed filaments.

Table 2. 3D-Printed Specimens.

Type of Test	Dimensions of 3D-Printed Specimens
Scanning Electron Microscopy (SEM)	$10 \times 2 \times 1$ mm
Tensile	ASTM D638 Standard, Type V [25]
Antibacterial	$\Phi\,5$ mm $\times$ 1 mm

The dimensions of the specimens were determined by the requirements of the testing equipment: $\Phi\,5$ mm $\times$ 1 mm for antibacterial testing and $10 \times 2 \times 1$ mm for Scanning Electron Microscopy (SEM). Figure 3 displays the filament reels for the Prusa 3D printer, used to produce these specific specimens in two different surface configurations. Notably, the $\Phi\,5$ mm $\times$ 1 mm specimens provided satisfactory results, making the larger surface specimens redundant for antibacterial testing.

PLA/ Joncryl/ TiO$_2$ Filament

PLA/ Joncryl Filament

Prusa i3 MK3S+ 3D printer

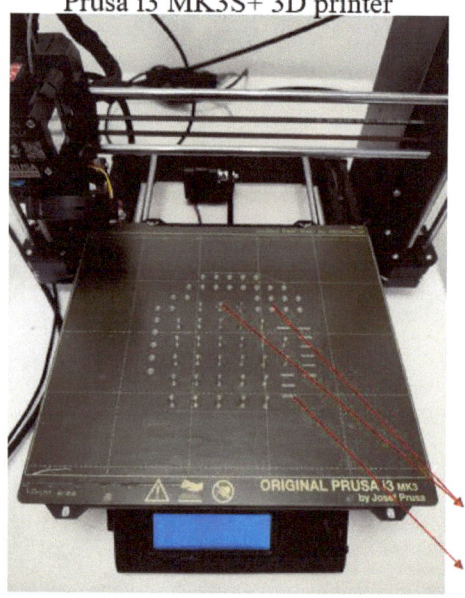

ASTM D638 Standard, Type V

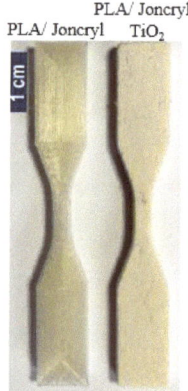

Specimens for Antibacterial Testing

Specimens for SEM

Figure 3. 3D Printer and Printed Specimens Visual Overview.

2.4. Coating Methodology: Dispersion Immersion Method

Between the two developed filaments, the PLA/Joncryl filament was selected for 3D printing the specimens that subsequently underwent a dispersion immersion coating process (as shown in Figure 4) to confer antimicrobial properties.

For the coating procedure, initially, the samples were immersed in a 1 M aqueous ammonia (NH$_3$) solution (pH = 11.5) and subjected to magnetic stirring at a speed of 435 rpm for 4 h at room temperature. Simultaneously, a 2 wt.% aqueous dispersion of TiO$_2$ was prepared. This dispersion underwent magnetic stirring for 1 h at a speed of

950 rpm, to prevent sedimentation in the stirring vessel. Around 30 min prior to removing the samples from the ammonia solution, an ultrasonication process was initiated using an ultrasonication probe. The process involved alternating cycles of 2 min of ultrasound treatment followed by 2 min of rest, repeated for a total of 6 cycles.

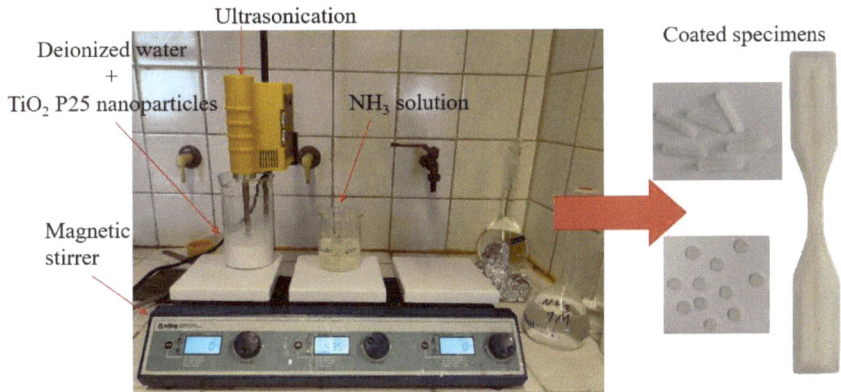

Figure 4. Coated specimens: materials and procedures used for dispersion immersion method.

After removal from the ammonia solution, the samples were thoroughly rinsed with deionized water and subsequently dried in an oven at 55 °C for 30 min. The dispersion containing deionized water and TiO_2 was heated at 70 °C and stirred magnetically for 30 min, with a stirring speed of 1200 rpm. Following this, the samples were immersed in this dispersion. The TiO_2 dispersion (with the specimens immersed) was heated to 70 °C and stirred magnetically at a speed of 875 rpm for a duration of 2 h. Afterwards, it was sonicated for 10 min, while maintaining the temperature at 70 °C. Finally, the samples were washed with ethanol and placed in an oven set at 75 °C for 10 min to facilitate drying. Upon completion of the aforementioned steps, the coating process concluded, resulting in the adhesion of TiO_2 powder particles to the samples (named as PLA TiO_2 coated).

2.5. Materials Characterization

2.5.1. Microscopy

The filaments' morphological features were examined with a stereoscope. Images were taken using a Jenoptik (Jena, Germany) ProgRes GRYPHAX Altair camera attached to a ZEISS (Oberkochen, Germany) SteREO Discovery V20 microscope and Gryphax image capturing software was used.

2.5.2. Scanning Electron Microscopy (SEM)

Scanning Electron Microscopy (SEM) images were captured using a JEOL (Tokyo, Japan) 2011 (JMS-840) electron microscope, equipped with an Oxford (Abingdon, UK) ISIS 300 energy-dispersive X-ray (EDX) micro-analytical system. Every specimen was positioned on the holder and coated with carbon to enhance the conductivity for the electron beam. The images were taken under an accelerating voltage of 2 kV, a probe current of 45 nA, and a counting time of 60 s.

2.5.3. X-ray Diffraction (XRD)

X-ray Diffraction (XRD) analyses of the polymers and copolymers were executed across a 2θ range of 5 to 80°, at intervals of 0.05°, and a scanning speed of 1.5 deg/min. The assessments were conducted using a MiniFlex II XRD system from Rigaku Co. (Tokyo, Japan) with Cu Kα radiation (λ = 0.154 nm).

2.5.4. Differential Scanning Calorimetry (DSC)

A PerkinElmer Pyris DSC-6 differential scanning calorimeter, which was calibrated with pure indium and zinc standards, was employed for the analysis. Samples of 5 ± 0.1 mg sealed in aluminum pans were used and all experiments were performed under N_2 atmosphere with a flow rate of 20 mL/min. Each specimen was subjected to a heating process from room temperature to 200 °C at a pace of 20 °C/min, then cooled down to 25 °C at the same rate of 20 °C/min and reheated to 200 °C at a 20 °C/min. The degree of crystallinity (X_c) was determined using Equation (1):

$$X_c(\%) = \left(\frac{\Delta H_m - \Delta H_{cc}}{\Delta H_f^0 - \frac{1 - \text{wt.\% additive}}{100}} \right) \times 100 \quad (1)$$

where ΔH_m, ΔH_{cc}, and ΔH_f^0 are the experimental melting enthalpy, the cold crystallization enthalpy, and the theoretical heat of fusion of 100% crystalline PLA ($\Delta H_f^0 = 93$ J/g), respectively.

2.5.5. Contact Angle Measurements

The water contact angle (WCA) was assessed with the Ossila (Sheffield, UK) Contact Angle Goniometer L2004A. The analysis of WCA for the samples was conducted through the sessile drop technique. A quantity of 25 µL of distilled water was delicately placed atop the surface of the 3D-printed plates ($n = 3$) and scaffolds. High-resolution images were captured within a span of 20 s and further analyzed using the Ossila Contact Angle Software v3.1.1.0. The statistical evaluation was conducted through a one-way ANOVA followed by a post hoc Tukey test, facilitated by the GraphPad Prism 6 software. A p-value of less than 0.05 was deemed as indicative of statistical significance.

2.5.6. Tensile Testing

Tensile testing evaluations were conducted utilizing a Shimadzu EZ Test Tensile Tester, Model EZ-LX, equipped a with a 2 kN load cell, following the ASTM D638 standards at a crosshead speed of 5 mm/min. For the testing, 3D-printed dumb-bell-shaped tensile type V test specimens were employed. Each sample underwent at least five separate assessments with the resulting data averaged to derive the mean values for Young's modulus, stress at break, and elongation at break. Data analysis was performed using one-way ANOVA, followed by a post hoc Tukey test, facilitated by GraphPad Prism 6 software. A p-value under 0.05 was established as the threshold for statistical significance.

2.5.7. Antibacterial Testing Methodology

Gram-positive Staphylococcus aureus (*S. aureus*, ATCC 25923), as well as Gram-negative Escherichia coli (*E. coli*, ATCC 25922) single colonies were inoculated in 10 mL of freshly prepared Nutrient Broth and incubated with agitation until reaching an OD600 measurement equal to 0.3–0.5. Then, 2 mL of the culture was moved to a microcentrifuge tube, spun for 1 min at 10.000 g, and the supernatant was removed. The cell pellet was resuspended in 2 mL PBS and spun for 1 min at 10.000 g. PBS washing was repeated twice. After the last spin, the pellet was resuspended in 2 mL PBS and 500 µL was transferred in four different glass flasks containing 4.5 mL PBS. Serial dilutions of each flask were transferred on Nutrient Agar plates using a bent glass pipette and incubated overnight at 37 °C to determine the initial cfu/mL for each flask. After transferring to the plates, a control filament was added in one of the flasks, a TiO_2-based filament was added to the second, a filament coated with TiO_2 was added to the third, whereas the last flask was used as a no filament control. The flasks were incubated for 1 h at 37 °C and 250 rpm, and then serial dilutions for each flask were transferred to fresh Nutrient Agar plates again. This procedure was repeated after one more hour of the flasks' incubation. The plates were left to incubate overnight at 37 °C. The number of colonies that represent the surviving bacteria was counted the following day and the possibility of antibacterial activity of the filament was determined.

The filaments' antibacterial effectiveness against *S. aureus* and *E. coli* is reported as the mean standard deviation (SD) after 60 and 120 min of contact. Each experimental procedure was replicated three times (n = 3) for each bacterium strain. For statistical analysis, two-way ANOVA with repeated measurements was performed.

3. Results and Discussion

3.1. Scanning Electron Microscopy (SEM) and Optical Microscopy

The morphological characteristics and the dispersion of the TiO_2 nanocomposites in the polymer matrices are examined via microscopic techniques. Figure 5 displays the side surface of randomly selected 3D-printed tensile test specimens, providing a quantitative assessment of interlayer fusion, interlayer defects, or possible inhomogeneities.

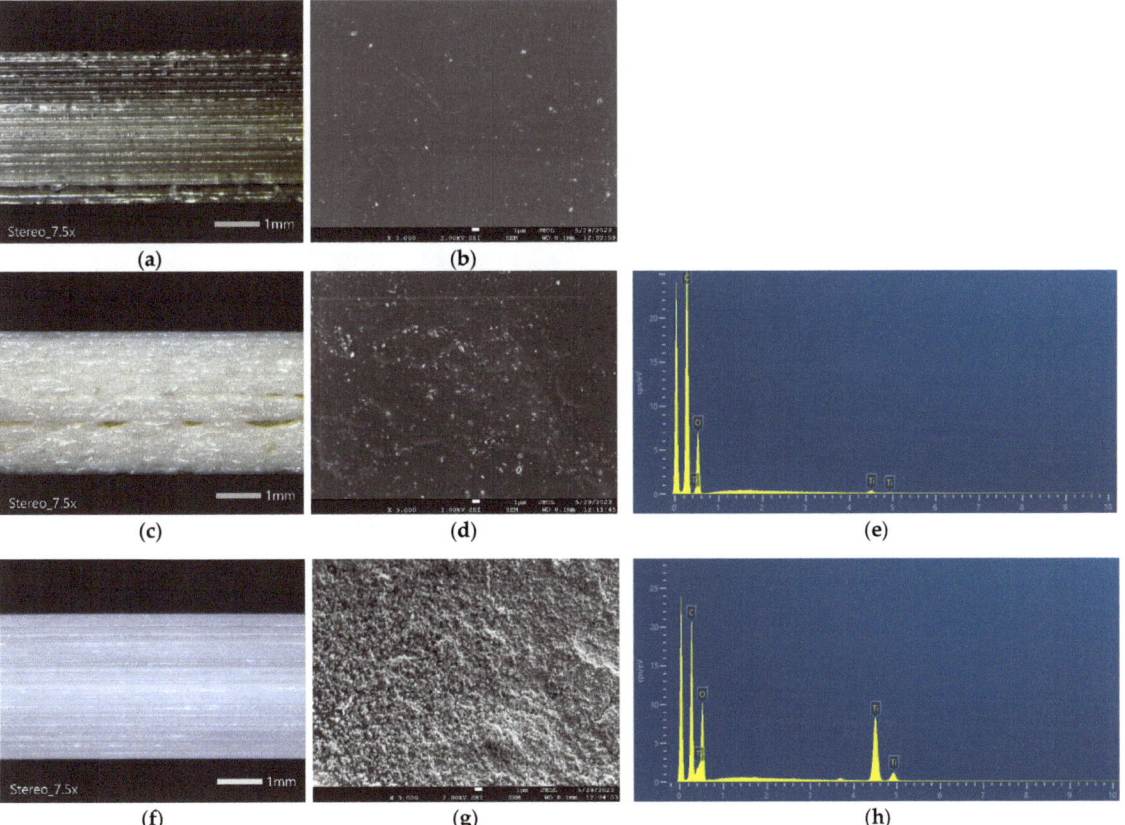

Figure 5. (**a,b**) Optical microscopy and SEM image of 3D-printed PLA (PLA/Joncryl filament), (**c,d**) optical microscopy and SEM image of 3D-printed PLA TiO_2 comp (PLA/Joncryl/TiO_2 filament), (**e**) EDX spectrum of 3D-printed PLA TiO_2 comp, (**f,g**) optical microscopy and SEM image of 3D-printed PLA TiO_2 coated, and (**h**) EDX spectrum of 3D-printed PLA TiO_2 coated.

It can be observed in Figure 5a,c,f that the 3D printing process utilizing the PLA/Joncryl/TiO_2 filament does not yield the same 3D printing quality as the PLA/Joncryl filament. This difference arises from the formation of TiO_2 nanoparticles' agglomerates, resulting in inconsistent material extrusion flow. This phenomenon is strongly associated with the dimension and the weight fraction of the inorganic additive in the polymer matrix. Specifically, it was found that >1 wt.%. of additive led to an increased agglomeration, and thus, intense surface roughness [26–28]. Furthermore, the processing method seems to affect the morphological features of the final specimen. Specifically, the dispersion of TiO_2 particles is observed to be more homogeneous in the case of the composite, resulting in a smooth surface. On the contrary, the coating procedure led to a surface with augmented roughness, due to the increased percentage of the TiO_2, as it was verified from the EDX analysis (Figure 5e,h).

3.2. X-ray Diffraction (XRD) Measurements

XRD patterns of the fabricated materials are presented in Figure 6. The characteristic diffraction peaks of PLA appeared at 2θ = 14.8°, 16.5°, 19°, and 22° resulting from the crystal planes (010), (200/110), (203), and (210) [29]. No diffraction peaks appeared in any sample, as thermal processing, such as extrusion and printing, usually leads to amorphous materials [30]. The small crystalline peaks that the coated material exhibits are probably due to the heating during the coating procedure, facilitating some cold crystallization.

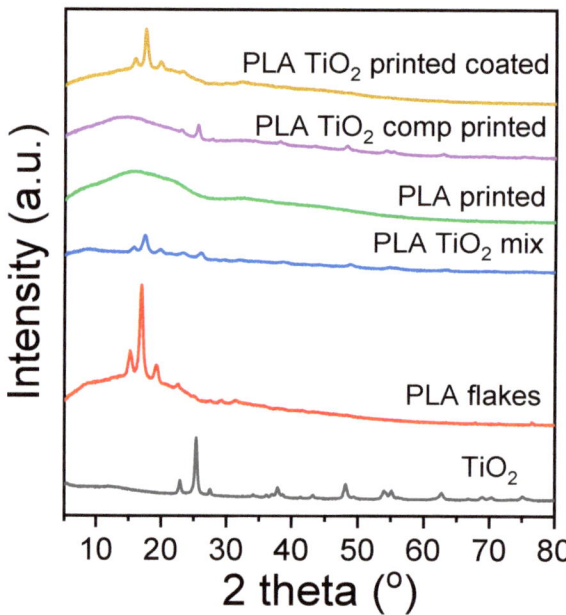

Figure 6. X-ray diffraction patterns of PLA, TiO_2, and the 3D-printed specimens.

3.3. Differential Scanning Calorimetry (DSC) Measurements

The thermal properties of the fabricated PLA/TiO_2 materials were determined using DSC analysis. The recorded DSC thermograms upon the first and the second heating scan are presented in Figure 7. The characteristic thermal transitions, including glass transition temperature (T_g), cold crystallization temperature (T_{cc}), and melting point (T_m), as well as the % degree of crystallinity of the materials, are summarized in Table 3. No crystallization peak was observed in any of the samples upon cooling from the melt. As can be observed, all the printed samples exhibited a very low degree of crystallinity, as processing methods, such as printing, tend to erase the matrix crystallinity [30]. All samples were amorphous, showing glass transition at 60–62 °C, cold crystallization, and subsequent melting. This observation is in agreement with the obtained XRD patterns. Furthermore, a small reduction in the T_g values of the printed samples in comparison to the PLA flakes can be attributed to some degradation during the thermal processing of 3D printing [31]. The coating process seems to also have an effect on the thermal transitions of the final sample, as T_g and T_m values decreased. However, the thermal transitions of the samples were not significantly affected.

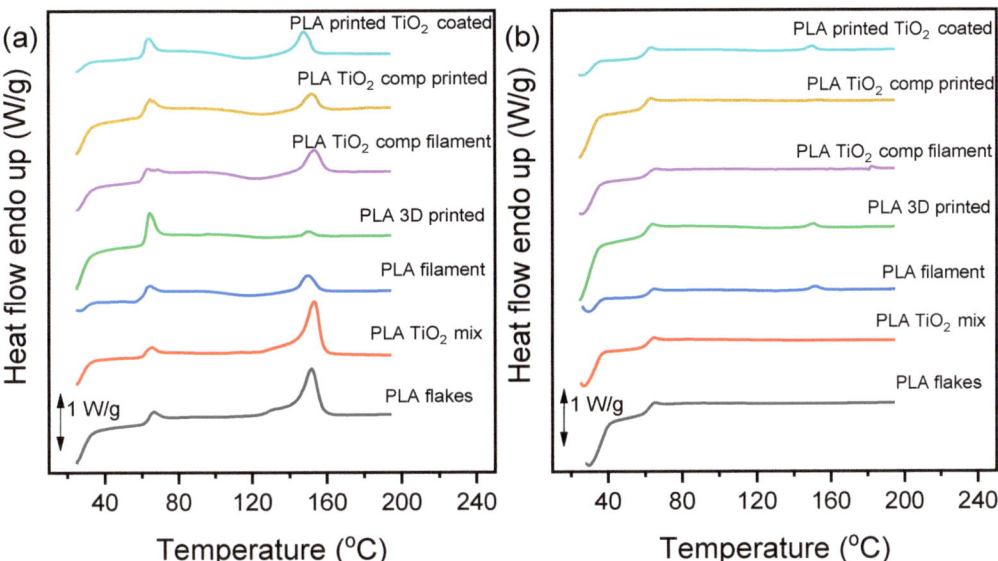

Figure 7. DSC graphs of the materials during heating with rate 20 °C/min, (**a**) first heating, and (**b**) second heating scan.

Table 3. Thermal characteristics of the samples as measured by DSC.

Sample	1st Heating				2nd Heating			
	T_g	T_{cc}	T_m	X_c	T_g	T_{cc}	T_m	X_c
PLA flakes	61.9	-	151.9	33.5	60.7	-	-	0
PLA TiO$_2$ mix	60.7	-	153.4	35.7	60.5	-	-	0
PLA filament	60.3	118.5	149.8	1.4	63.3	129.1	151.1	0
PLA 3D printed	61	127	150.7	0.0	59.7	127.7	150.7	0
PLA TiO$_2$ comp filament	59.9	121.7	153.3	3.0	60.5	-	-	0
PLA TiO$_2$ comp printed	61.8	124.7	152	1.1	58.9	-	152.7	0.1
PLA TiO$_2$ coated printed	60.1	116.8	148.1	0.5	58.9	125.8	143.1	0

3.4. Contact Angle Measurements

The contact angle of the surface with water plays a key role in the characterization of a material, as it can offer an insight into its absorption and its adhesion profile [32]. The water contact angle of polymeric materials is a function of their chemical composition and surface properties (roughness, heterogeneity, and preparation method), as well as temperature [33]. The hydrophilicity of the PLA/Joncryl (PLA), PLA/Joncryl/TiO$_2$ (PLA TiO$_2$ comp), and PLA/Joncryl coated (PLA TiO$_2$ coated) printed specimens was assessed through water contact angle measurements, and the methodology is illustrated in Figure 8. As can be observed, PLA/Joncryl specimens had a contact angle of ~54.2°. Neat PLA appeared less hydrophobic than expected because of its surface roughness. The addition of TiO$_2$ to PLA caused a statistically significant decrease in the contact angle values of both the composite and the coated specimen, and a decreasing trend in the values of the composite to the coated sample. This trend is a result of the increased free energy and the increased roughness of the specimen, as can be assumed from the corresponding SEM images [34]. Furthermore, the hydrophilicity of the incorporated TiO$_2$ nanoparticles, owing to the unsaturated reactive hydroxyl (-OH) groups, resulted in the significantly enhanced hydrophilicity of the final samples [35,36]. Although the hydrophilic character of a sample is related to its poor resistance to water, this may be beneficial for the contact between the microbial cells and the film which can facilitate the antimicrobial activity.

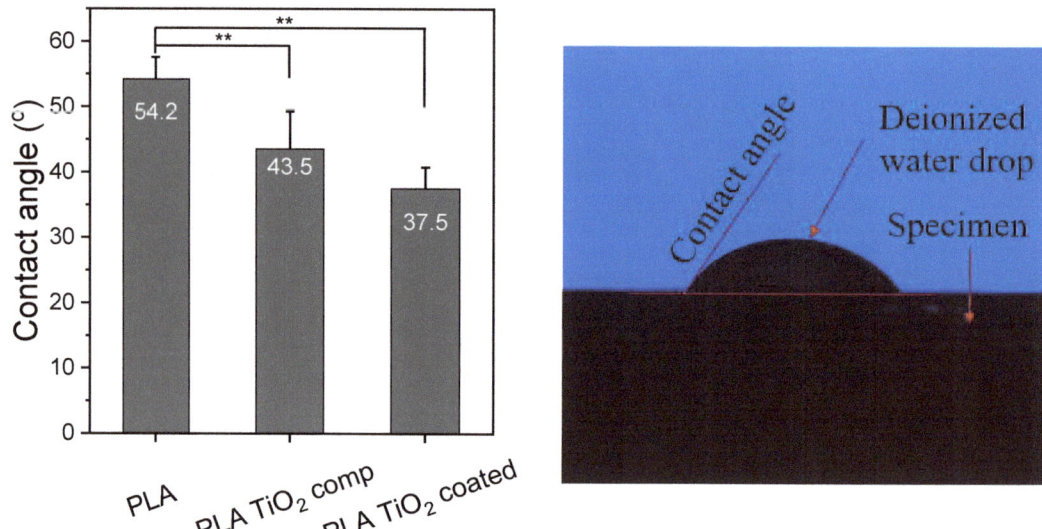

Figure 8. Water contact angle of 3D-printed PLA, PLA TiO$_2$ comp, and PLA TiO$_2$ coated. One-way ANOVA, ** p 0.001–0.01.

3.5. Tensile Testing Measurements

As the mechanical features of a polymer define its final applications, tensile tests were performed and the results are presented in Figure 9. In Figure 9a–c, stress–strain diagrams of the three specimen categories demonstrate differences in mechanical properties between specimens printed with PLA/Joncryl (PLA) filament, PLA/Joncryl/TiO$_2$ (PLA TiO$_2$ comp) filament, and specimens printed with PLA/Joncryl filament followed by the coating process (PLA TiO$_2$ coated). The PLA specimens exhibit the highest stress and strain at break, whereas the PLA TiO$_2$ coated specimens show the highest Young's modulus value. The stress–strain curves indicate that both the integration of TiO$_2$ nanoparticles and the coating process, aimed at providing antimicrobial properties to the final polymeric material, influence the mechanical properties of the developed filament. This is evident in the stress–strain curves, particularly in the ultimate tensile strength (UTS) or the maximum stress the material can endure before failure. The stress–strain curves of the PLA/Joncryl filament demonstrate the highest ultimate tensile strength. As Figure 9d–f illustrates, there is a statistically significant decrease in tensile strength and strain values among the PLA and either composite. The decreased values of the composite sample can be a result of defects on the printed structure (as shown in Figure 5c). The lower values of tensile strength and strain that the coated sample exhibits, in combination with its increased Young's modulus value, can be attributed to it being soaked in aqueous solution and ammonia, which could have caused some hydrolytic degradation to the material. However, the variation of the Young's modulus values between all samples was insignificant ($p > 0.05$).

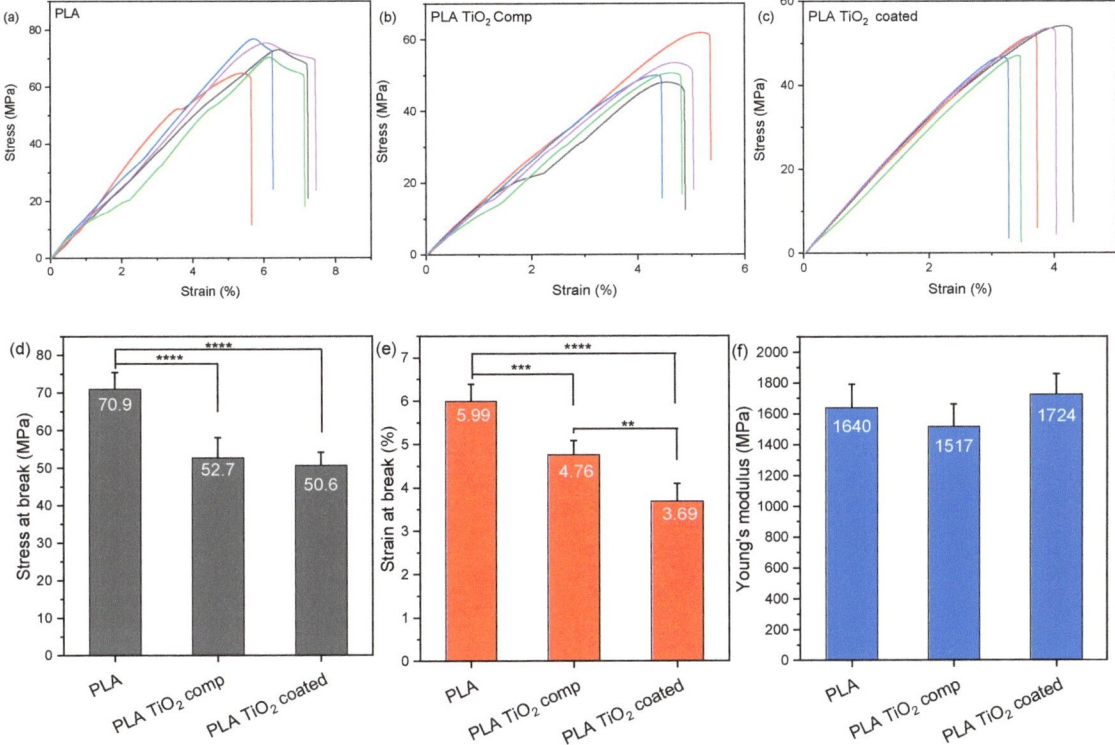

Figure 9. Stress–strain curves of (**a**) PLA and (**b**) PLA TiO$_2$ comp printed specimens, (**c**) stress at break, (**d**) strain at break, and (**e**) Young's modulus values. One-way ANOVA. ** $0.001 < p < 0.01$, *** $0.0001 < p < 0.001$, **** $p < 0.0001$.

Generally, the percentage and diameter of the particles incorporated in the matrix strongly affect its final features. It has been reported that the weight fractions of TiO$_2$ nanoparticles up to 1% increased the tensile strength and strain values, while at high TiO$_2$ concentrations, the self-networking of nanoparticles can take place [31]. This phenomenon may lead to the agglomeration and non-homogeneous dispersion of the particles [34]. This could potentially indicate that the agglomeration of TiO$_2$ nanoparticles at high loadings may contribute to the weakening of the mechanical properties, as evidenced in the stress–strain diagram, but it is necessary to achieve antimicrobial properties. Agglomerated nanoparticles restrict the interfacial area between themselves and the polymer matrix, resulting in a non-uniform particle distribution and a reduction in the nanoparticle concentration within the composite. Furthermore, it is found that the functionalization of TiO$_2$ or the addition of a plasticizer is essential for the improvement of the mechanical features of PLA/TiO$_2$ materials compared to the neat PLA [37–39]. Similar behavior has been observed with the presence of other metal-based nanoparticles, where concentrations above 1% wt. showed decreased mechanical properties [40–43].

All things considered, the diagrams in Figure 9 reveal that the stress at break (MPa) exhibits a decrease of 25.67% in the specimens printed with the PLA/Joncryl/TiO$_2$ filament compared to those printed with the PLA/Joncryl filament. Moreover, an additional decrease of 3.98% is observed in specimens printed with the PLA/Joncryl filament and subsequently subjected to a dispersion immersion coating process. A similar reduction is noticed in the strain at break (%) with a decrease of 20.53% when comparing the PLA/Joncryl to PLA/Joncryl/TiO$_2$ filament and an additional 22.47% decrease after the coating process.

3.6. Antibacterial Testing Measurements

In Figure 10, the x-axis represents the three time points when the cfu/mL was counted (i.e., 0 h, 1 h, and 2 h), while the y-axis represents the % average viability for each bacterium incubated with the different specimens.

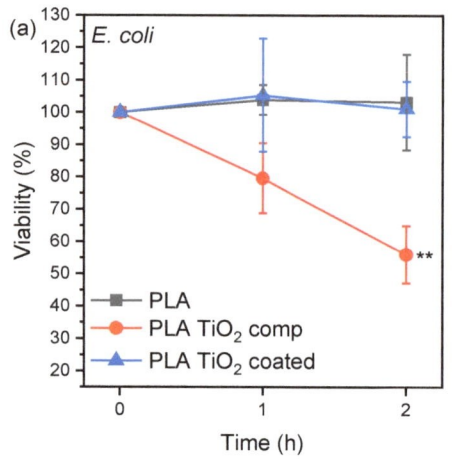
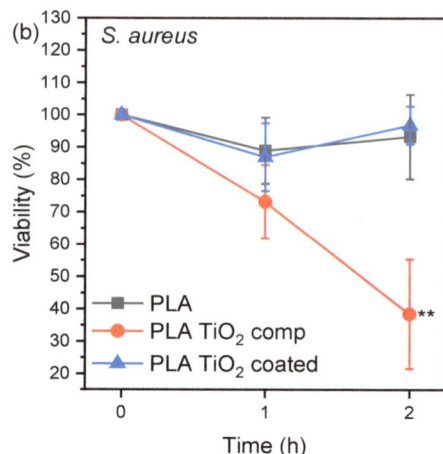

Figure 10. % Viability of bacteria (**a**) *E. coli* and (**b**) *S. aureus* after incubation with PLA, PLA TiO$_2$ comp, and PLA TiO$_2$ coated specimens. Two-way ANOVA with repeated measurements, ** $0.001 < p < 0.01$.

The average % viability is calculated as follows: First, the percentage of viability for each replicate is determined by dividing the cfu/mL of each hour by the cfu/mL at hour 0 and then multiplying by 100. Consequently, for each hour, three percentages of viability were obtained (one for each replicate), which were used to calculate the average % viability. These values were used to create the graph.

The results in Figure 10 indicate that the incubation of both bacteria species with the PLA TiO$_2$ comp specimen caused a statistically significant reduction over time in the microbes' viability compared to PLA and PLA TiO$_2$ coated specimens. There was no effect on the bacteria strain observed. TiO$_2$ is hydrophilic, and thus it decreased the water contact angle of PLA TiO$_2$ comp, but it is also antimicrobial, indicated by its generation of reactive oxygen species when it is exposed to light, which oxidize the cytoplasm of bacteria.

TiO$_2$ coatings imparted antimicrobial activity when dip-coated on PMMA [44], and their lack of efficiency herein could be due to the leaching of TiO$_2$ particles from the PLA TiO$_2$ coated specimen, resulting in a low ion concentration in the incubation medium. Concentration is the most important parameter that affects bacteria survival rate on plastic/metal oxide nanoparticles and the 5 wt.% of TiO$_2$ added in the PLA TiO$_2$ filament was effective [45]. Thus, directly adding metal oxides such as TiO$_2$ is an easy and efficient way of additive manufacturing antimicrobial objects.

4. Conclusions

In this study, we developed two different filaments for FFF feedstock material. One filament is developed for immediate use in 3D printing antibacterial parts, while the other is intended for 3D printing parts and subsequently providing them with antibacterial properties through a coating process. To ensure perfect repeatability in our experiments and the fabrication of all specimens under exactly the same parameters, the stability of all 3D printing parameters was maintained. Deviating from these parameters would lead to different results. To assess the mechanical performance and antibacterial activity of parts manufactured using these two methods, we conducted a series of tests to draw conclusions regarding their structure and properties. Although SEM and optical microscopy images confirmed the presence of TiO_2 nanoparticles, antibacterial tests demonstrated that only the parts manufactured by PLA/Joncryl/TiO_2 filament exhibited significant antibacterial properties. Interestingly, the coated parts exhibited non-antibacterial activity similar to those printed with the control PLA/Joncryl filament, while the PLA/Joncryl/TiO_2 filament specimens displayed notable antibacterial activity. Consequently, the coating methodology is not recommended, as it does not yield the desired results and involves a more complex procedure requiring access to chemical laboratories, substances, and equipment. In terms of the mechanical attributes of the printed components, the findings indicate that introducing TiO_2 nanoparticles undermines the mechanical properties, leading to the formation of agglomerates. Future studies should investigate the optimal pretreatment method for the materials and the ideal quantity (wt.%) and size of nanoparticles to achieve the best particle distribution and enhance mechanical properties while maintaining antibacterial characteristics.

Given access to the necessary equipment for filament development, this study proposes a DIY method for creating customized antibacterial parts suitable for production on any commercial FFF 3D printer. These parts could include plastic surfaces for various applications in hospitals, medical tools, laboratory plastic parts like Petri dishes, plastic centrifuge test tubes, and pipette tips for cases where reusable antibacterial plastics are applicable. They can also be used in research DIY devices such as bioreactors, orbital shakers, or other costly components that can be replaced with this cost-effective 3D printing solution. Finally, with further research and possible improvements, this proposed filament could be assessed for 3D printing implants or scaffolds to be used in bone reconstruction applications, potentially replacing existing materials used in these applications.

Author Contributions: Conceptualization, S.P. and Z.T.; methodology, E.M.P., D.N.B. and C.K.; software, S.P. and D.T.; validation, S.P., Z.T. and C.K.; formal analysis, S.P., E.X. and G.K.; investigation, S.P., Z.T., G.K. and E.X.; resources, D.N.B., D.T. and C.K.; data curation, S.P., G.K. and E.X.; writing—original draft preparation, S.P. and E.X.; writing—review and editing, Z.T. and E.M.P.; visualization, Z.T.; supervision, E.M.P.; project administration, E.M.P. and D.N.B.; funding acquisition, E.M.P. All authors have read and agreed to the published version of the manuscript.

Funding: Horizon 2020 PestNu project under the GA no. 101037128.

Institutional Review Board Statement: Not applicable.

Informed Consent Statement: Not applicable.

Data Availability Statement: All the data are included in the manuscript.

Acknowledgments: This work supported by the EU Green Deal project PestNu which has received funding from the European Union's Horizon 2020 research and innovation programme under Grant Agreement no. 101037128. The authors would like to thank candidates Chrysanthi Papoulia and Eleni Pavlidou for the SEM observations, as well as Michiel Noordam for the help with the statistical analysis. They would also like to thank Maria Eirini Grigora from I4byDesign Competence Center for her support with the 3devo Composer Series 350/450 filament maker operation.

Conflicts of Interest: The authors declare no conflict of interest.

References

1. Mastura, M.T.; Alkahari, M.R.; Syahibudil Ikhwan, A.K. Life Cycle Analysis of Fused Filament Fabrication: A Review. In *Design for Sustainability*; Elsevier: Amsterdam, The Netherlands, 2021; pp. 415–434. [CrossRef]
2. Gao, X.; Qi, S.; Kuang, X.; Su, Y.; Li, J.; Wang, D. Fused Filament Fabrication of Polymer Materials: A Review of Interlayer Bond. *Addit. Manuf.* **2021**, *37*, 101658. [CrossRef]
3. Pechlivani, E.M.; Pemas, S.; Kanlis, A.; Pechlivani, P.; Petrakis, S.; Papadimitriou, A.; Tzovaras, D.; Hatzistergos, K.E. Enhanced Growth of Bacterial Cells in a Smart 3D Printed Bioreactor. *Micromachines* **2023**, *14*, 1829. [CrossRef]
4. García Plaza, E.; Núñez López, P.; Caminero Torija, M.; Chacón Muñoz, J. Analysis of PLA Geometric Properties Processed by FFF Additive Manufacturing: Effects of Process Parameters and Plate-Extruder Precision Motion. *Polymers* **2019**, *11*, 1581. [CrossRef]
5. Moreno Nieto, D.; Alonso-García, M.; Pardo-Vicente, M.-A.; Rodríguez-Parada, L. Product Design by Additive Manufacturing for Water Environments: Study of Degradation and Absorption Behavior of PLA and PETG. *Polymers* **2021**, *13*, 1036. [CrossRef] [PubMed]

6. Pechlivani, E.M.; Papadimitriou, A.; Pemas, S.; Ntinas, G.; Tzovaras, D. IoT-Based Agro-Toolbox for Soil Analysis and Environmental Monitoring. *Micromachines* **2023**, *14*, 1698. [CrossRef]
7. Vidakis, N.; Petousis, M.; Velidakis, E.; Tzounis, L.; Mountakis, N.; Kechagias, J.; Grammatikos, S. Optimization of the Filler Concentration on Fused Filament Fabrication 3D Printed Polypropylene with Titanium Dioxide Nanocomposites. *Materials* **2021**, *14*, 3076. [CrossRef]
8. Ngo, T.D.; Kashani, A.; Imbalzano, G.; Nguyen, K.T.Q.; Hui, D. Additive Manufacturing (3D Printing): A Review of Materials, Methods, Applications and Challenges. *Compos. Part B Eng.* **2018**, *143*, 172–196. [CrossRef]
9. Ujfalusi, Z.; Pentek, A.; Told, R.; Schiffer, A.; Nyitrai, M.; Maroti, P. Detailed Thermal Characterization of Acrylonitrile Butadiene Styrene and Polylactic Acid Based Carbon Composites Used in Additive Manufacturing. *Polymers* **2020**, *12*, 2960. [CrossRef] [PubMed]
10. Ranakoti, L.; Gangil, B.; Bhandari, P.; Singh, T.; Sharma, S.; Singh, J.; Singh, S. Promising Role of Polylactic Acid as an Ingenious Biomaterial in Scaffolds, Drug Delivery, Tissue Engineering, and Medical Implants: Research Developments, and Prospective Applications. *Molecules* **2023**, *28*, 485. [CrossRef]
11. Bikiaris, N.D.; Koumentakou, I.; Samiotaki, C.; Meimaroglou, D.; Varytimidou, D.; Karatza, A.; Kalantzis, Z.; Roussou, M.; Bikiaris, R.D.; Papageorgiou, G.Z. Recent Advances in the Investigation of Poly(Lactic Acid) (PLA) Nanocomposites: Incorporation of Various Nanofillers and Their Properties and Applications. *Polymers* **2023**, *15*, 1196. [CrossRef]
12. Balla, E.; Daniilidis, V.; Karlioti, G.; Kalamas, T.; Stefanidou, M.; Bikiaris, N.D.; Vlachopoulos, A.; Koumentakou, I.; Bikiaris, D.N. Poly(Lactic Acid): A Versatile Biobased Polymer for the Future with Multifunctional Properties—From Monomer Synthesis, Polymerization Techniques and Molecular Weight Increase to PLA Applications. *Polymers* **2021**, *13*, 1822. [CrossRef] [PubMed]
13. Elsaka, S.E.; Hamouda, I.M.; Swain, M.V. Titanium Dioxide Nanoparticles Addition to a Conventional Glass-Ionomer Restorative: Influence on Physical and Antibacterial Properties. *J. Dent.* **2011**, *39*, 589–598. [CrossRef] [PubMed]
14. Vidakis, N.; Petousis, M.; Mountakis, N.; Maravelakis, E.; Zaoutsos, S.; Kechagias, J.D. Mechanical Response Assessment of Antibacterial PA12/TiO$_2$ 3D Printed Parts: Parameters Optimization through Artificial Neural Networks Modeling. *Int. J. Adv. Manuf. Technol.* **2022**, *121*, 785–803. [CrossRef]
15. Tekin, D.; Birhan, D.; Kiziltas, H. Thermal, Photocatalytic, and Antibacterial Properties of Calcinated Nano-TiO$_2$/Polymer Composites. *Mater. Chem. Phys.* **2020**, *251*, 123067. [CrossRef]
16. González, E.A.S.; Olmos, D.; Lorente, M.Á.; Vélaz, I.; González-Benito, J. Preparation and Characterization of Polymer Composite Materials Based on PLA/TiO$_2$ for Antibacterial Packaging. *Polymers* **2018**, *10*, 1365. [CrossRef]
17. Chen, W.; Nichols, L.; Brinkley, F.; Bohna, K.; Tian, W.; Priddy, M.W.; Priddy, L.B. Alkali Treatment Facilitates Functional Nano-Hydroxyapatite Coating of 3D Printed Polylactic Acid Scaffolds. *Mater. Sci. Eng. C* **2021**, *120*, 111686. [CrossRef]
18. Pechlivani, E.M.; Papadimitriou, A.; Pemas, S.; Giakoumoglou, N.; Tzovaras, D. Low-Cost Hyperspectral Imaging Device for Portable Remote Sensing. *Instruments* **2023**, *7*, 32. [CrossRef]
19. Reinhardt, M.; Kaufmann, J.; Kausch, M.; Kroll, L. PLA-Viscose-Composites with Continuous Fibre Reinforcement for Structural Applications. *Procedia Mater. Sci.* **2013**, *2*, 137–143. [CrossRef]
20. Trivedi, A.K.; Gupta, M.K.; Singh, H. PLA Based Biocomposites for Sustainable Products: A Review. *Adv. Ind. Eng. Polym. Res.* **2023**, *6*, 382–395. [CrossRef]
21. Grigora, M.-E.; Terzopoulou, Z.; Tsongas, K.; Bikiaris, D.N.; Tzetzis, D. Physicochemical Characterization and Finite Element Analysis-Assisted Mechanical Behavior of Polylactic Acid-Montmorillonite 3D Printed Nanocomposites. *Nanomaterials* **2022**, *12*, 2641. [CrossRef]
22. Kahraman, Y.; Alkan Goksu, Y.; Özdemir, B.; Eker Gümüş, B.; Nofar, M. Composition Design of PLA/TPU Emulsion Blends Compatibilized with Multifunctional Epoxy-based Chain Extender to Tackle High Impact Resistant Ductile Structures. *J. Appl. Polym. Sci.* **2022**, *139*, 51833. [CrossRef]
23. Standau, T.; Nofar, M.; Dörr, D.; Ruckdäschel, H.; Altstädt, V. A Review on Multifunctional Epoxy-Based Joncryl® ADR Chain Extended Thermoplastics. *Polym. Rev.* **2022**, *62*, 296–350. [CrossRef]
24. Grigora, M.-E.; Terzopoulou, Z.; Tsongas, K.; Klonos, P.; Kalafatakis, N.; Bikiaris, D.N.; Kyritsis, A.; Tzetzis, D. Influence of Reactive Chain Extension on the Properties of 3D Printed Poly(Lactic Acid) Constructs. *Polymers* **2021**, *13*, 1381. [CrossRef]
25. *ASTM D638-14*; Standard Test Method for Tensile Properties of Plastics. ASTM International: West Conshohocken, PA, USA, 2014.
26. Lopera, A.A.; Bezzon, V.D.N.; Ospina, V.; Higuita-Castro, J.L.; Ramirez, F.J.; Ferraz, H.G.; Orlando, M.T.A.; Paucar, C.G.; Robledo, S.M.; Garcia, C.P. Obtaining a Fused PLA-Calcium Phosphate-Tobramycin-Based Filament for 3D Printing with Potential Antimicrobial Application. *J. Korean Ceram. Soc.* **2023**, *60*, 169–182. [CrossRef]
27. Nájera, S.E.; Michel, M.; Kim, N.-S. 3D Printed PLA/PCL/TiO$_2$ Composite for Bone Replacement and Grafting. *MRS Adv.* **2018**, *3*, 2373–2378. [CrossRef]
28. Nájera, S.E.; Michel, M.; Kyung-Hwan, J.; Kim, J.N. Characterization of 3D Printed PLA/PCL/TiO$_2$ Composites for Cancellous Bone. *J. Mater. Sci. Eng.* **2018**, *7*, 417. [CrossRef]
29. Farid, T.; Herrera, V.N.; Kristiina, O. Investigation of Crystalline Structure of Plasticized Poly (Lactic Acid)/Banana Nanofibers Composites. *IOP Conf. Ser. Mater. Sci. Eng.* **2018**, *369*, 012031. [CrossRef]
30. Sangiorgi, A.; Gonzalez, Z.; Ferrandez-Montero, A.; Yus, J.; Sanchez-Herencia, A.J.; Galassi, C.; Sanson, A.; Ferrari, B. 3D Printing of Photocatalytic Filters Using a Biopolymer to Immobilize TiO$_2$ Nanoparticles. *J. Electrochem. Soc.* **2019**, *166*, H3239–H3248. [CrossRef]

31. Marra, A.; Silvestre, C.; Kujundziski, A.P.; Chamovska, D.; Duraccio, D. Preparation and Characterization of Nanocomposites Based on PLA and TiO_2 Nanoparticles Functionalized with Fluorocarbons. *Polym. Bull.* **2017**, *74*, 3027–3041. [CrossRef]
32. Hebbar, R.S.; Isloor, A.M.; Ismail, A.F. Contact Angle Measurements. In *Membrane Characterization*; Elsevier: Amsterdam, The Netherlands, 2017; pp. 219–255. [CrossRef]
33. Yekta-Fard, M.; Ponter, A.B. Factors Affecting the Wettability of Polymer Surfaces. *J. Adhes. Sci. Technol.* **1992**, *6*, 253–277. [CrossRef]
34. Feng, S.; Zhang, F.; Ahmed, S.; Liu, Y. Physico-Mechanical and Antibacterial Properties of PLA/TiO_2 Composite Materials Synthesized via Electrospinning and Solution Casting Processes. *Coatings* **2019**, *9*, 525. [CrossRef]
35. Nakamura, R.; Imanishi, A.; Murakoshi, K.; Nakato, Y. In Situ FTIR Studies of Primary Intermediates of Photocatalytic Reactions on Nanocrystalline TiO_2 Films in Contact with Aqueous Solutions. *J. Am. Chem. Soc.* **2003**, *125*, 7443–7450. [CrossRef]
36. Zhang, X.; Xiao, G.; Wang, Y.; Zhao, Y.; Su, H.; Tan, T. Preparation of Chitosan-TiO_2 Composite Film with Efficient Antimicrobial Activities under Visible Light for Food Packaging Applications. *Carbohydr. Polym.* **2017**, *169*, 101–107. [CrossRef] [PubMed]
37. Baek, N.; Kim, Y.T.; Marcy, J.E.; Duncan, S.E.; O'Keefe, S.F. Physical Properties of Nanocomposite Polylactic Acid Films Prepared with Oleic Acid Modified Titanium Dioxide. *Food Packag. Shelf. Life* **2018**, *17*, 30–38. [CrossRef]
38. Foruzanmehr, M.; Vuillaume, P.Y.; Elkoun, S.; Robert, M. Physical and Mechanical Properties of PLA Composites Reinforced by TiO_2 Grafted Flax Fibers. *Mater. Des.* **2016**, *106*, 295–304. [CrossRef]
39. Zhang, Q.; Li, D.; Zhang, H.; Su, G.; Li, G. Preparation and Properties of Poly(Lactic Acid)/Sesbania Gum/Nano-TiO_2 Composites. *Polym. Bull.* **2018**, *75*, 623–635. [CrossRef]
40. Zuo, M.; Pan, N.; Liu, Q.; Ren, X.; Liu, Y.; Huang, T.-S. Three-Dimensionally Printed Polylactic Acid/Cellulose Acetate Scaffolds with Antimicrobial Effect. *RSC Adv.* **2020**, *10*, 2952–2958. [CrossRef]
41. Wang, Y.; Wang, S.; Zhang, Y.; Mi, J.; Ding, X. Synthesis of Dimethyl Octyl Aminoethyl Ammonium Bromide and Preparation of Antibacterial ABS Composites for Fused Deposition Modeling. *Polymers* **2020**, *12*, 2229. [CrossRef]
42. Tylingo, R.; Kempa, P.; Banach-Kopeć, A.; Mania, S. A Novel Method of Creating Thermoplastic Chitosan Blends to Produce Cell Scaffolds by FDM Additive Manufacturing. *Carbohydr. Polym.* **2022**, *280*, 119028. [CrossRef]
43. Pušnik Črešnar, K.; Aulova, A.; Bikiaris, D.N.; Lambropoulou, D.; Kuzmič, K.; Fras Zemljič, L. Incorporation of Metal-Based Nanoadditives into the PLA Matrix: Effect of Surface Properties on Antibacterial Activity and Mechanical Performance of PLA Nanoadditive Films. *Molecules* **2021**, *26*, 4161. [CrossRef]
44. Su, W.; Wang, S.; Wang, X.; Fu, X.; Weng, J. Plasma Pre-Treatment and TiO_2 Coating of PMMA for the Improvement of Antibacterial Properties. *Surf. Coat. Technol.* **2010**, *205*, 465–469. [CrossRef]
45. Abdullah, T.; Qurban, R.O.; Bolarinwa, S.O.; Mirza, A.A.; Pasovic, M.; Memic, A. 3D Printing of Metal/Metal Oxide Incorporated Thermoplastic Nanocomposites With Antimicrobial Properties. *Front. Bioeng. Biotechnol.* **2020**, *8*, 568186. [CrossRef] [PubMed]

Disclaimer/Publisher's Note: The statements, opinions and data contained in all publications are solely those of the individual author(s) and contributor(s) and not of MDPI and/or the editor(s). MDPI and/or the editor(s) disclaim responsibility for any injury to people or property resulting from any ideas, methods, instructions or products referred to in the content.

Influence Analysis of Modified Polymers as a Marking Agent for Material Tracing during Cyclic Injection Molding

Tom Eggers [1,*], Sonja Marit Blumberg [1], Frank von Lacroix [1], Werner Berlin [2] and Klaus Dröder [2]

[1] Volkswagen AG Wolfsburg, Berliner Ring 2, 38440 Wolfsburg, Germany
[2] Institute of Machine Tools and Production Technology, Technische Universität Braunschweig, Langer Kamp 19b, 38106 Braunschweig, Germany
* Correspondence: tom.eggers1@volkswagen.de

Abstract: Injection molding (IM) is already an established technology for manufacturing polymer products. However, in the course of the increased use of recyclates for economic and ecological reasons, its application capability has been confronted with new requirements for reliability and reproducibility. In addition, the IM process is confronted with regulations regarding a verifiable recycling degree in polymers. With regard to the material identification and storage of manufacturer-, process- or product-related data in polymers, the implementation of a material-inherent marking technology forms a potential answer. The IM process combined with modified polymers (MP) as a marking technology turns out to be a feasible approach to manufacturing reproducibly and offers a high quality based on increased process awareness and fulfilling the required traceability. Therefore, this work focuses on the trial evaluation of MP within the IM process. The influence of MP on the material process behavior and mechanical and thermal component properties, as well as the influence of the IM process and recycling on MP traceability, are investigated. No discernible influences of MP on the investigated properties could be identified, and the traceability from the initial material to a recyclate could be confirmed. MP is suitable for monitoring the aging state of polymers in IM.

Keywords: injection molding; tracing; traceability; polymers; marking agent; modified polymers; process optimization; process predictability; recycling; circular economy

1. Introduction

Injection molding (IM) is among the most prominent representatives of polymer-based manufacturing processes and is characterized by a causal flow from the initial material to the end product. In the IM process, the plastification of granular or powdered polymers into a flowable state occurs using dissipation and heat conduction along a temperature-controlled cylinder. The plastified material is injected into the closed mold via the axial movement of the injection unit and subsequently solidifies. In the case of thermoplastic materials, the mold is temperature-controlled to a temperature below the solidification temperature of the melt so that the melt cools and solidifies as it enters the mold. Part quality is influenced by the IM process parameters, such as the temperatures from the hopper to the nozzle, the holding pressure and the back pressure, as well as the screw rotation speed, holding time and mold temperature [1–4].

The widespread use of thermoplastics and their composites in the present is a cause of growing concern because of their adverse effects on the environment [5]. Although the use of recyclates in the manufacturing of new polymer products is an important contribution to a circular economy, as it enables material cycles to be closed, the amount of recyclates in new products is still low due to technical barriers to reprocessing [6]. The remaining fraction ends up in landfills or in the oceans and becomes microplastics [7,8]. According to Yang et al. [9], a current barrier to the use of recycled polymers is the lower quality of recyclates compared to virgin materials. Even though mechanical recycling is less energy-intensive than chemical recycling [10], it is capable of having a significant impact on the quality of the recyclates [11,12]. The material is subject to various material aging processes [13–15]. Besides the rheological properties [16–18], the mechanical properties are also affected due to the reduction in molecular weight [10,11,18–29]. As a result, the recycling of polymers tends to decrease the stability of the IM process and decrease part quality [14,21,30]. A common method of reusing recycled polymers is to mix them with virgin materials to obtain materials with average properties [21,31–36]. A ratio of approximately 30% [21] to 50% [19] recycled to virgin material was reported in previous studies.

As different batches and material streams have different material properties, the identification of the different classes of recyclates is required [5,9,32,36,37]. In Auer et al. [37], the current problems of recycling and their causes were summarized. Although the use of methods to determine material composition is a solution to reduce quality problems, Auer et al. [37] evaluated the current sorting technology as inappropriate. For this, Auer et al. [37] mentioned the use of tracing technologies that can sort the previously unsortable materials. Modified polymers were evaluated as a suitable material-inherent tracing technology for use in polymer processing manufacturing [38]. This tracing technology has already been successfully investigated in the selective laser sintering of polyamide 12 in a previous study [39]. Modified polymers are sequence-defined polymers, and they are encodable via targeted polymerization so that information can be stored at the molecular level [40].

IM combined with modified polymers as a marking technology turns out to be a promising way to manufacture reproducibly and with high quality, based on an advanced process awareness [41]. Therefore, it is based on the concept of the tracer-based optimization of the process [38] via a material marking. Material traceability in injection molding is achievable due to the incremental addition of one or more different modified polymers at each new recycling step the material passes through. The markers are analyzed at defined analysis points in the recycling process and before the IM process. It is intended to determine the mixture of the material on the basis not only of either the qualitative or the quantitative analysis of the markers but also via a combination of both variants. Thus, an increasing amount of different modified polymers propagates through the continuous material in a manner analogous to the recyclate classes formed. The composition of the markers can, in turn, be used to infer the different recyclate proportions. Once the history of the material or its composition is identified, it is possible to draw conclusions about the quality of the material. The respective marker codes are linked to the material history in a database, which is expanded via new codes with each new recycling of the material. If the recyclate proportions are known, an overall material quality can be defined from the respective material qualities of the individual recyclate proportions, and the process parameters can be specifically adjusted, as a result of which the scattering material properties are compensated for. As the material is refreshed via virgin material, the proportions of the frequently recycled materials, as well as the associated markers, decrease. When the respective marker concentration falls below a certain limit, traceability is no longer guaranteed. The associated amount of recyclate is then already so small that it can be neglected. In addition to the stepwise material marking, the mechanical properties of marked products must also be examined to maintain the functionality of products.

Consequently, the aim of the present study is to validate modified polymers for application in IM. Investigations are carried out to determine the influence of the modified polymer on the processing behavior of the material and the component properties and to determine the influence of the IM process and recycling on the traceability of the modified polymer. Thermal and rheological material properties, as well as mechanical and thermal component properties, are analyzed. Since repeated extrusion and IM are evaluated as suitable methods to investigate recyclability [29,42–46], these methods are used in the present study. The influence of the cyclic processing of the marked material, in particular, thermal stress and stress due to shear forces [1–4,47], on the functionality and detectability of the marker is investigated. Mass spectroscopy is used to determine the traceability of the marker.

2. Materials and Methods

In order to minimize disturbance variables and avoid aging processes due to humidity, temperature and ultraviolet radiation [13,14,21,48], materials and test specimens were stored, and test procedures were applied under controlled ambient conditions of 23 °C and 50% relative humidity. The storage of materials and test specimens was airtight and protected from ultraviolet radiation.

2.1. Injection Molding Material

The Daplen EF155AE elastomer-modified polypropylene compound with 10% mineral filling from Borealis AG (Vienna, Austria) was used. This material is among the thermoplastic polyolefins [13,15,49] and can be easily processed with commercially available injection molding machines. Pre-drying the material at 80 °C for approximately 2 h was recommended in order to avoid residual moisture. Referring to the manufacturer's data, the specific material density was 0.95 g/cm^3. The material choice was based on the fact that polypropylene is one of the most commonly used semi-crystalline thermoplastics in injection molding [13,50–52] and the most widely used polymer in the automotive industry [53].

2.2. Modified Polymer

The modified polymer used in this study was supplied by Polysecure GmbH (Freiburg, Germany) and branded as POLTAG® technology [41,54–56]. A previous study [38] listed further information about the marking agent. Compared to the investigated modified polymer used in a previous study [39], this modified polymer contained a different main molar mass, as well as different mass numbers of the sequences (Table 1).

Table 1. Main molar mass and mass numbers of the sequences of the used modified polymers.

Main Molar Mass [Da]	Sequences [Da]
918.5	131.0 | 246.2 | 347.3 | 434.3 | 535.4 | 650.4 | 781.4 | 882.5 | 981.5

2.3. Production of the Master Batch

On the basis of the used injection molding material and modified polymer, a master batch was produced. Via a spray-drying process, the modified polymer was added to the injection molding material via mixing. In the master batch, a concentration of 30 ppm of the modified polymer was used. The method of producing the master batch based on the modified polymer was confirmed in a previous study [39]. Through weighing the material fractions, the concentrations of the modified polymer in the material were determined to be 1 ppm and 10 ppm. For the quantification of the material mass, an EW 4200-2NM scale from KERN & Sohn GmbH (Balingen-Frommern, Germany) was applied.

2.4. Injection Molding Processing and Recycling Procedure

In Figure 1, the injection molding and mechanical recycling process used are illustrated. The presented process flow of recycling was in accordance with Tamrakar et al. [20]. A horizontal injection molding machine [1] of the Allrounder 470 E Golden Electric type from ARBURG GmbH + Co KG (Loßburg, Germany) was used for injection molding. In the initial process cycle (R0), unmarked and marked granulates were fed into the injection molding machine to produce test samples. All samples were produced via injection molding using the parameters listed in Table 2. The parameters were in accordance with the DIN EN ISO 19069-2:2020-01 [57] standard, the parameters listed in Isayev et al. [1] and Dominghaus et al. [51] and the specifications of the manufacturer of the injection molding machine. The cylinder temperature was gradually increased from the hopper to the nozzle end to ensure a smooth transition from a solid to a molten polymer and to reduce the screw wear [1,2,20,47]. A defined number of specimens were taken for further characterization.

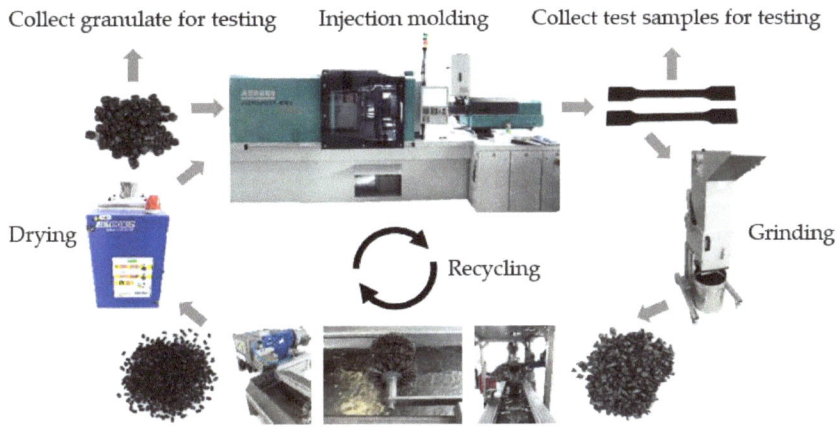

Figure 1. Process flow of the used recycling steps to process of the used injection molding material. The process flow was divided into injection molding, collecting of the specimens for testing (tensile bars), grinding of the remaining specimens, extruding and regranulation of the shredded material, drying of the granulate, collecting of the granulate for testing and renewed injection molding.

Table 2. Selected parameters of the injection molding machine used.

Option	Selected Setting
Temperatures (hopper to nozzle) [°C]	35 | 180 | 190 | 195 | 200 | 200
Mold temperature [°C]	40
Dosing quantity [cm^3]	36.5
Decompression [cm^3]	3
Screw speed [m/s]	0.25
Back pressure [bar]	80
Residual mass cushion [m^3]	4.6
Volume flow [cm^3/s]	16
Max. injection pressure [bar]	280
Switching volume [cm^3]	7
Max. hold pressure [bar]	200
Hold time [s]	40
Residual cooling time [s]	18

In Table 3, the investigated samples of each injection molding process are listed. The rest of the test samples were ground using a granulator of the C17.26sv SE type from Wanner Technik GmbH (Wertheim-Reicholzheim, Germany) and a grinding sieve with a mesh size of 6 mm.

Table 3. Breakdown of the test samples used for each process cycle.

Type	Dimension	Number	Usage
Tensile bar [1]	1A [2]	5	Tensile test
Tensile bar [1]	1A [2]	10	Vicat softening temperature test, Charpy impact test
Tensile bar [1]	1A [2]	2	Traceability

[1] Tensile bars presented in Figure 1. [2] DIN EN ISO 527-2 [58].

Then, the ground material was extruded by a twin screw extruder of the ZE25A-40D-UTX-UG type from KraussMaffei Berstorff (Laatzen, Germany) with gravimetric dosing from Scholz Dosiertechnik GmbH (Großostheim, Germany). All materials were extruded according to the parameters listed in Table 4. The extruded strands were cooled in a water bath using a strand pelletizing unit of the ips-SG-E 30 Kombi type from IPS Intelligent Pelletizing Solutions GmbH & Co KG (Niedernberg, Germany) with an integrated strand cooling tank, dried via blasting with an integrated strand dewatering system and processed to a granulate with a granulate size of 3 mm. A cold cutting process [1,47] was applied in this case. Next, the granulates were dried at 80 °C for a duration of 2 h in a JETBOXX granulate dryer from HELIOS GmbH (Rosenheim, Germany).

Table 4. Selected parameters of the extrusion and pelleting machine used.

Option	Selected Setting
Temperature (hopper to nozzle) [°C]	50	190	190	190	190	190	195	195	200	200	200
Temperature at the output [°C]	209
Screw speed [1/min]	160
Nozzle head pressure [bar]	12
Volume flow [kg/h]	8
Capacity utilization [%]	30
Cut-off speed [m/min]	33

Before the next cycle (R1), a granulate quantity of 20 g was taken for further tests. The remaining amount of granulates was further processed. Then, the pre-dried granulates were injection-molded, and the recycling process was repeated up to ten times (R10). The number of recycling steps was based on Aurrekoetxea et al. [29], and it indicated an asymptotic behavior of the investigated thermal and mechanical properties after ten recycling steps. For each investigated material, an initial amount of 5 kg of the respective material was provided. Before each cycling step, the hopper, screw and mold were cleaned using compressed air and a vacuum cleaner, as well as dry wipes.

2.5. Thermogravimetric Analysis

Thermogravimetric analysis (TGA) was applied using a TGA/DSC-1 measuring system from Mettler Toledo (Gießen, Germany). Measurements were performed according to the DIN EN ISO 11358-1:2022-07 [59] standard. The change in the mass of a sample over a defined, time-related temperature curve under the influence of a purge gas was determined. Whether marking with the used modified polymer led to a different thermally induced degradation behavior of the polymer was examined [13,60]. The sample was heated from room temperature to 900 °C at a heating rate of 20 K/min under a nitrogen atmosphere (flow: 70 mL/min) and was stored at this temperature and atmosphere for 10 min. Afterwards, the sample was cooled to 400 °C at a cooling rate of 20 K/min under a nitrogen atmosphere (flow: 70 mL/min). Next, the sample was heated to 900 °C under the presence of oxygen (flow: 70 mL/min) at a heating rate of 20 K/min and was stored at this temperature and atmosphere for 15 min. The sample weight for each measurement was 12 mg to 14 mg. The analyzed TGA measurement results included the maximum pyrolytic degradation temperature and the residue as mean values. No standard deviation was provided by the TGA measurement.

2.6. Differential Scanning Calorimetry Testing

Differential scanning calorimetry (DSC) was performed using a DSC-3$^+$ measuring system from Mettler Toledo (Gießen, Germany). Measurements were obtained according to the DIN EN ISO 11357-1:2023-06 [61] standard and were realized under the presence of a nitrogen atmosphere. Each measurement was carried out using a sample weight of 10 mg ± 2 mg. The heating and cooling cycles were performed in the temperature interval between 50 °C and 200 °C. The temperature rate was 10 °C/min. Among other data from the DSC measurement, both the melting and crystallization temperatures and the onset temperatures were output as mean values. Whether marking with the used modified polymer tightened or widened the thermal process window of the polymer was considered. The DSC testing did not provide any standard deviation.

2.7. Melt Flow Index Test

Melt flow index testing was realized with an Mflow measuring device from Zwick/Roell GmbH & Co. KG (Ulm, Germany) and was performed according to the DIN EN ISO 1133-1:2012-03 [62] standard. In order to determine the melt flow rate (MFR), a testing load of 2.16 kg at a temperature of 230 °C and a filling quantity of 4 g was used. Melt flow index measurement is a single-point test method that represents a single measuring point of the viscosity curve and gives an initial indication of a change in the rheological behavior of the melt [63]. Whether the modified polymer used had an influence on the MFR value of the polymer was considered. To avoid moisture, the material was pre-dried at 80 °C for 2 h using a JETBOXX granulate dryer from HELIOS GmbH. Before testing, the nozzle was cleaned with cleaning tools and cotton cloths. The test started at a position of the piston of 50 mm. Thereby, five sections with a measured length of 5 mm were recorded. The mass of each of the five sections was quantified using an AB-100 scale from PCE Deutschland GmbH (Meschede, Germany). The extruded mass within the defined interval described the MFR value, which was expressed in the unit g/10 min.

2.8. Tensile Test

A tensile test was performed on a tensile testing machine of the Zwick/Roell Z100 type from Zwick/Roell GmbH&Co. KG. The equipment of the testing machine included a makroXtens mechanical extensometer and two wedge clamping jaws designed for a normal force of up to 10 kN. Furthermore, the equipment included an Xforce K load cell determined for the same load limit and a testControl II control unit. According to the DIN EN ISO 19069-2:2020-01 [57] standard, the tensile test specimens (Table 3) were conditioned in a constant climate chamber of the KBF 240 type from BINDER GmbH (Tuttlingen, Germany) for 96 h at a temperature of 23 °C and 50% relative humidity. During the transfer from the constant climate chamber to the tensile testing machine, the samples were stored in airtight containers together with silicate pellets for moisture absorption. Tensile testing was performed from the conditioned state of the samples and was done within ten days after manufacturing the samples. The test specimens were evaluated according to the DIN EN ISO 527-1 [64] standard. According to the DIN EN ISO 19069-2:2020-01 [57] standard, the tensile strength and elongation at break were measured at a test speed of 50 mm/min. In contrast, Young's modulus was measured at a speed of 1 mm/min. In addition, Young's modulus was determined as the secant modulus, according to the DIN EN ISO 527-1 [64] standard, in the interval of elongation from 0.05% to 0.25%. Thereby, the specimens were exposed to a preload of 0.1 MPa at a test speed of 1 mm/min. Five specimens of each process step were tested, and the mean value, as well as the standard deviation, was reported.

2.9. Charpy Impact Test

A Charpy impact test was performed on a pendulum impact tester of the Zwick/Roell HIT25P type from Zwick/Roell GmbH & Co. KG using a testControl II control unit. According to the DIN EN ISO 179-1:2010-11 [65] standard, the test specimens for Charpy impact testing were conditioned in a constant climate chamber of the KBF 240 type from BINDER GmbH for 96 h at a temperature of 23 °C and 50% relative humidity. For Charpy impact testing, a test specimen of type 1 was used. Due to the toughness of the material used, a Charpy impact test was performed according to the DIN EN ISO 179-1/1fA [65] standard. The specimens were prepared from the parallel part of the specimens presented in Table 3 and corresponded to the multipurpose test for specimen type A, according to the DIN EN ISO 3167:2014-11 [66] standard. During the transfer from the constant climate chamber to the pendulum impact tester, the specimens were stored in airtight containers together with silicate pellets for moisture absorption. The Charpy impact test was performed with the conditioned state of the samples and was done within ten days after manufacturing the samples. A 7.5 J pendulum was used in this test. As a result of the Charpy impact test, among other data, Charpy double-edge-notched impact strength (dNIS) was analyzed. Ten specimens of each process step were tested, and the mean value, as well as the standard deviation, was reported.

2.10. Vicat Softening Temperature Test

The Vicat softening temperature (VST) of the specimens was measured according to the DIN EN ISO 306:2023-03 [67] standard using a VST/HDT Compact 3 system from Coesfeld GmbH & Co. KG (Dortmund, Germany). The system used consisted of a bar with a support disk for the test weights and a fixture for the indenter tip, as well as a calibrated dial gauge for determining the indentation depth. Before the measurement, the surface of the specimen was cleaned of burrs and irregularities. Based on the B 50 VST method, the specimens were loaded with a force of 50 N, and the heating rate was 50 °C/h. During measurement, the specimens were positioned on a specimen fixture and heated at the defined heating rate in a silicone bath. The specimens were prepared from the end pieces of the specimens presented in Table 3, and the size of the specimens was 10 mm × 10 mm × 4 mm. The temperature at which a tip penetrated 1 mm deep into the surface of the test specimen was determined. The indenter tip has a circular cross section of 1 mm^2. The measured temperature was defined as the VST. Three specimens were used for each measurement. The mean value and the standard deviation were used to determine the VST. The objective is to quantitatively characterize the thermal stability of a polymer [68]. Since the VST responds to a change in molecular size, the measurement can be used to indicate the processing-induced thermal damage of the material [68,69].

2.11. Tandem Mass Spectroscopy

Tandem mass spectroscopy (MS/MS) was used to detect, sequence and measure the traceability of the modified polymer [40,70–72]. An MS/MS device of the API 4000 MS/MS type was used, which was supplied by AB SCIEX (Darmstadt, Germany). The method for detection, sequencing and determining the traceability of the modified polymer was analogous to that of a previous study [39]. With the exception of the extraction methods listed in Table 5, the same procedure and parameters were used. The extraction methods depended on the investigated specimens. According to a previous study [39], the analyte was delivered to the atmospheric pressure chemical ionization source using a syringe pump of the Model 100 type from kd Scientific Inc. (Holliston, MA, USA). The syringe pump flow rate was 0.06 mL/h [39]. Detection and sequencing were both recorded for at least 5 min. Between each measurement, the MS/MS device was cleaned. Therefore, a liquid consisting of methanol and 3 mmol ammonium acetate was injected for at least 10 min.

Table 5. Extraction methods used for the investigated samples [39].

Sample	Analyzed Quantity	Extraction Method
Granulate	5 g	Mixed with 10 mL of ethanol
		Placed in an ultrasonic bath at 40 °C for a duration of 30 min
		Filtration with a 22 μm filter
Component [1]		Placed in a rotary vacuum evaporator
		Dilution with 2 mL of methanol and 3 mmol of ammonium acetate

[1] Shredded pieces of tensile bar listed in Table 2 for analyzing traceability.

Based on a Python tool used in a previous study [39], the results of the MS/MS measurements were evaluated. For this purpose, the sum of the determined intensities within a measurement interval of ±1 Da was determined for all mass numbers of the modified polymer (Table 1). The respective maximum peak of the mass spectrum was used to calibrate the molar mass. The modified polymer was considered to be detected if an intensity greater than 0 was determined for the mass number of the main mass, as well as for the majority of the mass number of all sequences within the measurement interval, and peaks were clearly visible [39].

3. Results
3.1. Influence of the Modified Polymer Used on the Material Properties
3.1.1. Initial Material Properties

In Table 6, the investigated initial material properties for the injection molding material and master batch are presented. The determined crystallization and melting temperatures, as well as their onset temperatures, deviated between the injection molding material and the master batch by less than approximately 1%. Regarding the thermogravimetric analysis, the master batch shows an approximately 0.34% higher maximum pyrolytic degradation temperature and an approximately 0.55% higher residue than the injection molding material. The values of the melt flow rate (MFR) differ by approximately 2.4%. Considering the standard deviation, the deviation of the MFR values is approximately 0.1%.

Table 6. Investigated material properties for the injection molding material and master batch used based on the injection molding material used and the modified polymer used.

Properties	Injection Molding Material	Master Batch
PE [1] onset crystallization temperature [°C]	113.16	113.09
PE crystallization temperature [°C]	110.02	109.95
PE onset melting temperature [°C]	107.86	107.01
PE melting temperature [°C]	121.31	120.95
PP [2] onset crystallization temperature [°C]	133.30	133.02
PP crystallization temperature [°C]	129.69	129.28
PP onset melting temperature [°C]	158.43	159.11
PP melting temperature [°C]	168.62	167.75
Max. pyrolytic degradation temperature [°C]	473.95	475.56
Residue [%]	9.1925	9.2435
Melt flow rate [g/10 min]	17.73 ± 0.19	17.31 ± 0.21

[1] Polyethylene-specific peak. [2] Polypropylene-specific peak.

3.1.2. Thermal Process Window

The influence of the modified polymer on the thermal process window, characterized by melting and crystallization temperature and depending on the concentration of the modified polymer and the recycling step, is presented in Figure 2. While there was an increase in melting temperature (T_M) between the first and fifth steps for all concentrations considered, there was a decrease in T_M from R0 to R10 for all concentrations observed. From R0 to R10, the T_M for 0 ppm decreased by approximately 0.36%. In contrast, for 1 ppm, the T_M decreased by approximately 0.09% from R0 to R10, and for 30 ppm, it decreased by approximately 1.31%. At R0, the deviations of the T_M between the unmarked material and the marked materials amounted to a maximum of approximately 0.34%. At R1, the maximum deviation of T_M was approximately 0.8%. At R10, the deviations of the T_M between the unmarked material and the marked materials amounted to a maximum of approximately 1.47%.

Regarding crystallization temperature (T_C), there was an increase from R0 to R10 for nearly all concentrations considered. From R0 to R10, the T_C for 0 ppm increased by approximately 0.53%. For 1 ppm, the T_C increased by approximately 1.54% from R0 to R10, and for 30 ppm, it increased by approximately 1.42%. At R0, the deviations of the T_C between the unmarked material and the marked materials amounted to a maximum of approximately 0.32%. At R1, the maximum deviation of T_C was approximately 1.17%. At R10, the deviations of the T_C between the unmarked material and the marked materials amounted to a maximum of approximately 0.96%.

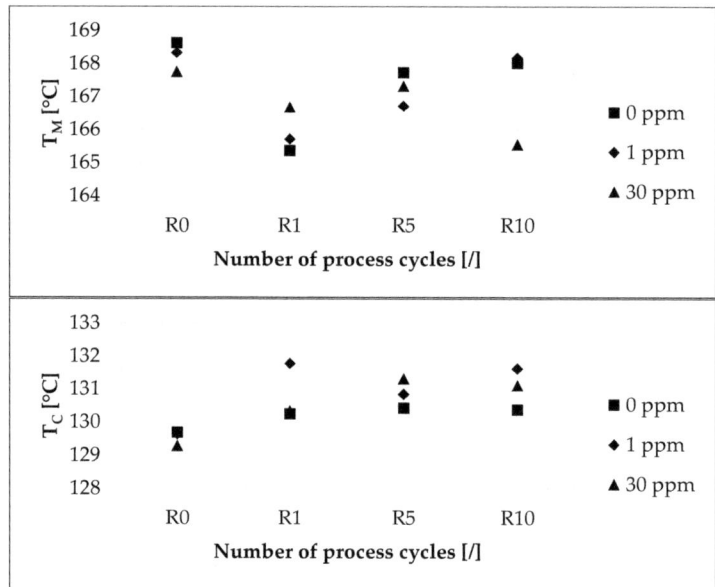

Figure 2. Influence of the modified polymer used and the recycling on the melting (T_M) and crystallization temperature (T_C). The material samples are based on the polypropylene compound injection molding material used and were collected in a dry state before each process step. Concentrations of the modified polymer used of 0 ppm, 1 ppm and 30 ppm were studied. The recycling steps are listed on the X-axis. Process step R0 describes the initial injection molding process step. Steps R1, R5 and R10 describe the first, fifth and tenth process steps of the recycling.

3.1.3. Melt Flow Rate

The influence of the modified polymer on the mean value of the melt flow rate (MFR), depending on the concentration of the modified polymer and the recycling step, is presented in Figure 3. While the mean value of the MFR decreased for individual concentrations of the respective process cycles, the mean value of the MFR increased in the overall view from R0 to R10.

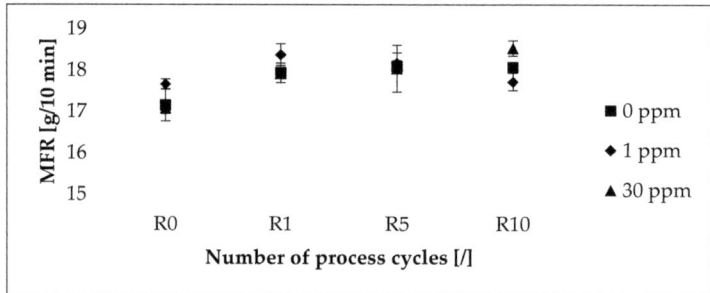

Figure 3. Influence of the modified polymer used and the recycling on the melt flow rate (MFR). The material samples are based on the polypropylene compound injection molding material used and were collected in a dry state before each process step. Concentrations of the modified polymer used of 0 ppm, 1 ppm and 30 ppm were studied. The recycling steps are listed on the X-axis. Process step R0 describes the initial injection molding process step. Steps R1, R5 and R10 describe the first, fifth and tenth process steps of the recycling.

The mean value of the MFR for 0 ppm increased by approximately 4.61% from R0 to R1. In contrast, for 1 ppm, the mean value of the MFR increased by approximately 4.08% from R0 to R1, and for 30 ppm, it increased by approximately 4.92%. From R1 to R10, the mean value of the MFR for 0 ppm increased by approximately 0.72%. In contrast, for 1 ppm, the mean value of the MFR decreased by approximately 3.43% from R0 to R10, and for 30 ppm, it increased by approximately 3.40%. At R0, the deviations of the mean value of the MFR between the unmarked material and the marked materials amounted to a maximum of approximately 2.97%. At R1, the maximum deviation of the mean value of the MFR was approximately 2.45%. At R10, the deviations of the mean value of the MFR between the unmarked material and the marked materials amounted to a maximum of approximately 2.55%. The deviations of the mean values of the MFR as a function of the concentration of the modified polymer were also within the standard deviation across the recycling steps.

3.2. Influence of the Modified Polymer Used on the Component Properties

3.2.1. Tensile Test

The influence of the modified polymer used on the mechanical properties, depending on the concentration of the modified polymer and the recycling step, is presented in Figure 4. With regard to the mean value of Young's modulus for 0 ppm, there was a decrease of approximately 3.3% from R0 to R10. In contrast, for 1 ppm, from R0 to R10, an increase of approximately 7.92% was recorded, and for 30 ppm, from R0 to R10, an increase of approximately 1.59% was recorded. With regard to the mean value of Young's modulus at R0, the deviation between 0 ppm and 30 ppm was approximately 0.93%. At R10, this deviation amounted to approximately 4.08%. With regard to the mean value of the tensile strength for 0 ppm, there was a decrease of approximately 1.02% from R0 to R10. In contrast, for 1 ppm, from R0 to R10, a decrease of approximately 0.32% was recorded, and for 30 ppm, from R0 to R10, a decrease of approximately 0.33% was recorded. With regard to the mean value of the tensile strength at R0, the deviation between 0 ppm and 30 ppm was approximately 1.76%. At R10, this deviation amounted to approximately 1.03%. Regarding the mean value of elongation at break for 0 ppm, there was an increase of approximately 34% from R0 to R10. In contrast, for 1 ppm, from R0 to R10, a decrease of approximately 46.2% was recorded, and for 30 ppm, from R0 to R10, a decrease of approximately 25.1% was recorded. With regard to the mean value of elongation at break at R0, the deviation between 0 ppm and 30 ppm was approximately 13.6%. At R10, this deviation amounted to approximately 36.5%. The deviations of the mean values of Young's modulus, the tensile strength and the elongation at break, depending on the concentration of the modified polymer, were within the standard deviation even across the recycling steps.

3.2.2. Charpy Impact Test

The influence of the modified polymer on the double-edge-notched impact strength (dNIS), depending on the concentration of the modified polymer and the recycling step, is presented in Figure 5.

The mean value of the dNIS for 0 ppm increased by approximately 5.13% from R0 to R10. In contrast, for 1 ppm, the mean value of the dNIS increased by approximately 4.33% from R0 to R10, and for 30 ppm, it increased by approximately 5.52%. At R0, the maximum deviation of the mean value of the dNIS between the unmarked material and marked materials was approximately 2.63%. At R10, the maximum deviation of the mean value of the dNIS between the unmarked material and marked materials was approximately 3.38%. The deviations of the mean values of the dNIS, depending on the concentration of the modified polymer, were within the standard deviation even across the recycling steps, with the exception of recycling step R1.

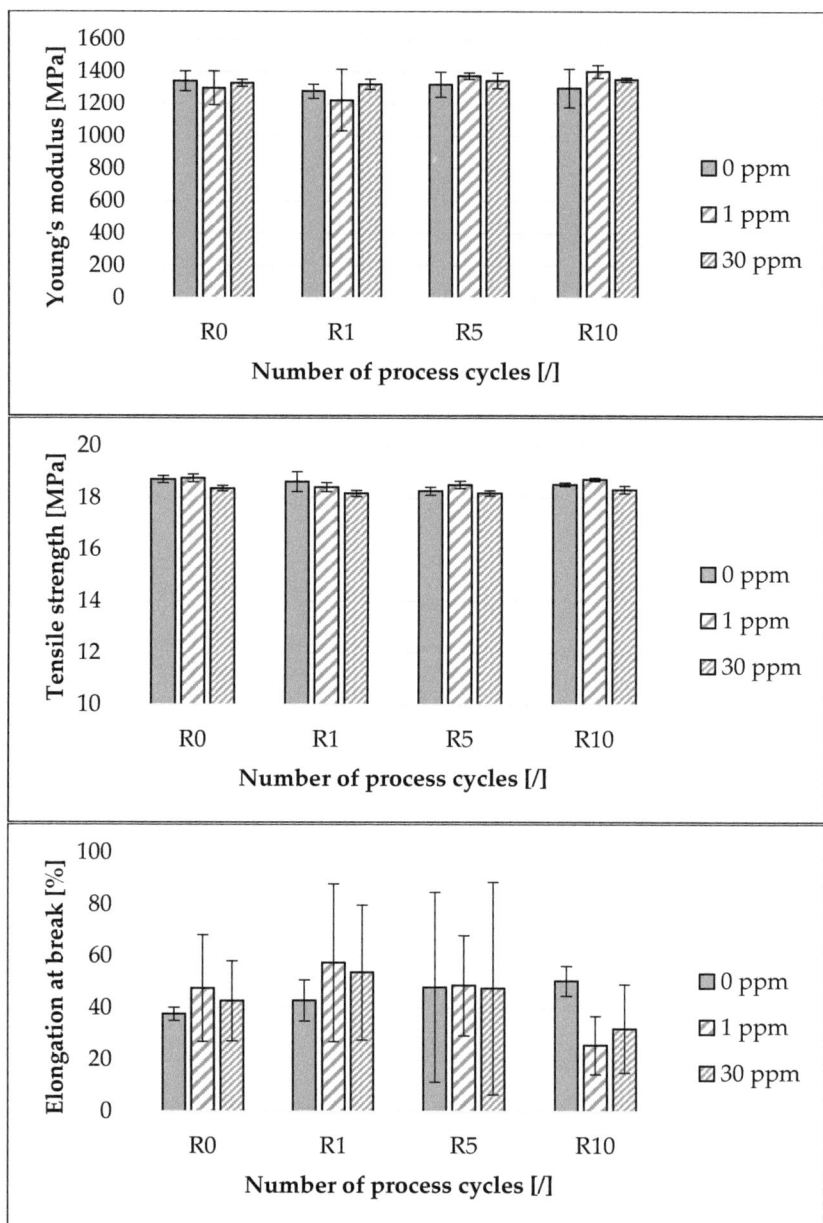

Figure 4. Influence of the modified polymer used and the recycling on the investigated mechanical properties. The specimens were manufactured from the polypropylene compound injection molding material used. Concentrations of the modified polymer used of 0 ppm, 1 ppm and 30 ppm were examined. The recycling steps are listed on the X-axis. Process step R0 describes the initial injection molding process step. Steps R1, R5 and R10 describe the first, fifth and tenth process steps of the recycling. In each case, five tensile bars were analyzed.

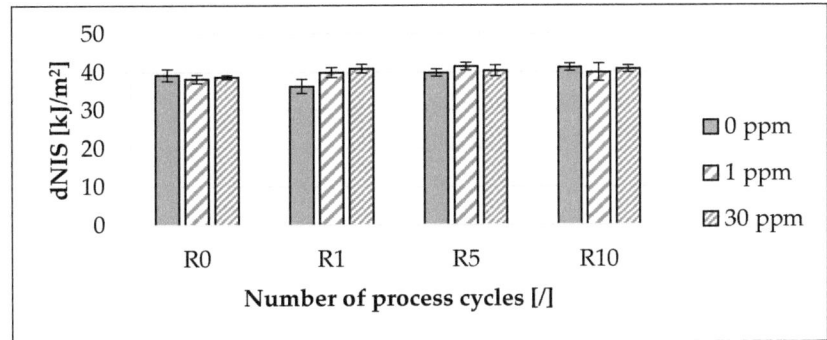

Figure 5. Influence of the modified polymer used and the recycling on the double-edge-notched impact strength (dNIS). The specimens were manufactured from the polypropylene compound injection molding material used. Concentrations of the modified polymer used of 0 ppm, 1 ppm and 30 ppm were examined. The recycling steps are listed on the X-axis. Process step R0 describes the initial injection molding process step. Steps R1, R5 and R10 describe the first, fifth and tenth process steps of the recycling. In each case, ten specimens were analyzed.

3.2.3. Vicat Softening Temperature

The influence of the modified polymer on the Vicat softening temperature (VST), depending on the concentration of the modified polymer and the recycling step, is presented in Figure 6. The mean value of the VST for 0 ppm increased by approximately 1.9% from R0 to R10. In contrast, for 1 ppm, the mean value of VST increased by approximately 0.74% from R0 to R10, and for 30 ppm, it increased by approximately 0.2%.

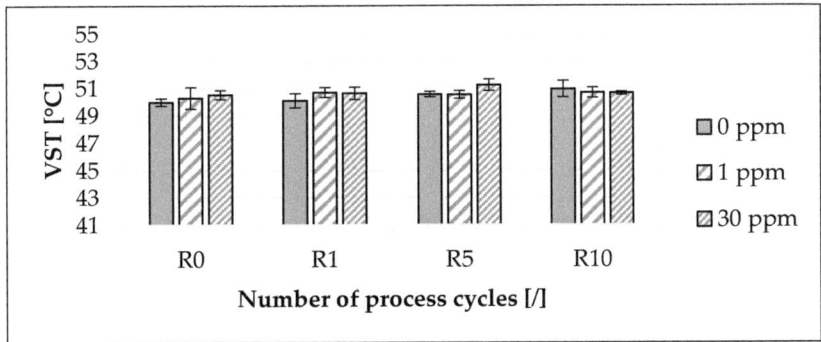

Figure 6. Influence of the modified polymer used and the recycling on the Vicat softening temperature (VST). The specimens were manufactured from the polypropylene compound injection molding material used. Concentrations of the modified polymer used of 0 ppm, 1 ppm and 30 ppm were examined. The recycling steps are listed on the X-axis. Process step R0 describes the initial injection molding process step. Steps R1, R5 and R10 describe the first, fifth and tenth process steps of the recycling. In each case, ten specimens were analyzed.

The maximum deviation of the mean value of the VST between the unmarked material and marked materials at R0 was approximately 1.06%. At R10, the maximum deviation of the mean value of the VST between the unmarked material and marked materials was approximately 0.59%. The deviations of the mean values of the VST as a function of the concentration of the modified polymer were also within the standard deviation across the recycling steps.

3.3. Traceability of the Modified Polymer Used

The traceability of the modified polymer at different recycling steps is presented in Figure 7. The modified polymer was detectable at all investigated concentrations down to a concentration of 1 ppm, and the main mass, as well as all individual sequences, was detectable in all recycling steps (Figure 8).

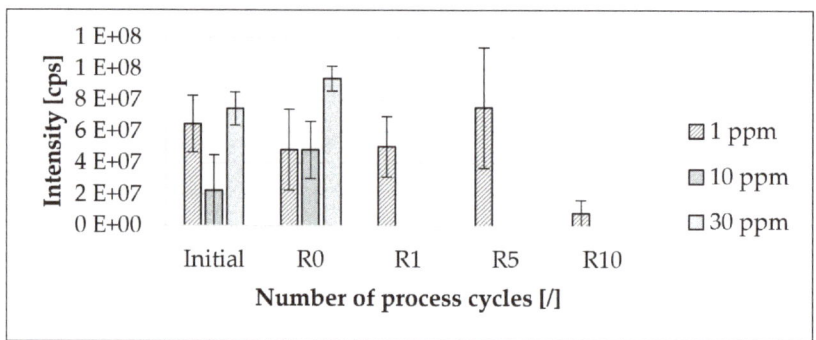

Figure 7. Traceability of the modified polymer used at different process steps. The recycling steps are listed on the X-axis. The initial step describes the initial granulate before injection molding processing. Process step R0 describes the initial injection molding process step. Steps R1, R5 and R10 describe the first, fifth and tenth process steps of the recycling. The specimens are based on the polypropylene injection molding material used. The arithmetic mean of the sum of the intensities within the considered interval of ±1 Da around the main molar mass of three MS/MS measurements each was formed. For the initial step and step R0, concentrations of the modified polymer used of 1 ppm, 10 ppm and 30 ppm were considered. For steps R1, R5 and R10, a concentration of 1 ppm of the modified polymer used was assumed.

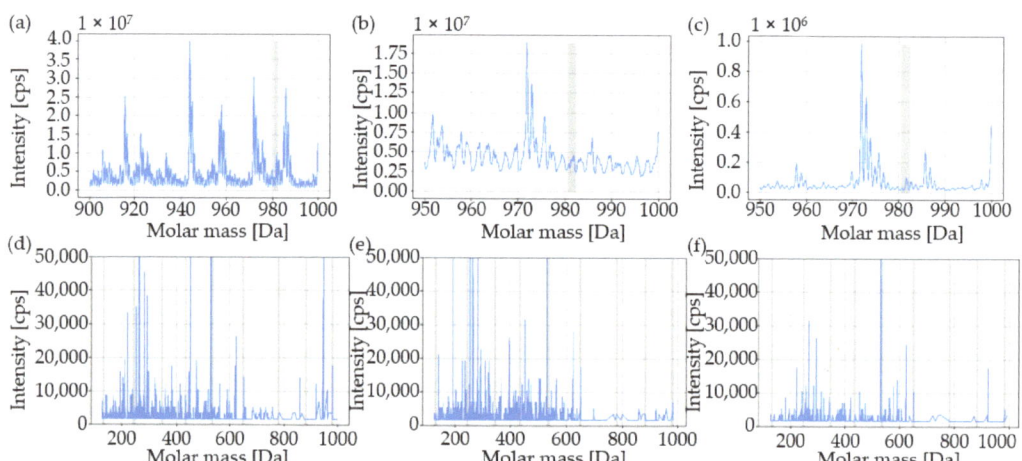

Figure 8. MS/MS measurements (detection and sequencing) of the selected process steps for the concentration of the modified polymer used of 1 ppm. The intensity is plotted over the molar mass: (**a,d**) initial injection molding material; (**b,e**) initial injection molding process step R0; and (**c,f**) tenth process step R10 of recycling. The intervals in which the main masses and individual sequences were detected are highlighted with gray bars. The specimens are based on the polypropylene injection molding material used.

The determined intensities did not increase linearly with the selected concentrations. Furthermore, there was no apparent correlation between the concentration of the modified polymer and the mean value of the detected intensity of the main mass (Figures 7 and 8). If the standard deviation is taken into account, there was no change in intensity as a result of the initial injection molding process for all the investigated concentrations of the modified polymer used. In the initial material, from 1 ppm to 30 ppm, there was an increase in the mean value of the intensity of the modified polymer of approximately 14.6%. In contrast, the mean value of the intensity of the modified polymer decreased from 1 ppm to 10 ppm by approximately 65.2%. From the initial material to R0, the mean value of the intensity of the modified polymer used at a concentration of 1 ppm decreased by approximately 25.4%. In contrast, the mean value of the intensity of the modified polymer at 30 ppm increased by approximately 25.9%. For R0, the deviation between 1 ppm and 30 ppm relative to the mean value of the intensity of the modified polymer used was about 93.6%. From the initial material to R5, the mean value of the intensity of the modified polymer at 1 ppm increased by about 15.9%. In contrast, the average intensity of the modified polymer at 1 ppm decreased by 89.2% from R5 to R10. At R10, eight out of nine sequences were detectable.

4. Discussion

The investigated thermal and rheological material properties (Table 6, Figures 2 and 3), as well as the mechanical (Figures 4 and 5) and thermal component properties (Figure 6), were not significantly influenced by the modified polymer. On the one hand, this observation is confirmed by the marginal deviations of the mean values of the respective properties with increasing concentrations. If the minimum and maximum standard deviations are taken into account, the variation in the respective properties due to the presence of the modified polymer used was negligible. This observation can be attributed to the selected concentrations of the modified polymer used. These were too low to identify any influence on the investigated material and component properties via the chosen testing methods [39]. On the other hand, this observation is confirmed by a previous study [39] that investigated modified polymers as marking agents in a polyamide 12 material via selective laser sintering. Taking into account the concentrations of the modified polymer used in the base material used and the selected scope of the investigation, no discernible influence on the material and component properties was observed. Since comparable material and component properties, as well as comparable concentrations of the marking agent, were used in this study, the influence of the marking agent on the material and component properties identified in the present study was also evaluated as not discernible. As a result, the processing properties of the marked material in injection molding remained the same as those of the unmarked material. This is confirmed by the fact that the chosen concentration of 30 ppm in the master batch was already at the maximum and, thus, lower concentrations behave analogously in the investigated material [39]. Furthermore, any influence attributable to the modified polymer could also be masked by other additives in the investigated material [13,24,36,73–81].

The selected concentrations of the marking agent are in accordance with an earlier study [39]. In addition to the easier master batch handling to set lower concentrations, the choice was based on the use of minimal concentrations for a future marking agent application in injection molding. Furthermore, the concentration of the master batch was selected in consultation with the manufacturer of the modified polymer. With regard to possible stepwise material marking and process optimization [38], at an initial concentration of the modified polymer of 1 ppm per process cycle, a maximum concentration of no more than 30 ppm can be expected after 30 process cycles. This total results from the stepwise addition of modified polymers. Due to the usual recycling rates in injection molding [19,21], the respective concentration of the specific modified polymer (initially 1 ppm) after more than 30 process cycles is so low that this concentration, as well as the associated amount of material, can be neglected. This assumption is based on a theoretical consideration. A concentration of 1 ppm is already enough to provide full functionality and traceability of the modified polymer in the material used even after multiple processing steps (Figures 7 and 8). Moreover, various studies [41,54,55,70,82] have confirmed the chosen concentration of modified polymers in different materials. Choosing higher concentrations of the modified polymer does not add value in terms of functionality or traceability [39]. The modified polymer is traceable in the initial material, as well as in the recycled material, down to a concentration of 1 ppm even after ten recycling steps (Figures 7 and 8).

As there is no apparent correlation between the investigated concentration of the modified polymer and the determined mean value of the intensity of the main mass, only a qualitative analysis is available. This observation is confirmed by a previous study [39]. Additives in the injection molding material, like carbon black particles, are extracted from the injection molding material during the used extraction process (Table 5) and are deposited in the measuring chamber of the mass spectroscope during the MS/MS measurement [39]. Consequently, no precise quantification of the used modified polymer is possible. Although no quantification of the mean value of the intensity of the main mass is possible, the decrease in intensity recorded in recycling step R10, combined with the absence of a single sequence, could indicate a possible degradation of the modified polymer. Nevertheless, the functionality of the modified polymer is still present after ten recycling steps. Considering the manufacturing process of the used master batch, it can be expected that all granulates are marked and, thus, that there is sufficient dispersion of the modified polymer in the material [39].

Regarding the melting temperature, a small decrease occurred from R0 to R10 for all concentrations (Figure 2). This is in contrast to the observations of a previous study [29]. However, if only the change in melting temperature from R1 to R10 is considered, a small increase in melting temperature with increasing recycling steps can be observed. The melting temperature of the sample with the 30 ppm marker at R10 is considered to be an anomaly, based on the previous results. The increasing melting temperature is due to a higher crystallinity of the polymer via multiple processing. The reduction in molecular weight during recycling increases the mobility and folding ability of the chains, which results in the formation of thicker lamellae and, consequently, a higher degree of crystallinity [29]. However, crystallinity was not quantified in the investigation. The observed increase in the crystallization temperature from R0 to R10 (Figure 2) also resulted from the increasing degree of crystallinity and is confirmed by an earlier study [29]. It should be taken into account that the recorded changes in the respective melting and crystallization temperatures are based on a single measurement.

The investigated increasing mean value of the melt flow rate (MFR) from R0 to R10, illustrated in Figure 3, is in accordance with previous studies [18,19,29,83] that observed an increasing MFR with increasing numbers of process cycles. The observed small changes in the MFR up to R5 are in accordance with Aurrekoetxea et al. [29], who attributed this observation to the presence of stabilizers in the material. The MFR value reaches its maximum after the first recycling step. Further recycling no longer significantly shortens the mean chain lengths. The increasing MFR with increasing recycling steps is due to the decreasing molecular weight. Therefore, a decreasing melt viscosity of a polymer decreases with a decreasing molecular weight, and polymers with a decreasing molecular weight have an increasing melt flow rate [18,29,83,84]. A more robust statement would be provided via recording flow curves. Since the measurement is carried out below the relevant processing range, no direct conclusions can be drawn about the processing behavior [85,86].

The resulting investigated mechanical component properties at R0 are in accordance with previous studies [1,87]. With regard to Young's modulus (Figure 4), there are different tendencies with increasing recycling steps, depending on the concentration of the marking agent. However, the changes were all within the respective standard deviation, and no clear tendency was observed. In contrast, previous studies have recorded both a decrease [18,20,84,88–90] and an increase [29,91] in Young's modulus with increasing steps of recycling. For example, Tamrakar et al. [20] investigated a decrease in Young's modulus of 12% after five recycling steps. Meneghetti et al. [88] indicated a decrease in Young's modulus for polypropylene with increasing recycling content in the material by up to 20%. In contrast, Aurrekoetxea et al. [29] and Brostow and Corneliussen [91] attributed the increasing crystallinity in the material during recycling as the cause of the increasing Young's modulus. The crystalline structures are much stiffer than amorphous structures and prevent rotations of the chain segments [29].

With regard to tensile strength (Figure 4), there are similar tendencies with increasing recycling steps, depending on the concentration of the marking agent. The decreasing tensile strength with recycling is in accordance with previous studies [18,20,84,88,90,92–97]. Multiple recycling steps lead to a change in material structure, which, due to degradation, results in a reduction in viscosity due to the scission of the chains and a significant loss of mechanical properties [18,84,90,92–97]. Tamrakar et al. [20] investigated a decrease in tensile strength of 4.6% after five recycling steps, and Meneghetti et al. [88] indicated a decrease in the tensile strength of polypropylene with increasing recycling content in the material by up to 20%. In contrast, Aurrekoetxea et al. [29] indicated increasing tensile strength with increasing numbers of recycling. Although it has not been investigated in this work, the decrease in tensile strength with increasing recycling steps could result from a thermomechanical degradation of the additives in the material during recycling [98].

Regarding elongation at break (Figure 4), there are different tendencies with increasing recycling steps, depending on the concentration of the marking agent. Previous studies [68,99] have mentioned that elongation at break exhibits a high statistical variation, which can be confirmed in the present study. The large spread of the results is probably due to a heterogeneous degradation in the material. Studies have shown that the oxidation of polyolefins (like polypropylene) is a heterogeneous process [5,100–105]. Due to the high standard deviations of the mean values, no discernible influence of either the recycling or the marking agent could be identified. Previous studies have recorded a decrease in elongation at break [5,18,29,83,84,89,106–113] with increasing steps of recycling. Decreasing elongation at break is due to the higher crystallinity of the recycled material [107] and to the reduction in the molecular weight with further recycling steps [29]. As a result of the reduced molecular weight, the density of bonding molecules incorporated into at least two crystalline lamellae, and also the number of those bonding the spherulites, decreases. In addition, the probability of chain entanglement in the amorphous phase decreases. As a result, the structure is less connected [106,109]. The concentration of binding molecules is also influenced by the crystallization temperature, as the density of the binding molecules decreases with increasing recycling due to shorter molecules and higher crystallization temperatures [29,106,110]. In addition, shorter molecules are less entangled and have fewer C-C bonds to stretch than long molecules of fresh material [29]. This assumption can be confirmed in this study (Figure 2).

Although no clear increase in Young's modulus or tensile strength could be observed, as indicated by Aurrekoetxea et al. [29], the increasing melting and crystallization temperature indicate an increase in crystallinity. In contrast, the elongation at break decreases with increasing recycling steps, which could be partially confirmed in this work. The reason why the expected increases in Young's modulus and tensile strength were not recorded in this study may be due to an insufficient number of recycling stages. The number of recycling steps was based on Aurrekoetxea et al. [29], who indicated an asymptotic behavior of the investigated mechanical properties after 10 recycling steps. Furthermore, even when the marking agent is recycled up to ten times, there are no marker-related degradation effects that result in an influence on the material and component properties. This is confirmed by the fact that the material and component properties at all recycling steps indicate no discernible influence of the marking agent. Even though the material properties of the investigated material and master batch differed slightly, and the component properties changed a little with the increase in the concentration of the modified polymer used, the variation of the investigated properties was within the standard deviation of the mean values.

Regarding Charpy impact strength (Figure 5), there are similar tendencies with increasing recycling steps, depending on the concentration of the marking agent. In contrast with earlier studies [83,88], Charpy impact strength increases with increasing recycling steps, increasing crystallinity (Figure 2) and increasing the melt flow rate (Figure 3). The result is, nevertheless, taken as a given since the complete breakage of the specimen was recorded in all Charpy impact tests [114]. One possible reason for the deviation from the previous study [83] is that a double-edge-notched specimen shape [65] was used. A possible explanation of why the marking agent had no discernible influence on Charpy impact strength is that polypropylene with a melt flow rate greater than 10 g/10 min (Table 6 and Figure 3) shows no change in notched impact strength with the addition of additives [83,115].

With regard to the Vicat softening temperature (VST) (Figure 6), there are similar tendencies with increasing recycling steps, depending on the concentration of the marking agent. Grellmann and Seidler [68] revealed that the softening behavior of polypropylene composites is mainly determined by the matrix. Compared to other studies [69,116], the VST is lower. This finding could be due to the elastomer-modified polypropylene compound used since the literature values refer to pure substances. Although the results of this study indicate a change in molecular weight due to recycling, this thermally induced damage to the material could not also be reflected in the VST [68,69]. Even though Arndt and Lechner [69] indicated the influence of additives on VST, this influence could not be confirmed in this study. This could be attributed to the reason that the chosen concentrations of the marking agent used were too low to influence the studied material and component properties using the selected testing methods [39].

5. Conclusions and Outlook

The identification and traceability of materials and components is a major challenge in the injection molding process. This challenge can be overcome through the use of a material-inherent marking technology. This allows the aging state of polymers and the composition of individual material mixtures to be monitored, as a result of which scatter-reducing process parameters can be set. Based on a systematic selection of modified polymers as a proper marking technology for application in polymer processing manufacturing, as well as the experimental validation of this marking agent in selective laser sintering in previous studies [38,39], this study focused on the investigation of the suitability of this marking agent in injection molding. In particular, the impact of the modified polymer used on the thermal and rheological material properties, as well as the mechanical and thermal component properties, during multiple processing steps of the marked material was investigated. In addition, the influence of multiple processing steps of the marked material on the presence and functionality of the used modified polymer was analyzed via mass spectroscopy. The key findings can be summarized as follows:

- Considering the applied investigations and concentrations of the modified polymer used in the injection molding material used, the marking technology does not have any discernible influence on the investigated properties of the material and components;
- During injection molding, the marked material shows analogous processing behavior to the unmarked material;
- The modified polymer used can be reliably detected in the material down to a concentration of 1 ppm, even after ten recycling steps;
- The concentration of the modified polymer and the measured intensity of the marking agent in the injection molding material used do not correlate;
- The dispersion of the modified polymer used in the injection molding material is sufficient;
- The modified polymer is suitable as a marking agent in injection molding;
- The modified polymer allows encoding of the material used at the molecular level.

The applicability of modified polymers for use in injection molding can be confirmed. This demonstrates the thermal stability of modified polymers [54,55] for application in the injection molding of polypropylene. Within the considered concentration ranges of the modified polymer, the marked injection molding material is substituted for the previously used injection molding material. While the modified polymer used has no identifiable impact on the investigated material and component properties, the modified polymer could conceivably be present in the material as a heterogeneous extraneous germ, influencing the crystallization and morphology of the polymer [117,118]. Moreover, it is possible that the modified polymer might become enriched in the amorphous phase of the polymer. This is because segregation occurs during crystallization, resulting in a segregation reaction [119]. An analysis of the injection molding material and master batch morphology could provide information about the possible effects of the modified polymer [13,15]. Furthermore, the use of a high-purity polymer could be useful to identify possible effects [39].

Based on this study's findings, an injection-molding-specific coding strategy can be developed. The tracer-based process optimization [38] must be elaborated upon for the present application. This involves analyzing the materials at defined stages in the injection molding process and adjusting the process parameters in a targeted manner. For the implementation of the marking technology in injection molding, the extraction method has to be optimized. The avoidance of material-specific additives such as carbon black particles during MS/MS analysis could enable the quantification of modified polymers [39]. In addition, another method of extracting and transferring the modified polymers to the MS/MS device is conceivable. The application of the desorption electrospray-ionization MS/MS analysis method enables the possibility of measuring the marking agent in situ [56,120–122]. As a result, the traceability effort is minimized. In addition, the analytical equipment can be included in the injection molding process. Moreover, there are other potentially suitable marking technologies listed in an earlier study [38]. Their suitability for injection molding has to be evaluated as well.

Author Contributions: Conceptualization, T.E., S.M.B. and F.v.L.; methodology, T.E.; validation, T.E., S.M.B. and F.v.L.; formal analysis, T.E. and S.M.B.; investigation, T.E. and S.M.B.; resources, T.E., F.v.L. and K.D.; data curation, T.E.; writing—original draft preparation, T.E.; writing—review and editing, T.E., S.M.B., F.v.L., W.B. and K.D.; visualization, T.E.; supervision, F.v.L. and W.B.; project administration, K.D.; funding acquisition, T.E. All authors have read and agreed to the published version of the manuscript.

Funding: We acknowledge support from the Open Access Publication Funds of Technische Universität Braunschweig.

Institutional Review Board Statement: Not applicable.

Informed Consent Statement: Not applicable.

Data Availability Statement: The raw/processed data required to reproduce these findings cannot be shared at this time, as the data also form part of an ongoing study.

Acknowledgments: The authors would like to thank the companies Borealis AG and Polysecure GmbH for providing information and technical support.

Conflicts of Interest: The authors declare no conflict of interest. The funders played no role in the design of the study, in the collection, analyses or interpretation of data, in the writing of the manuscript or in the decision to publish the results. The results, opinions and conclusions expressed in this publication are not necessarily those of Volkswagen Aktiengesellschaft.

References

1. Isayev, A.I.; Kamal, M.R.; Liu, S.-J. (Eds.) *Injection Molding: Technology and Fundamentals*; Hanser: Munich, Germany; Cincinnati, OH, USA, 2009; ISBN 978-3-446-41685-7.
2. Yang, Y.; Gao, F.; Lu, N.; Chen, X. *Injection Molding Process Control, Monitoring, and Optimization*; Hanser Publishers: Munich, Germany; Cincinnati, OH, USA, 2016; ISBN 978-1-56990-593-7.
3. Rosato, D.V. (Ed.) *Injection Molding Handbook*, 3rd ed.; Kluwer Academic: Boston, MA, USA, 2000; ISBN 978-0-7923-8619-3.

4. Osswald, T.A.; Turng, L.-S.; Gramann, P.J.; Beaumont, J. *Injection Molding Handbook*, 2nd ed.; Hanser Gardner: Munich, Germany; Cincinnati, OH, USA, 2008; ISBN 978-3-446-40781-7.
5. Jansson, A.; Möller, K.; Gevert, T. Degradation of post-consumer polypropylene materials exposed to simulated recycling—Mechanical properties. *Polym. Degrad. Stab.* **2003**, *82*, 37–46. [CrossRef]
6. Schyns, Z.O.G.; Shaver, M.P. Mechanical Recycling of Packaging Plastics: A Review. *Macromol. Rapid Commun.* **2021**, *42*, e2000415. [CrossRef] [PubMed]
7. Lehner, R.; Weder, C.; Petri-Fink, A.; Rothen-Rutishauser, B. Emergence of Nanoplastic in the Environment and Possible Impact on Human Health. *Environ. Sci. Technol.* **2019**, *53*, 1748–1765. [CrossRef] [PubMed]
8. Ölund, G.; Eriksson, E. *Resthandteringsalternativ för Plastförpackningar–en Miljöpåverknadsbedömmning*; CIT Ekologik Forsuring Tabell: Göteborg, Sweden, 1998.
9. Yang, Y.; Boom, R.; Irion, B.; van Heerden, D.-J.; Kuiper, P.; Wit, H.d. Recycling of composite materials. *Chem. Eng. Process. Process Intensif.* **2012**, *51*, 53–68. [CrossRef]
10. Gonçalves, R.M.; Martinho, A.; Oliveira, J.P. Recycling of Reinforced Glass Fibers Waste: Current Status. *Materials* **2022**, *15*, 1596. [CrossRef]
11. Al-Salem, S.M.; Lettieri, P.; Baeyens, J. Recycling and recovery routes of plastic solid waste (PSW): A review. *Waste Manag.* **2009**, *29*, 2625–2643. [CrossRef]
12. Ragaert, K.; Delva, L.; van Geem, K. Mechanical and chemical recycling of solid plastic waste. *Waste Manag.* **2017**, *69*, 24–58. [CrossRef]
13. Ehrenstein, G. *Polymer-Werkstoffe: Struktur; Eigenschaften; Anwendung*; Carl Hanser Fachbuchverlag: Munich, Germany, 2011; ISBN 978-3-446-42283-4.
14. Dahlmann, R.; Haberstroh, E.; Menges, G. *Menges Werkstoffkunde Kunststoffe, Vollständig neu Bearbeitete Auflage*, 7th ed.; Hanser: Munich, Germany, 2022; ISBN 978-3-446-45801-7.
15. Bonnet, M. *Kunststoffe in der Ingenieuranwendung: Verstehen und Zuverlässig Auswählen*; Vieweg + Teubner Verlag/GWV Fachverlage GmbH Wiesbaden: Wiesbaden, Germany, 2009; ISBN 978-3-8348-0349-8.
16. Dostál, J.; Kašpárková, V.; Zatloukal, M.; Muras, J.; Šimek, L. Influence of the repeated extrusion on the degradation of polyethylene. Structural changes in low density polyethylene. *Eur. Polym. J.* **2008**, *44*, 2652–2658. [CrossRef]
17. da Costa, H.M.; Ramos, V.D.; Rocha, M.C. Rheological properties of polypropylene during multiple extrusion. *Polym. Test.* **2005**, *24*, 86–93. [CrossRef]
18. González-González, V.A.; Neira-Velázquez, G.; Angulo-Sánchez, J.L. Polypropylene chain scissions and molecular weight changes in multiple extrusion. *Polym. Degrad. Stab.* **1998**, *60*, 33–42. [CrossRef]
19. Martins, M.H.; de Paoli, M.-A. Polypropylene compounding with post-consumer material. *Polym. Degrad. Stab.* **2002**, *78*, 491–495. [CrossRef]
20. Tamrakar, S.; Couvreur, R.; Mielewski, D.; Gillespie, J.W.; Kiziltas, A. Effects of recycling and hygrothermal environment on mechanical properties of thermoplastic composites. *Polym. Degrad. Stab.* **2023**, *207*, 110233. [CrossRef]
21. Kaiser, W. *Kunststoffchemie für Ingenieure: Von der Synthese bis zur Anwendung*, 5th ed.; Hanser: München, Germany, 2021; ISBN 978-3-446-45191-9.
22. Abad, M.J.; Ares, A.; Barral, L.; Cano, J.; Díez, F.J.; García-Garabal, S.; López, J.; Ramírez, C. Effects of a mixture of stabilizers on the structure and mechanical properties of polyethylene during reprocessing. *J. Appl. Polym. Sci.* **2004**, *92*, 3910–3916. [CrossRef]
23. Hamad, K.; Kaseem, M.; Deri, F. Effect of recycling on rheological and mechanical properties of poly(lactic acid)/polystyrene polymer blend. *J. Mater. Sci.* **2011**, *46*, 3013–3019. [CrossRef]
24. Marrone, M.; La Mantia, F.P. Re-stabilisation of recycled polypropylenes. *Polym. Recycl.* **1996**, *2*, 17–26.
25. Catalina, F. Degradación y estabilización de polipropileno. *Rev. Plásticos Mod.* **1998**, *502*, 391–397.
26. Tiganis, B.E.; Shanks, R.A.; Long, Y. Effects of processing on the microstructure, melting behavior, and equilibrium melting temperature of polypropylene. *J. Appl. Polym. Sci.* **1996**, *59*, 663–671. [CrossRef]
27. Ying, Q.; Zhao, Y.; Liu, Y. A study of thermal oxidative and thermal mechanical degradation of polypropylene. *Makromol. Chem.* **1991**, *192*, 1041–1058. [CrossRef]
28. Hinsken, H.; Moss, S.; Pauquet, J.-R.; Zweifel, H. Degradation of polyolefins during melt processing. *Polym. Degrad. Stab.* **1991**, *34*, 279–293. [CrossRef]
29. Aurrekoetxea, J.; Sarrionandia, M.A.; Urrutibeascoa, I.; Maspoch, M.L. Effects of recycling on the microstructure and the mechanical properties of isotactic polypropylene. *J. Mater. Sci.* **2001**, *36*, 2607–2613. [CrossRef]
30. Chanda, M.; Roy, S.K. *Plastics Fabrication and Recycling*; CRC Press: Boca Raton, FL, USA, 2016; ISBN 9781420080636.
31. Marrone, M.; La Mantia, F.P. Monopolymers blends of virgin and recycled polypropylene. *Polym. Recycl.* **1996**, *2*, 9–15.
32. Steigemann, U. Werkstoff- und ressourcenschonende Recyclingstrategien. *Lightweight Des.* **2012**, *5*, 26–31. [CrossRef]
33. Wenguang, M.; La Mantia, F.P. Processing and mechanical properties of recycled PVC and of homopolymer blends with virgin PVC. *J. Appl. Polym. Sci.* **1996**, *59*, 759–767. [CrossRef]
34. Wenguang, M.A.; La Mantia, F.P. Recycling of Post-consumer Polyethylene Greenhouse Films: Monopolymer Blends of recycled and virgin Polyethylene. *Polym. Netw. Blends* **1995**, *5*, 173–180.
35. Valenza, A.; La Mantia, F.P. Recycling of polymer waste: Part II—Stress degraded polypropylene. *Polym. Degrad. Stab.* **1988**, *20*, 63–73. [CrossRef]

36. Pfaendner, R.; Herbst, H.; Hoffmann, K.; Sitek, F. Recycling and restabilization of polymers for high quality applications. An Overview. *Angew. Makromol. Chem.* **1995**, *232*, 193–227. [CrossRef]
37. Auer, M.; Schmidt, J.; Diemert, J.; Gerhardt, G.; Renz, M.; Galler, V.; Woidasky, J. Quality Aspects in the Compounding of Plastic Recyclate. *Recycling* **2023**, *8*, 18. [CrossRef]
38. Eggers, T.; von Lacroix, F.; van de Kraan, F.; Reichler, A.-K.; Hürkamp, A.; Dröder, K. Investigations for Material Tracing in Selective Laser Sintering: Part I: Methodical Selection of a Suitable Marking Agent. *Materials* **2023**, *16*, 1043. [CrossRef]
39. Eggers, T.; von Lacroix, F.; Goede, M.F.; Persch, C.; Berlin, W.; Dröder, K. Investigations for Material Tracing in Selective Laser Sintering: Part II: Validation of Modified Polymers as Marking Agents. *Materials* **2023**, *16*, 2631. [CrossRef]
40. Lutz, J.-F. Coding macromolecules: Inputting information in polymers using monomer-based alphabets. *Macromolecules* **2015**, *48*, 4759–4767. [CrossRef]
41. Lutz, J.-F. Les Polymères, Messagers à l'Échelle Moléculaire. *IT Ind. Technol.* **2021**, 1–11.
42. Fearon, P.K.; Marshall, N.; Billingham, N.C.; Bigger, S.W. Evaluation of the oxidative stability of multiextruded polypropylene as assessed by physicomechanical testing and simultaneous differential scanning calorimetry-chemiluminescence. *J. Appl. Polym. Sci.* **2001**, *79*, 733–741. [CrossRef]
43. Incarnato, L.; Scarfato, P.; Acierno, D. Rheological and mechanical properties of recycled polypropylene. *Polym. Eng. Sci.* **1999**, *39*, 749–755. [CrossRef]
44. Incarnato, L.; Scarfato, P.; Gorrasi, G.; Vittoria, V.; Acierno, D. Structural modifications induced by recycling of polypropylene. *Polym. Eng. Sci.* **1999**, *39*, 1661–1666. [CrossRef]
45. Kartalis, C.N.; Papaspyrides, C.D.; Pfaendner, R.; Hoffmann, K.; Herbst, H. Mechanical recycling of postused high-density polyethylene crates using the restabilization technique. I. Influence of reprocessing. *J. Appl. Polym. Sci.* **1999**, *73*, 1775–1785. [CrossRef]
46. Dintcheva, N.; Jilov, N.; La Mantia, F.P. Recycling of plastics from packaging. *Polym. Degrad. Stab.* **1997**, *57*, 191–203. [CrossRef]
47. Hopmann, C.; Michaeli, W. *Einführung in die Kunststoffverarbeitung*; Hanser: München, Germany, 2017; ISBN 978-3-446-45355-5.
48. Moalli, J. (Ed.) *Plastics Failure: Analysis and Prevention*; Plastics Design Library: Norwich, NY, USA, 2010; ISBN 978-1-884-20792-1.
49. Eyerer, P.; Elsner, P.; Hirth, T. *Polymer Engineering: Technologien und Praxis*; Springer: Berlin/Heidelberg, Germany, 2008; ISBN 978-3-540-72402-5.
50. Baur, E.; Brinkmann, S.; Osswald, T.A.; Schmachtenberg, E. *Saechtling Kunststoff Taschenbuch*; Hanser: München, Germany, 2013; ISBN 978-3-446-43442-4.
51. Domininghaus, H.; Elsner, P.; Eyerer, P.; Hirth, T. (Eds.) *Kunststoffe: Eigenschaften und Anwendungen*; Springer: Berlin/Heidelberg, Germany, 2012; ISBN 9783642161728.
52. Pasquini, N.; Addeo, A. (Eds.) *Polypropylene Handbook*, 2nd ed.; Hanser: Munich, Germany, 2005; ISBN 3-446-22978-7.
53. Gruden, D. *Umweltschutz in der Automobilindustrie: Motor, Kraftstoffe, Recycling*; Vieweg + Teubner/GWV Fachverlage GmbH Wiesbaden: Wiesbaden, Germany, 2008; ISBN 978-3-8348-0404-4.
54. Dost, G.; Kummer, B.; Matloubi, M.; Moesslein, J.; Treick, A. Produkt- und Materialpässe Nützen der Kreislaufwirtschaft Nur, Wenn sie Tatsächlich Robust mit Produkten und Materialien Verknüpft Sind! Kurzfassung, Freiburg, Germany. 2022. Available online: https://polysecure.eu/fileadmin/main/Unternehmen/Media-Files/220425_Produkt_Materialpass_Kurzfassung_Dt.pdf (accessed on 20 August 2022).
55. Dost, G.; Matloubi, M.; Treick, A.; Kummer, B. Booster für eine gelingende Kreislaufwirtschaft. *Recycl. Mag. Sonderh* **2022**, 84–86.
56. Youssef, I.; Carvin-Sergent, I.; Konishcheva, E.; Kebe, S.; Greff, V.; Karamessini, D.; Matloubi, M.; Ouahabi, A.A.; Moesslein, J.; Amalian, J.-A.; et al. Covalent Attachment and Detachment by Reactive DESI of Sequence-Coded Polymer Taggants. *Macromol. Rapid Commun.* **2022**, *43*, 2200412. [CrossRef]
57. DIN EN ISO 19069-2:2020-01; Kunststoffe—Polypropylen (PP)-Formmassen—Teil 2: Herstellung von Probekörpern und Bestimmung von Eigenschaften (ISO 19069-2:2016). Beuth Verlag GmbH: Berlin, Germany, 2020; Deutsche Fassung EN ISO 19069-2:2016.
58. DIN EN ISO 527-2:2012-06; Kunststoffe—Bestimmung der Zugeigenschaften—Teil 2: Prüfbedingungen für Form und Extrusionsmassen (ISO 527-2:2012). Beuth Verlag GmbH: Berlin, Germany, 2012; Deutsche Fassung EN ISO 527-2:2012.
59. DIN EN ISO 11358-1:2022-07; Kunststoffe—Thermogravimetrie (TG) von Polymeren—Teil 1: Allgemeine Grundsätze (ISO 11358-1:2022). Beuth Verlag GmbH: Berlin, Germany, 2022; Deutsche Fassung EN ISO 11358-1:2022.
60. Frick, A.; Stern, C. *Einführung in die Kunststoffprüfung: Prüfmethoden und Anwendungen*; Hanser: München, Germany, 2017; ISBN 978-3-446-44351-8.
61. DIN EN ISO 11357-1:2023-06; Kunststoffe—Dynamische Differenzkalorimetrie (DSC)—Teil 1: Allgemeine Grundlagen (ISO 11357-1:2023). Beuth Verlag GmbH: Berlin, Germany, 2023; Deutsche Fassung EN ISO 11357-1:2023.
62. DIN EN ISO 1133-1:2012-03; Kunststoffe—Bestimmung der Schmelze-Massefließrate (MFR) und der Schmelze-Volumenfließrate (MVR) von Thermoplasten—Teil 1: Allgemeines Prüfverfahren (ISO 1133-1:2011). Beuth Verlag GmbH: Berlin, Germany, 2012; Deutsche Fassung EN ISO 1133-1:2011.
63. Mielicki, C. Prozessnahes Qualitätsmanagement beim Lasersintern von Polyamid 12. Ph.D. Thesis, Universität Duisburg-Essen, Duisburg/Essen, Germany, 2014.
64. DIN EN ISO 527-1:2019-12; Kunststoffe—Bestimmung der Zugeigenschaften—Teil 1: Allgemeine Grundsätze (ISO 527-1:2019). Beuth Verlag GmbH: Berlin, Germany, 2019; Deutsche Fassung EN ISO 527-1:2019.

65. DIN EN ISO 179-1:2010-11; Kunststoffe—Bestimmung der Charpy-Schlageigenschaften—Teil 1: Nicht Instrumentierte Schlagzähigkeitsprüfung (ISO 179-1:2010). Beuth Verlag GmbH: Berlin, Germany, 2010; Deutsche Fassung EN ISO 179-1:2010.
66. DIN EN ISO 3167:2014-11; Kunststoffe—Vielzweckprobekörper (ISO 3167:2014). Beuth Verlag GmbH: Berlin, Germany, 2014; Deutsche Fassung EN ISO 3167:2014.
67. DIN EN ISO 306:2023-03; Kunststoffe—Thermoplaste—Bestimmung der Vicat-Erweichungstemperatur (VST) (ISO 306:2022). Beuth Verlag GmbH: Berlin, Germany, 2023; Deutsche Fassung EN ISO 306:2022.
68. Grellmann, W.; Seidler, S. (Eds.) *Kunststoffprüfung*; Hanser-Verlag: München, Germant, 2011; ISBN 978-3-446-42722-8.
69. Arndt, K.-F.; Lechner, M.D. *Polymer Solids and Polymer Melts–Mechanical and Thermomechanical Properties of Polymers*; Springer Berlin Heidelberg: Berlin/Heidelberg, Germany, 2014; ISBN 978-3-642-55165-9.
70. Al Ouahabi, A.; Amalian, J.-A.; Charles, L.; Lutz, J.-F. Mass spectrometry sequencing of long digital polymers facilitated by programmed inter-byte fragmentation. *Nat. Commun.* 2017, *8*, 967. [CrossRef]
71. Gunay, U.S.; Petit, B.E.; Karamessini, D.; Al Ouahabi, A.; Amalian, J.-A.; Chendo, C.; Bouquey, M.; Gigmes, D.; Charles, L.; Lutz, J.-F. Chemoselective synthesis of uniform sequence-coded polyurethanes and their use as molecular tags. *Chem* 2016, *1*, 114–126. [CrossRef]
72. Lutz, J.-F.; Ouchi, M.; Liu, D.R.; Sawamoto, M. Sequence-controlled polymers. *Science* 2013, *341*, 1238149. [CrossRef] [PubMed]
73. Guerrica-Echevarría, G.; Eguiazábal, J.I.; Nazábal, J. Effects of reprocessing conditions on the properties of unfilled and talc-filled polypropylene. *Polym. Degrad. Stab.* 1996, *53*, 1–8. [CrossRef]
74. Setnescu, R.; Barcutean, C.; Jipa, S.; Setnescu, T.; Negoiu, M.; Mihalcea, I.; Dumitru, M.; Zaharescu, T. The effect of some thiosemicarbazide compounds on thermal oxidation of polypropylene. *Polym. Degrad. Stab.* 2004, *85*, 997–1001. [CrossRef]
75. Jipa, S.; Setnescu, R.; Setnescu, T.; Zaharescu, T. Efficiency assessment of additives in thermal degradation of i-PP by chemiluminescence I. Triazines. *Polym. Degrad. Stab.* 2000, *68*, 159–164. [CrossRef]
76. Gregorová, A.; Cibulková, Z.; Košíková, B.; Šimon, P. Stabilization effect of lignin in polypropylene and recycled polypropylene. *Polym. Degrad. Stab.* 2005, *89*, 553–558. [CrossRef]
77. Tocháček, J. Effect of secondary structure on physical behaviour and performance of hindered phenolic antioxidants in polypropylene. *Polym. Degrad. Stab.* 2004, *86*, 385–389. [CrossRef]
78. Gugumus, F. Aspects of the impact of stabilizer mass on performance in polymers 2. Effect of increasing molecular mass of polymeric HALS in PP. *Polym. Degrad. Stab.* 2000, *67*, 299–311. [CrossRef]
79. Chmela, Š.; Hrdlovič, P. The influence of substituents on the photo-stabilizing efficiency of hindered amine stabilizers in polypropylene. *Polym. Degrad. Stab.* 1990, *27*, 159–167. [CrossRef]
80. Hamid, S.H. *Handbook of Polymer Degradation*; CRC Press: Boca Raton, FL, USA, 2000; ISBN 978-0-429-08258-0.
81. Gijsman, P.; Gitton, M. Hindered amine stabilisers as long-term heat stabilisers for polypropylene. *Polym. Degrad. Stab.* 1999, *66*, 365–371. [CrossRef]
82. Buttitta, A. "Tracer-Based-Sorting"—Die Zukunft der Kunststoffverwertung. *EU-Recycl. Fachmag. Eur. Recycl.* 2021, *38*, 16–18.
83. Kausch, H.-H. (Ed.) *Intrinsic Molecular Mobility and Toughness of Polymers II*; Springer-Verlag GmbH: Berlin/Heidelberg, Germany, 2005; ISBN 978-3-540-26162-9.
84. Canevarolo, S.V. Chain scission distribution function for polypropylene degradation during multiple extrusions. *Polym. Degrad. Stab.* 2000, *70*, 71–76. [CrossRef]
85. Wortberg, J. *Qualitätssicherung in der Kunststoffverarbeitung: Rohstoff-, Prozess- und Produktqualität; Tabellen*; Hanser: München, Germany; Wien, Austria, 1996; ISBN 9783446171336.
86. Pahl, M.; Gleißle, W.; Laun, H.-M. (Eds.) *Praktische Rheologie der Kunststoffe und Elastomere*; VDI-Verlag: Düsseldorf, Germany, 1995; ISBN 978-3-18-234192-5.
87. World Encyclopedia. Modern Plastics Worldwide. 2007, pp. 160–217. Available online: https://www.plasticstoday.com/resin-pricing/ (accessed on 17 June 2023).
88. Meneghetti, G.; Ricotta, M.; Sanità, M.; Refosco, D. Fully Reversed Axial Notch Fatigue Behaviour of Virgin and Recycled Polypropylene Compounds. *Procedia Struct. Integr.* 2016, *2*, 2255–2262. [CrossRef]
89. Elloumi, A.; Pimbert, S.; Bourmaud, A.; Bradai, C. Thermomechanical properties of virgin and recycled polypropylene impact copolymer/CaCO$_3$ nanocomposites. *Polym. Eng. Sci.* 2010, *50*, 1904–1913. [CrossRef]
90. Oblak, P.; Gonzalez-Gutierrez, J.; Zupančič, B.; Aulova, A.; Emri, I. Processability and mechanical properties of extensively recycled high density polyethylene. *Polym. Degrad. Stab.* 2015, *114*, 133–145. [CrossRef]
91. Brostow, W.; Corneliussen, R.D. (Eds.) *Failure of Plastics*; Hanser: Munich, Germany, 1986; ISBN 9780029475102.
92. Coulier, L.; Orbons, H.G.; Rijk, R. Analytical protocol to study the food safety of (multiple-)recycled high-density polyethylene (HDPE) and polypropylene (PP) crates: Influence of recycling on the migration and formation of degradation products. *Polym. Degrad. Stab.* 2007, *92*, 2016–2025. [CrossRef]
93. Scaffaro, R.; La Mantia, F.P.; Botta, L.; Morreale, M.; Tz. Dintcheva, N.; Mariani, P. Competition between chain scission and branching formation in the processing of high-density polyethylene: Effect of processing parameters and of stabilizers. *Polym. Eng. Sci.* 2009, *49*, 1316–1325. [CrossRef]
94. Mendes, A.A.; Cunha, A.M.; Bernardo, C.A. Study of the degradation mechanisms of polyethylene during reprocessing. *Polym. Degrad. Stab.* 2011, *96*, 1125–1133. [CrossRef]
95. Sun, L.; Zhao, X.Y.; Sun, Z.Y. Study on the Properties of Multi-Extruded Recycled PE and PP. *AMR* 2014, *1003*, 96–99. [CrossRef]

96. Yin, S.; Tuladhar, R.; Shi, F.; Shanks, R.A.; Combe, M.; Collister, T. Mechanical reprocessing of polyolefin waste: A review. *Polym. Eng. Sci.* **2015**, *55*, 2899–2909. [CrossRef]
97. Baltes, L.; Costiuc, L.; Patachia, S.; Tierean, M. Differential scanning calorimetry—A powerful tool for the determination of morphological features of the recycled polypropylene. *J. Therm. Anal. Calorim.* **2019**, *138*, 2399–2408. [CrossRef]
98. Gupta, V.B.; Mittal, R.K.; Sharma, P.K.; Mennig, G.; Wolters, J. Some studies on glass fiber-reinforced polypropylene. Part II: Mechanical properties and their dependence on fiber length, interfacial adhesion, and fiber dispersion. *Polym. Compos.* **1989**, *10*, 16–27. [CrossRef]
99. Battisti, M. Spritzgießcompoundieren von Polymer-Nanocomposites auf Basis von Schichtsilikaten. Ph.D. Thesis, University of Leoben, Leoben, Austria, 2015.
100. Celina, M.; George, G.A. A heterogeneous model for the thermal oxidation of solid polypropylene from chemiluminescence analysis. *Polym. Degrad. Stab.* **1993**, *40*, 323–335. [CrossRef]
101. Celina, M.; George, G.A. Heterogeneous and homogeneous kinetic analyses of the thermal oxidation of polypropylene. *Polym. Degrad. Stab.* **1995**, *50*, 89–99. [CrossRef]
102. Gugumus, F. Thermooxidative degradation of polyolefins in the solid state—6. Kinetics of thermal oxidation of polypropylene. *Polym. Degrad. Stab.* **1998**, *62*, 235–243. [CrossRef]
103. Gugumus, F. Thermooxidative degradation of polyolefins in the solid state—7. Effect of sample thickness and heterogeneous oxidation kinetics for polypropylene. *Polym. Degrad. Stab.* **1998**, *62*, 245–257. [CrossRef]
104. Knight, J.B.; Calvert, P.D.; Billingham, N.C. Localization of oxidation in polypropylene. *Polymer* **1985**, *26*, 1713–1718. [CrossRef]
105. Richters, P. Initiation Process in the Oxidation of Polypropylene. *Macromolecules* **1970**, *3*, 262–264. [CrossRef]
106. Ibhadon, A.O. Fracture mechanics of polypropylene: Effect of molecular characteristics, crystallization conditions, and annealing on morphology and impact performance. *J. Appl. Polym. Sci.* **1998**, *69*, 2657–2661. [CrossRef]
107. Greco, R.; Coppola, F. Influence of crystallization conditions on the mechanical properties of isotactic polypropylene. *Plast. Rubber Process. Appl.* **1986**, *6*, 35–41.
108. Seiffert, S.; Kummerlöwe, C.; Vennemann, N. (Eds.) *Makromolekulare Chemie: Ein Lehrbuch für Chemiker, Physiker, Materialwissenschaftler und Verfahrenstechniker*; Springer Spektrum: Berlin, Germany, 2020; ISBN 978-3-662-61108-1.
109. Karger-Kocsis, J. *Polypropylene Structure, Blends and Composites: Volume 3 Composites*; Springer: Dordrecht, The Netherlands, 1995; ISBN 978-94-010-4233-8.
110. Greco, R.; Ragosta, G. Isotactic polypropylenes of different molecular characteristics: Influence of crystallization conditions and annealing on the fracture behaviour. *J. Mater. Sci.* **1988**, *23*, 4171–4180. [CrossRef]
111. Sugimoto, M.; Ishikawa, M.; Hatada, K. Toughness of polypropylene. *Polymer* **1995**, *36*, 3675–3682. [CrossRef]
112. Tjong, S.C.; Shen, J.S.; Li, R. Morphological behaviour and instrumented dart impact properties of β-crystalline-phase polypropylene. *Polymer* **1996**, *37*, 2309–2316. [CrossRef]
113. Jancar, J.; DiAnselmo, A.; DiBenedetto, A.T.; Kucera, J. Failure mechanics in elastomer toughened polypropylene. *Polymer* **1993**, *34*, 1684–1694. [CrossRef]
114. DIN EN ISO 179-2:2020-09; Kunststoffe—Bestimmung der Charpy-Schlageigenschaften—Teil 2: Instrumentierte Schlagzähigkeitsprüfung (ISO 179-2:2020). Beuth Verlag GmbH: Berlin, Germany, 2020; Deutsche Fassung EN ISO 179-2:2020.
115. Moerl, M. Steigerung der Zähigkeit von Isotaktischem Polypropylen durch Kontrolle der Morphologie Mittels 1,3,5-Benzoltrisamiden. Ph.D. Thesis, Universität Bayreuth, Bayreuth, Germany, 2017.
116. Carlowitz, B. *Tabellarische Übersicht über die Prüfung von Kunststoffen*; Giesel-Verl. für Publizität: Isernhagen, Germany, 1992; ISBN 978-3980294201.
117. Bastian, M.; Hochrein, T. *Einfärben von Kunststoffen: Produktanforderungen—Verfahrenstechnik—Prüfmethodik*; Hanser: Munich, Germany, 2018; ISBN 978-3-446-45398-2.
118. Johnsen, U.; Spilgies, G.; Zachmann, H.G. Abhängigkeit der heterogenen Keimbildung in Polypropylen von der Kristallisationstemperatur und von der Art der Beimengung der Fremdsubstanz. *Kolloid Z.U.Z.Polym.* **1970**, *240*, 762–765. [CrossRef]
119. Hornbogen, E.; Eggeler, G.; Werner, E. *Werkstoffe: Aufbau und Eigenschaften von Keramik-, Metall-, Polymer- und Verbundwerkstoffen*; Springer Vieweg: Berlin, Germany, 2017; ISBN 978-3-642-53866-7.
120. Amalian, J.-A.; Mondal, T.; Konishcheva, E.; Cavallo, G.; Petit, B.E.; Lutz, J.-F.; Charles, L. Desorption electrospray ionization (DESI) of digital polymers: Direct tandem mass spectrometry decoding and imaging from materials surfaces. *Adv. Mater. Technol.* **2021**, *6*, 2001088. [CrossRef]
121. Takats, Z.; Wiseman, J.M.; Gologan, B.; Cooks, R.G. Mass spectrometry sampling under ambient conditions with desorption electrospray ionization. *Science* **2004**, *306*, 471–473. [CrossRef]
122. Cooks, R.G.; Ouyang, Z.; Takats, Z.; Wiseman, J.M. Ambient mass spectrometry. *Science* **2006**, *311*, 1566–1570. [CrossRef]

Disclaimer/Publisher's Note: The statements, opinions and data contained in all publications are solely those of the individual author(s) and contributor(s) and not of MDPI and/or the editor(s). MDPI and/or the editor(s) disclaim responsibility for any injury to people or property resulting from any ideas, methods, instructions or products referred to in the content.

Article

Blown Composite Films of Low-Density/Linear-Low-Density Polyethylene and Silica Aerogel for Transparent Heat Retention Films and Influence of Silica Aerogel on Biaxial Properties

Seong Baek Yang [1], Jungeon Lee [1], Sabina Yeasmin [1], Jae Min Park [1], Myung Dong Han [2], Dong-Jun Kwon [3,*] and Jeong Hyun Yeum [1,*]

1. Department of Biofibers and Biomaterials Science, Kyungpook National University, Daegu 41566, Korea; ysb@knu.ac.kr (S.B.Y.); dlwjddjs2@gmail.com (J.L.); yeasminsabina44@yahoo.com (S.Y.); woawoa9025@naver.com (J.M.P.)
2. Hans Intech Co., Ltd., Daegu 41243, Korea; mdhan@hanschemtek.com
3. Department of Materials Engineering and Convergence Technology, Research Institute for Green Energy Convergence Technology, Gyeongsang National University, Jinju 52828, Korea
* Correspondence: rorrir@empas.com (D.-J.K.); jhyeum@knu.ac.kr (J.H.Y.); Tel.: +82-55-772-1656 (D.J.K.); +82-53-950-5739 (J.H.Y.)

Citation: Yang, S.B.; Lee, J.; Yeasmin, S.; Park, J.M.; Han, M.D.; Kwon, D.-J.; Yeum, J.H. Blown Composite Films of Low-Density/Linear-Low-Density Polyethylene and Silica Aerogel for Transparent Heat Retention Films and Influence of Silica Aerogel on Biaxial Properties. *Materials* 2022, 15, 5314. https://doi.org/10.3390/ma15155314

Academic Editors: Eduard-Marius Lungulescu, Radu Setnescu and Cristina Stancu

Received: 16 June 2022
Accepted: 27 July 2022
Published: 2 August 2022

Publisher's Note: MDPI stays neutral with regard to jurisdictional claims in published maps and institutional affiliations.

Copyright: © 2022 by the authors. Licensee MDPI, Basel, Switzerland. This article is an open access article distributed under the terms and conditions of the Creative Commons Attribution (CC BY) license (https://creativecommons.org/licenses/by/4.0/).

Abstract: Blown films based on low-density polyethylene (LDPE)/linear low-density polyethylene (LLDPE) and silica aerogel (SA; 0, 0.5, 1, and 1.5 wt.%) were obtained at the pilot scale. Good particle dispersion and distribution were achieved without thermo oxidative degradation. The effects of different SA contents (0.5–1.5 wt.%) were studied to prepare transparent-heat-retention LDPE/LLDPE films with improved material properties, while maintaining the optical performance. The optical characteristics of the composite films were analyzed using methods such as ultraviolet–visible spectroscopy and electron microscopy. Their mechanical characteristics were examined along the machine and transverse directions (MD and TD, respectively). The MD film performance was better, and the 0.5% composition exhibited the highest stress at break. The crystallization kinetics of the LDPE/LLDPE blends and their composites containing different SA loadings were investigated using differential scanning calorimetry, which revealed that the crystallinity of LDPE/LLDPE was increased by 0.5 wt.% of well-dispersed SA acting as a nucleating agent and decreased by agglomerated SA (1–1.5 wt.%). The LDPE/LLDPE/SA (0.5–1.5 wt.%) films exhibited improved infrared retention without compromising the visible light transmission, proving the potential of this method for producing next-generation heat retention films. Moreover, these films were biaxially drawn at 13.72 MPa, and the introduction of SA resulted in lower draw ratios in both the MD and TD. Most of the results were explained in terms of changes in the biaxial crystallization caused by the process or the influence of particles on the process after a systematic experimental investigation. The issues were strongly related to the development of blown nanocomposites films as materials for the packaging industry.

Keywords: blends; composites; blown film extrusion; silica aerogel; biaxial properties; morphology; thermal properties; mechanical properties

1. Introduction

Polymer blending is defined as a process in which at least two polymers are blended to produce a new material with different physical characteristics. This is mainly performed to improve and increase certain characteristics such as the thermal barrier [1]. Blends of linear low-density polyethylene (LLDPE) and low-density polyethylene (LDPE) are of considerable importance in industrial applications. Their good processability and excellent mechanical properties make LDPE/LLDPE films suitable for packaging applications. LLDPE is added to LDPE because of its superior mechanical characteristics such as higher tensile strength, impact properties, and elongation at break. The LDPE/LLDPE films are characterized by low haze and better bubble stability. Furthermore, manufacturers can use conventional LDPE film-blowing devices to blend LLDPE with LDPE without modification [2]. Despite the benefits of LDPE/LLDPE blends in film applications, blend miscibility, and the miscibility of LDPE/LLDPE blends have certain effects on their properties, and few studies have been conducted on the miscibility of LDPE/LLDPE blends [3–6]. Most researchers have reported that LLDPE/LDPE blends are miscible at low LDPE contents and show immiscibility at higher LDPE [7]. In this study, the LDPE concentrations were kept low, and the differential scanning calorimetry (DSC) thermograms exhibited two overlapping peaks, which may be due to the phase separation of the LDPE and LLDPE components in the blended film after crystallization [8]. Although LDPE and LLDPE resins are the most versatile polymers, their applications are restricted by drawbacks such as low strength, stiffness, and poor heat resistance [9]. To solve these problems and prepare materials with improved properties, the preparation of polyethylene (PE) nanocomposites with different inorganic nanofillers has been reported [10,11].

Currently, energy cost and availability are important concerns, and heat energy thermal insulation is an efficient strategy to address these issues. Low thermal conductivity is considered to be an essential feature of materials that can be reinforced by incorporating fillers into the main matrix [12]. Nanoporous networks of aerogels filled with gas (above 90% of aerogels are composed of air) show excellent characteristics including high specific surface area (500–1200 m^2/g), low thermal conductivity (0.013–0.04 W/m K), low dielectric constant (1.1–2), low density (0.003–0.1 g/cm^3), high optical transmission in the visible range (90%), and high insulating ability. These properties make them good candidates as insulators in different applications [13]. Low thermal conductivity is one of the major characteristics of silica aerogels (SAs), making them applicable in the field of insulation. However, SAs have poor mechanical properties, which restrict their application. Therefore, SA-based composites are generally used [14,15].

Blown film extrusion is the main processing method for producing a biaxial melt-drawn film. This method requires the use of air pressure for initiating a transverse direction (TD) draw, in addition to a higher haul-off roll speed for delivering a machine direction (MD) draw. Billions of pounds of polymer are processed annually by using this technique. Blown film extrusion is used to produce agricultural and construction films, industrial films and bags, stretch films, polyvinyl chloride cling films, liners, high barriers, and small tube systems [16,17].

The purpose of this study was to improve both the thermal and mechanical properties of the LDPE/LLDPE blend films; hence, LDPE/LLDPE/SA composites were prepared using a twin-screw extruder to prepare an LDPE/LLDPE/SA (0–1.5 wt.%) masterbatch, and then an LDPE/LLDPE/SA (0–1.5 wt.%) film was prepared using the blown method [18,19]. The effects of various SA contents on the morphology, draw ratio, and mechanical and thermal characteristics of the prepared composite films were studied. Our proposed blown film extrusion of composites based on LDPE/LLDPE and SA could widen their application as thermal insulating films in the packaging field while preserving the biaxial film properties initiated by the processing method or the impact of the particles on the processing method.

2. Experimental Section

2.1. Materials

Polymer samples of LDPE and LLDPE were purchased from Equate and Seongji Industrial Co., Ltd. (Apryang-myeon, Gyeongsan-si, Korea), and the material characteristics of both polymers are listed in Table 1 (Part a). The SA powder was provided by EM-POWER Co., Ltd. (Asan-si, Chungnam, Korea), and its properties are listed in Table 1 (Part b).

Table 1. (**a**). The material properties of the low-density polyethylene (LDPE)/linear low-density polyethylene samples. (**b**). The technical data of the silica aerogel (SA).

				a.				
Name	Quantity (g) Used for Blown Film Preparation	Grade	Melt Flow Index	Density	Melting Point (°C)	Haze (%)		Gloss (GU)
LDPE	15,000	LDPE 150E	0.25 g/10 min	0.921 g/cm^3	96	-		-
LLDPE	35,000	CEFOR 1221P	2.0 g/10 min	0.918 g/cm^3	116	0.56		151
				b.				
Particle Size	Pore Diameter	BULK DENSITY	Surface Chemistry	BET(Brunauer–Emmett–Teller) Surface Area		Porosity		Heavy Metal
20–30 µm	20–30 µm	100 kg/m^2	Hydrophobic	500 m^2/g		Less than 90%		N/A

2.2. Preparation of LDPE/LLDPE/SA (0–1.5 wt.%) Extruded Blown Composite Film

Figure 1 displays the preparation method of the LDPE/LLDPE/SA masterbatch and film. Before the preparation of the LDPE/LLDPE masterbatch (30%/70%) with various SA contents (0–1.5 wt.%), the feed rate was set by calculating the weight of the LDPE, LLDPE, and SA exiting through the feeder. The screw rate was fixed at 480 rpm, at 150–160 °C, and extruded by mixing LDPE, LLDPE, and SA (0–1.5 wt.%) through the feeder. LDPE/LLDPE/SA was fed into the twin-screw extruder via the hopper and discharged through the feeder at a fixed rate. The extrudate was appropriately cooled through the water-cooling zone and cut using a pelletizer to obtain a masterbatch chip. The prepared masterbatch of LDPE/LLDPE/SA with various SA contents (0–1.5 wt.%) was fed to a blown film maker with LDPE/LLDPE in a constant ratio to prepare the LDPE/LLDPE/SA composite film. For the air-blown pure-blend LDPE/LLDPE film, any further addition of LDPE/LLDPE to the LDPE/LDPE masterbatch chip was not necessary.

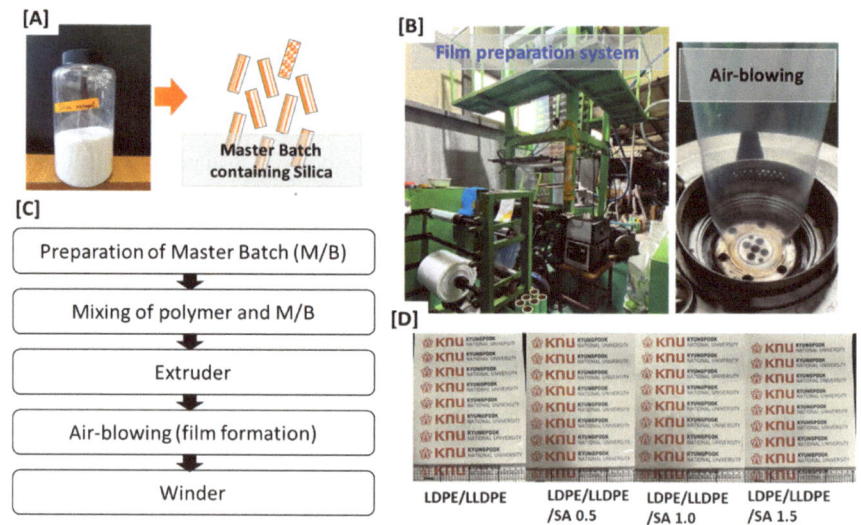

Figure 1. A demonstration of the experiment. (**A**) SA powder and masterbatch containing SA; (**B**) film preparation system at the pilot scale; (**C**) diagram showing the preparation of the air blown film; and (**D**) a macro photograph of the prepared films.

2.3. Instrumental Analysis

The dispersion states and morphologies of the LDPE/LLDPE/SA composites with various SA contents (0–1.5 wt.%) were tested by applying scanning electron microscopy (SEM,

SU8220, Hitachi, Japan) at an accelerating voltage of 10.0 kV. A gold coating was applied to each sample before the analysis. The crystallization behavior of the LDPE/LLDPE/SA (0–1.5 wt.%) composites were studied by performing X-ray diffraction (XRD, D/Max–2500, Rigaku, Tokyo, Japan) using Cu-Kα radiation under operational conditions of 40 kV with 2θ in the range of 2°–40°, having a step interval of 0.02°. Differential scanning calorimetry (DSC) (Q 2000, TA Instruments, New Castle, DE, USA) was performed to analyze the thermal characteristics and crystallinity of the LDPE/LLDPE/SA (0–1.5 wt.%) composite films. First, all of the samples were exposed to a temperature from 30 °C to 300 °C at a heating rate of 10 °C min^{-1}, then kept for 10 min at 300 °C to remove the thermal history, and subsequently cooled down to 30 °C at a cooling rate of 10 °C min^{-1}. A second heating method was performed at 300 °C at a similar scanning rate. The degree of crystallinity (X_c) was calculated from the peak area of the DSC thermograms [17]. The mechanical properties were evaluated using an Instron 5567 material testing system at 25 °C, as per the ASTM D638-96 type II requirements [18]. All data were estimated based on the average of three sample measurements. The light transmittance (T%) of the film was measured by employing ultraviolet–visible (UV–Vis) spectroscopy (K Lab Co., Ltd., Optizen O 2120UV, Daejeon, Korea) in the 200–800 nm wavelength range. The infrared thermal images were captured by a FLIR system AB (Täby, Sweden) infrared thermal imager.

3. Results and Discussion

Microstructural FE-SEM micrographs of the prepared samples with various SA contents are shown in Figure 2. The FE-SEM images clearly show the three-dimensional structure of SA and the effect of SA content on the surface morphology of the LDPE/LLDPE blend film (Figure 2A–D). These results show that the smoothest surfaces were obtained for the LDPE/LLDPE blend film without SA (Figure 2A), and all blown composite films exhibited a good SA distribution in both MD and TD. These results are very significant when considering the mechanical properties of the composite films. However, spherical particles were observed, and the surface roughness increased due to aggregation as the concentration of SA increased (Figure 2B–D) [19]. The optical properties of the LDPE/LLDPE blend films containing different SA contents were determined by measuring the transmittance in the 200–1100 nm range, and the results are presented in Figure 3. The transparency of all of the polymer composite films (0–1.5 wt.%) was good in the visible light range (380–700 nm), in spite of a slight decrease in the transparency with an increase in the content of SA. SA may scatter and absorb ultraviolet light, indicating that it has a UV-blocking function. Photographs of the blended films with different SA contents are exhibited in Figure 1D. All of them showed good visible light transmittance capacity. Similar results were obtained for the LDPE/silica nanocomposite films, showing that various silica contents (0.5–1.5 wt.%) had no significant influence on the transmittance [20].

Some polymers remain in the crystalline state because their molecular chains can be stretched and narrowly arranged in parallel. PE is an orthorhombic crystalline polymer. The crystallization characteristics have an important effect on the physical properties, melting point, and mechanical strength of the polymers. Therefore, it is crucial to investigate the changes in the crystallinity of PE [21]. The XRD patterns of the LDPE/LLDPE composite films with various SA contents are shown in Figure 4. From these scans, it can be observed that the two major peaks of PE were mainly presented in the 2θ range of 10° to 30°.

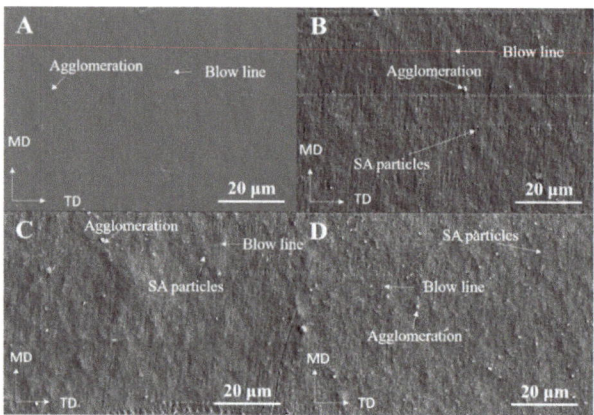

Figure 2. The field-emission scanning electron microscopy (FE-SEM) images of the LDPE/LLDPE blend polymer film with various SA contents: (**A**) 0 wt.%, (**B**) 0.5 wt.%, (**C**) 1 wt.%, and (**D**) 1.5 wt.% (MD = machine direction and TD = transverse direction).

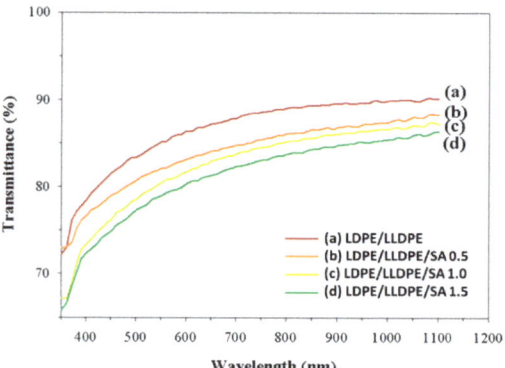

Figure 3. The UV–Vis spectra of the LDPE/LLDPE blend polymer film with various SA contents: (a) 0 wt.%, (b) 0.5 wt.%, (c) 1 wt.%, and (d) 1.5 wt.%.

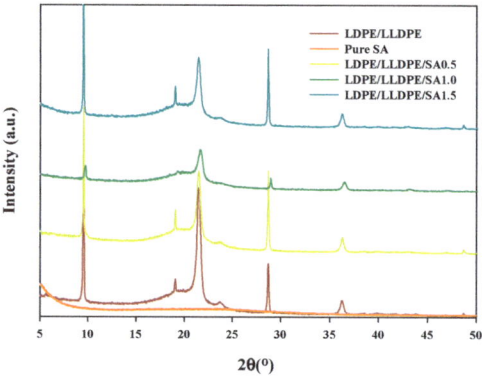

Figure 4. The XRD data of the pure SA and LDPE/LLDPE blend polymer film with various SA contents (0–1.5 wt.%).

The XRD peak at 21.6° was ascribed to the 110 reflections of PE [22]. No obvious peaks of SA appeared in the XRD pattern of the LDPE/LLDPE/SA composite film, indicating that the SA was fully exfoliated. Silica is an amorphous solid, and the amorphous nature of silica was confirmed by a broad peak (2θ = 22°, (101)) in the XRD pattern of pure silica [23]. A large peak at 22° also occurred in our investigation for the pure SA. When a small amount (0.5–1.5 wt.%) of SA was added to the LDPE/LLDPE blend, the absence of the main peaks of SA in the spectra of the composite films may have been caused by the masking effect of the LDPE/LLDPE blend matrix due to the small content of SA [21]. However, it clearly had an impact on the peak intensity of PE. Additionally, after the addition of SA (0.5–1.5 wt.%), a modest shift in the peak location of PE, particularly in the range of 20° to 22°, was seen. These findings can be quite convincingly explained by the various material properties that are brought about by adding SA (0.5–1.5 wt.%), either as a consequence of the process being affected by particles or as a result of the variation in SA dispersion in the extrusion blown LDPE/LLDPE blend matrix with an increase in the SA concentration. The corresponding SEM images of the samples support the XRD measurements [22].

Figure 5 displays the stress–strain curves for the pure LDPE/LLDPE 70/30 blend and the LDPE/LLDPE/SA blown film with various SA contents (0–1.5 wt.%) in the MD, and the tensile properties of the studied films are summarized in Table 2. According to Figure 5B, the stress at break value of the pure blend film increased after adding the SA content (0.5–1.5 wt.%) and the composite with 0.5 wt.% of SA loading showed the highest stress at break (37.96 MPa). In this case, the enhancement of the tensile properties can be clarified by considering the possible influence of SA on the molecular orientation during the extrusion blowing along the MD as well as the good matrix–particle adhesion [24], consequently, promoting the increase in the tensile strength of the composite film with the addition of SA particles (0.5–1.5 wt.%). However, if the SA amount exceeded 0.5%, the stress at break decreased gradually from 37.96 to 27.72 MPa. In contrast, the stress–strain curves illustrated that the strain at break values of the LDPE/LLDPE blend film gradually decreased from 882.18% to 349.01% with the incorporation of SA (0.5–1.5 wt.%). The SA particles may be stuck inside the entanglements, thus resulting in a restriction in the polymer's total chain mobility [20].

The gradual decrease in mechanical properties above a 0.5 wt.% SA loading may deteriorate the dispersion in the LDPE/LLDPE blend solution, as mentioned earlier [25]. In addition, agglomerated SA particles exhibit poor tensile properties. In this study, the SA particles were well-dispersed at lower loading (0.5 wt.%), showing excellent reinforcing efficiency. Moreover, as shown in Figure 5A, the stress increased gradually with filler loading (0.5–1.5 wt.%) within a lower (90%) strain. This behavior is likely to be related to the stiff layers of silicate with a high aspect ratio, which produce a high degree of interaction and appropriate interfacial adhesion properties. Moreover, this tendency restricts the free movement of the polymer chains, increasing the tensile strength value [26]. However, when the strain values increased, a strain hardening mechanism developed, which may be a result of an orientated crystalline structure of polymer in both MD (Figure 5B) and TD (Figure 6B) [25]. Nanocomposites containing various SA concentrations (0.5–1.5%) showed different strain hardening mechanisms than the pure blend film, as was expected by considering the influence of SA dispersion on the polymer chain orientation along the blow direction during the film-blowing [27], which was also demonstrated by the yield point at low strain values (Figures 5A and 6A) and that, with an increase in SA content (0.5–1.5%), the yield stress and yield strain of the blown film, as shown in Tables 2 and 3, increased from 6.34–7.13%, and 9.92–11.49 MPa, respectively, for MD and by 7.77–8.28%, 6.11–8.25 MPa, respectively, for TD (i.e., the mechanical strength and flexibility increased as the SA content increased). The crystallinity of the blown film (Table 4) was also influenced by the effect of SA on the molecular chain, as the molecular orientation decreased the polymer fractional free volume and molecular flexibility, and induced crystallinity [28].

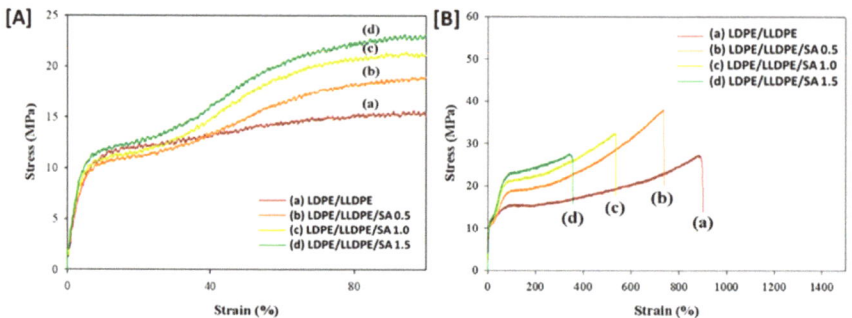

Figure 5. The stress–strain curve for the LDPE/LLDPE blend polymer films with various SA contents: (a) 0 wt.%, (b) 0.5 wt.%, (c) 1 wt.%, and (d) 1.5 wt.%. (**A**) from 0 to 90% and (**B**) from 0 to 1500%. The samples were cut along the MD.

Table 2. The mechanical properties of the LLDPE/LDPE/SA (0–1.5%) air blown composite films along the MD.

Specimen	Thickness (μm)	Yield Strain (%)	Yield Stress (MPa)	Stress at Break (MPa)	Strain at Break (%)	Young Modulus (MPa)
LDPE/LLDPE	20 ± 10	8.04	10.75	27.56	882.18	139.55
LDPE/LLDPE/SA 0.5 wt.%	20 ± 11	6.34	9.92	37.96	742.53	152.44
LDPE/LLDPE/SA 1 wt.%	20.8 ± 12	6.60	10.40	32.32	526.72	198.55
LDPE/LLDPE/SA 1.5 wt.%	21.5 ± 13	7.13	11.49	27.73	349.014	222.01

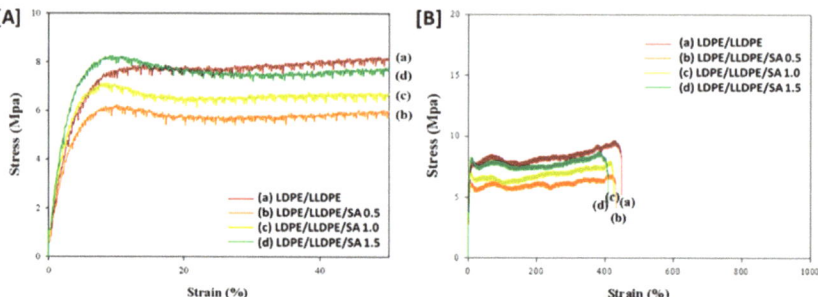

Figure 6. The stress–strain curve for the LDPE/LLDPE blend polymer film with various SA contents: (a) 0 wt.%, (b) 0.5 wt.%, (c) 1 wt.%, and (d) 1.5 wt.%. (**A**) from 0 to 90% and (**B**) from 0 to 1000%. The samples were cut along the TD.

Table 3. The mechanical properties of the LLDPE/LDPE/SA (0–1.5 wt.%) air blown composite films along the TD.

Specimen	Thickness (μm)	Yield Strain (%)	Yield Stress (MPa)	Stress at Break (MPa)	Strain at Break (%)	Young Modulus (MPa)
LDPE/LLDPE	20 ± 10	10.42	7.75	8.61	448.202	64.84
LDPE/LLDPE/SA 0.5 wt.%	20 ± 11	7.77	6.11	6.31	430.64	66.13
LDPE/LLDPE/SA 1 wt.%	20.8 ± 12	7.80	7.10	7.88	417.76	88.98
LDPE/LLDPE/SA 1.5 wt.%	21.5 ± 13	8.28	8.25	8.05	403.52	94.93

Table 4. The DSC analysis data for the LDPE/LLDPE binary blend and LLDPE/LDPE/SA (0–1.5%) air blown composite films.

Sample Type	T_{onset} (°C)	T_{endset} (°C)	T_{peak} (°C)	ΔT (°C)	T_{m1} (°C)	T_{m2} (°C)	ΔH_c (J/g)	X_c (%)
LDPE/LLDPE	110.48	63.87	107.21	3.27	108.58	118.05	95.75	32.67
LDPE/LLDPE/SA(0.5)	110.55	63.90	107.25	3.3	108.54	118.02	97.03	33.11
LDPE/LLDPE/SA(1.0)	110.38	63.42	106.74	3.64	108.19	117.86	94.26	32.17
LDPE/LLDPE/SA(1.5)	110.51	63.75	106.91	3.6	108.03	117.86	94.02	32.02

Because of the effect of the film-blowing ratio and traction ratio in the film-blowing procedure, the mechanical behavior of the film in different directions was different, and the MD film performance (Figure 5) was often higher than the TD film performance (Figure 6) [29]. In the composite films, the nanoparticles spread in a preferential position during blown film extrusion, which could result in changes in the mechanical properties [30]. To determine the influence of processing and particle existence on the mechanical behavior of the blown films, a stress–strain test was also performed in the TD, as presented in Figure 6, and the mechanical properties are summarized in Table 3. The mechanical properties of the LDPE/LLDPE/SA (0.5–1.5 wt.%) films in both directions, MD and TD, had a different behavior than those of the pure blend LDPE/LLDPE film. Unlike the MD samples (27.73–37.96 MPa) presented in Figure 5, the samples in the TD showed lower stress at break (6.31–8.05 MPa) compared to the unfilled blend matrices; this can be related to the low ductility of the composite films compared to the unfilled blend matrix, resulting in early failure of the samples during the test [31]. The difference in molecular chain orientation and the effect of SA particle dispersion on the molecular chain orientation along the blow direction (MD) may also be responsible for this difference between the MD and TD film samples. This difference may allow for a differential increase in the composite film stiffness (Tables 2 and 3) [32]. Moreover, according to Figures 5A and 6A, the stress–strain curves of the LDPE/LLDPE/SA (0.5–1.5 wt.%) composite films in both directions showed the same characteristic ductile deformation behavior of semicrystalline polymers [33] and the reinforcing effect of SA particles.

The crystallization behaviors of the LDPE/LLDPE/SA composite films with various SA contents (0–1.5 wt.%) were studied by DSC, and the resulting DSC cooling and heating curves are shown in Figure 7.

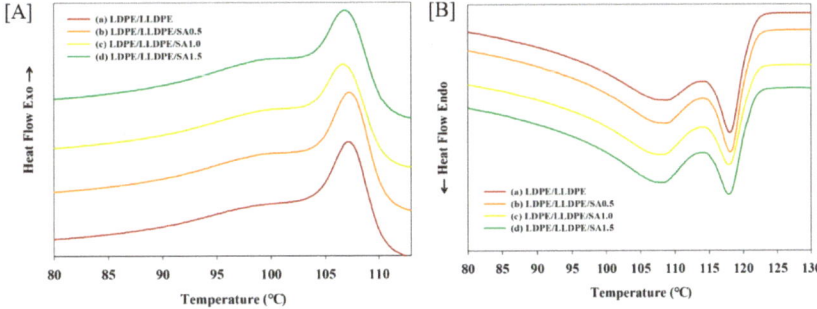

Figure 7. The DSC cooling curves (**A**) and heating curves (**B**) for the LDPE/LLDPE blend film with various SA contents: (a) 0 wt.% (b) 0.5 wt.%, (c) 1 wt.%, and (d) 1.5 wt.%.

The experimental results in terms of the melting temperature (T_m), crystallization temperature (T_c), crystallinity (X_c), and heat of crystallization (ΔH_c) are listed in Table 4. The crystallization temperature, enthalpy of crystallization, and X_c were obtained from the cooling cycle (exothermal peak) (Figure 7A). The second heating run for the neat LDPE/LLDPE blend composite film and those with various loadings of SA (0.5–1.5 wt.%)

are shown in Figure 7B and the data are listed in Table 4. The T_m was calculated based on the second heating cycle (Figure 7B) and the two melting temperatures (T_{m1} and T_{m2}) for all of the studied films are listed in Table 4, corresponding to the first and second endotherms, respectively.

According to Table 4, the addition of SA, irrespective of its content, widened the crystallization window (increasing $\Delta T = T_{onset} - T_{peak}$) and delayed the crystallization progression. This is because SA increases the viscosity of the LDPE/LLDPE matrix, hampers LDPE/LLDPE chain movements, and slows down crystallization development. The composite with 0.5 wt.% SA loading yielded the highest crystallinity (X_c%, 33.1%). However, if the SA content exceeded 0.5%, the crystallinity decreased, and the heat enthalpy of the crystallization (ΔH_c) results showed the same tendency (Table 4). This result suggests that at higher SA contents (1–1.5 wt.%), it did not act as a nucleating agent but hampered polymer chain movements by absorbing polymer segments on its surface. Similar findings were reported in a previous study [34]. As expected, the DSC endotherm of the binary blend LDPE/LLDPE sample showed two distinct major peaks (T_{m1}, T_{m2}) (Figure 7B). A similar feature was shown in the endotherms of the LDPE/LLDPE samples reported earlier [8]. Furthermore, the effect of SA content on the LDPE/LLDPE melting point was carefully observed. As the amount of SA increased in the LDPE/LLDPE blend, the T_{m1} and T_{m2} values decreased, and its intensity changed, which can be attributed to the effect of the filler as it influences the thermal motion of the polymer [35]. It should be mentioned that the MD film was used for the crystallization process and the crystallization curves should not only reflect the general filler effect [36] of SA particles on LDPE/LLDPE crystallization, but also the effect of SA particle dispersion on the molecular chain orientation along the blow direction (MD), as molecular orientation significantly affects the molecular flexibility, which induces the crystallinity [28]. Thus, the variation in the crystalline peak (107.21–106.91 °C) for the extrusion blown LDPE/LLDPE/SA (0–1.5%) system may arise from the strong effect of blown extrusion on the crystal degree orientation [37] and SA aggregates with the increase in the SA concentration as observed through SEM (Figure 2A–D). Additionally, the possible influence of the embedded SA particles on the molecular orientation along the flow direction is also reflected in the mechanical behavior of the film in different directions (Figures 5 and 6) [22].

The thermal insulation performance of the LDPE/LLDPE/SA (0–1.5 wt.%) composite films was confirmed by the IR camera and are presented in Figure 8B. The graph in Figure 8B(f) shows the temperature profiles obtained by placing the LDPE/LLDPE/SA (0–1.5 wt.%) composite films on a hot plate and exposing them to heating for 6 s and subsequent cooling, and after 6 s of heating, the hot plate was turned off, and the temperature was collected from the circle-area marked spots of each film sample, as shown by the temperature–time curve in Figure 8B(a–e). The thickness of all of the studied films is summarized in Table 2. The results show that the film surface temperature decreased, and the heat stored in the film was emitted. After 15 min, all of the studied films showed almost the same temperature; however, after 30 min, the pure blend LDPE/LLDPE film showed a faster cooling rate than the LDPE/LLDPE/SA composite film with various SA contents, and the heat retention capacity increased with the addition of SA (0.5–1 wt.%).

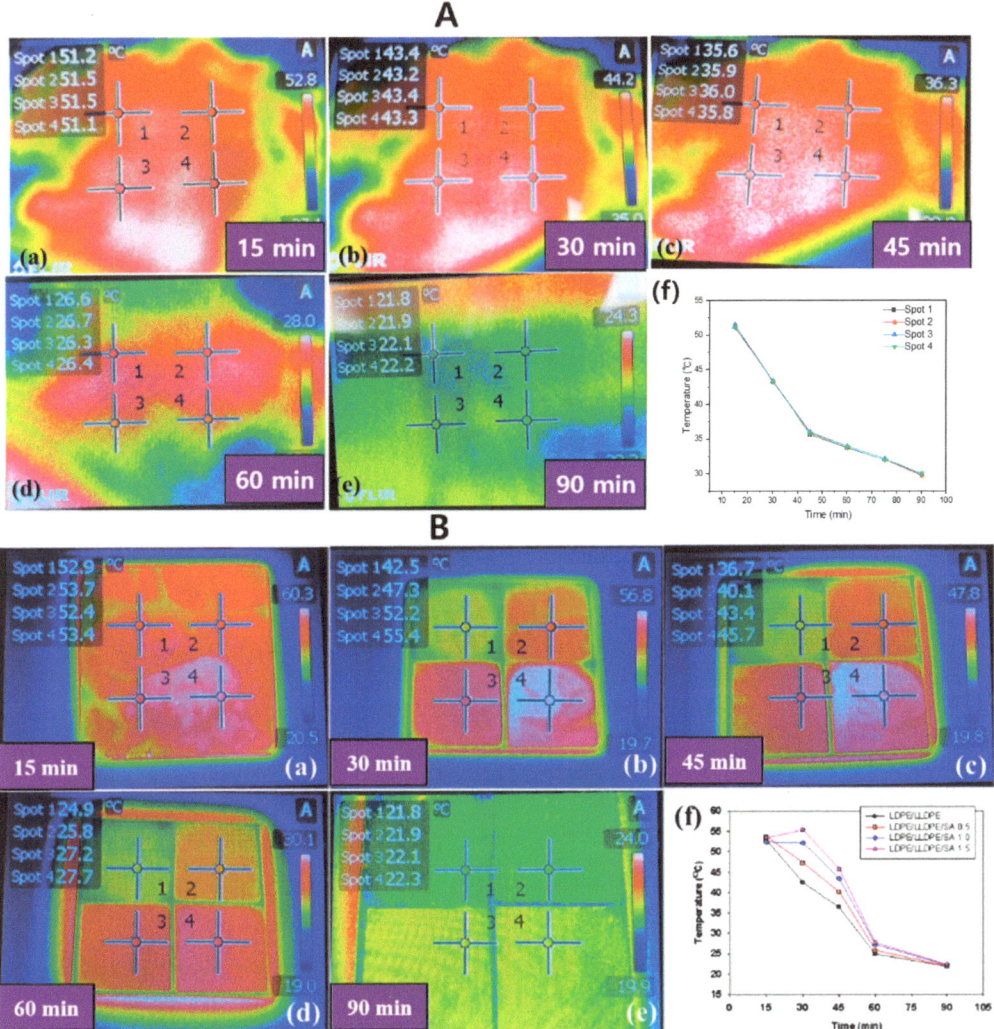

Figure 8. The infrared thermal images and temperature-time curve of (**A**) only the hot plate and (**B**) the prepared LDPE/LLDPE/SA (0–1.5%) films obtained by the hot plate. (**a**–**e**) of (**A**,**B**) are measuring time such as (**a**) 15 min, (**b**) 30 min, (**c**) 45 min, (**d**) 60 min, and (**e**) 90 min. (**f**) represented the temperature according to time.

A similar observation was found in a previous report, where it was shown that the addition of SA reduced the thermal conductivity by approximately 30% with 1–2 wt.% of SA loading [22]. Thus, the prepared composite LDPE/LLDPE/SA (0.5–1.5 wt.%) film had a lower thermal conductivity than the neat LDPE/LLDPE blend film, and was suitable for application as a thermally insulating material. A control experiment was also conducted simultaneously (Figure 8A(a–f)), and in contrast to the composite films, the temperature was measured from the circle-area marked spots of only the hot plate surface without any films. The results showed a similar time–temperature relationship (Figure 8A(f)) and were consistent with the results of all of the examined films (Figure 8B(f)), which showed the same time range (15–90 min) to reach room temperature from the starting temperature (51 °C).

Figure 9 depicts the draw ratio at two different directions: MD and TD. As can be observed, a linear decrease in the draw ratio was obtained as a function of the SA content for both directions. However, the main aspect to note in these curves is that the highest maximum draw ratio was achieved for the pure LDPE/LLDPE blend film. Another interesting observation is that all curves showed a similar dependence on the SA content (0–1.5%). This indicates that the decreasing tendency of the draw ratio when increasing the SA content is the result of the direct SA reinforcement effect. As indicated above, the main effect of the addition of SA particles is the decreased drawability of the LDPE/LLDPE matrix. This decreased draw ratio resulted in a restriction in the molecular orientation and chain extension when the SA content increased, which in turn was responsible for the subsequent decrease in the draw ratio in both directions.

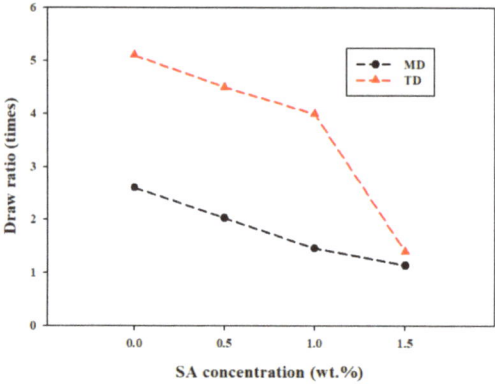

Figure 9. The draw ratio results in the MD and TD of the LDPE/LLDPE blend film as a function of SA content.

Clearly, all of the investigated films, LDPE/LLDPE and the composites incorporating 0.5 wt.%, 1 wt.%, or 1.5 wt.% SA, revealed the same unique relationship between the draw ratio and fillers, revealing that the increase in draw ratio was mainly related to the oriented polymer matrix rather than an additional effect of the filler's reinforcement. Moreover, the obtained draw ratio for the TD was higher than that for the MD, which may be due to a variation in the orientation of the polymer chain during processing of the blown films.

4. Conclusions

Heat-retention LDPE/LLDPE blend films containing highly insulating SA particles (0.5–1.5 wt.%) were prepared successfully by employing blown extrusion at a pilot scale, and the influence of various amounts of SAs on the material properties of composite films was investigated. Macroscopically, the almost homogeneous appearance of the blown films was the result of a good SA dispersion within the LDPE/LLDPE blend, and this was verified by performing SEM, XRD, UV–Vis spectroscopy, and DSC. The SA nucleating capacity was revealed by the increase in the crystallinity degree and crystallization temperature when 0.5 wt.% of SA was incorporated in the composite films. Therefore, the particle dispersion, nucleating character, and strong effect of blown extrusion played a significant role in the introduction of the preferential LDPE/LLDPE crystalline orientation and the properties of the final film. Based on this logic, differences in the tensile behavior of the composite films were observed between the MD and TD. Almost all of the films (LDPE/LLDPE/SA (0–1 wt.%)) showed a higher elongation at break in the MD than in the TD. However, films with 1.5% SA showed almost the same value in both directions. This may be due to the agglomeration at higher SA content, which reduces mobility and elongation. These results confirmed that along with the processing method, the SA particles were also responsible for initiating changes in the biaxial nature of the blown films. Based on the orientation of the dual mechanical characteristics, these composite films can be used as packaging materials in many applications. Moreover, the resulting material showed enhanced infrared retention, and the incorporation of the filler material did not hamper visible light transmission; thus, these films have the potential to be applied in greenhouses with good optical performance.

Author Contributions: Conceptualization, S.B.Y. and J.L.; Methodology, J.L.; Software, J.M.P.; Validation, J.M.P. and D.-J.K.; Investigation, S.Y. and M.D.H.; Data curation, S.Y.; Writing—original draft preparation, S.B.Y.; Writing—review and editing, S.B.Y. and S.Y.; Visualization, S.Y. and D.-J.K.; Supervision, J.H.Y.; Project administration, M.D.H. and J.H.Y. All authors have read and agreed to the published version of the manuscript.

Funding: This research was supported by the Technology Commercialization Support Program (2022-DG-RD-0092) funded by the Ministry of Science and ICT of Korea.

Institutional Review Board Statement: Not applicable.

Informed Consent Statement: Not applicable.

Data Availability Statement: Not applicable.

Conflicts of Interest: The authors declare no conflict of interest.

References

1. Campuzano, J.; Lopez, I. Study of the effect of dicumyl peroxide on morphological and physical properties of foam injection molded poly (lactic acid)/poly (butylene succinate) blends. *Express Polym. Lett.* **2020**, *14*, 673–684. [CrossRef]
2. Zhang, Q.; Chen, W.; Zhao, H.; Ji, Y.; Meng, L.; Wang, D.; Li, L. In-situ tracking polymer crystallization during film blowing by synchrotron radiation X-ray scattering: The critical role of network. *Polymer* **2020**, *198*, 122492. [CrossRef]
3. Delgadillo-Velázquez, O.; Hatzikiriakos, S.; Sentmanat, M. Thermorheological properties of LLDPE/LDPE blends: Effects of production technology of LLDPE. *J. Polym. Sci. Part B Polym. Phys.* **2008**, *46*, 1669–1683. [CrossRef]
4. Fang, Y.; Carreau, P.J.; Lafleur, P.G. Thermal and rheological properties of mLLDPE/LDPE blends. *Polym. Eng. Sci.* **2005**, *45*, 1254–1264. [CrossRef]
5. Hussein, I.A.; Williams, M.C. Rheological study of the influence of branch content on the miscibility of octene m-LLDPE and ZN-LLDPE in LDPE. *Polym. Eng. Sci.* **2004**, *44*, 660–672. [CrossRef]
6. Schlund, B.; Utracki, L. Linear low density polyethylenes and their blends: Part 3. Extensional flow of LLDPE's. *Polym. Eng. Sci.* **1987**, *27*, 380–386. [CrossRef]
7. Al-Attar, F. Thermal, Mechanical and Rheological Properties of Low Density/Linear Low Density Polyethylene Blend for Packing Application. *J. Mater. Sci. Chem. Eng.* **2018**, *6*, 32. [CrossRef]
8. Monwar, M.; Yu, Y. Determination of the Composition of LDPE/LLDPE Blends via 13C NMR. In *Macromolecular Symposia*; Wiley: Hoboken, NJ, USA, 2020; p. 1900013.
9. Busu, W.N.W.; Chen, R.S.; Shahdan, D.; Yusof, M.J.M.; Saad, M.J.; Ahmad, S. Statistical Optimization Using Response Surface Methodology for Enhanced Tensile Strength of Polyethylene/Graphene Nanocomposites. *Int. J. Integr. Eng.* **2021**, *13*, 109–117.
10. Al Sheheri, S.Z.; Al-Amshany, Z.M.; Al Sulami, Q.A.; Tashkandi, N.Y.; Hussein, M.A.; El-Shishtawy, R.M. The preparation of carbon nanofillers and their role on the performance of variable polymer nanocomposites. *Des. Monomers Polym.* **2019**, *22*, 8–53. [CrossRef]
11. Deng, J.; Ding, Q.M.; Li, W.; Wang, J.H.; Liu, D.M.; Zeng, X.X.; Liu, X.Y.; Ma, L.; Deng, Y.; Su, W. Preparation of nano-silver-containing polyethylene composite film and Ag Ion migration into food-simulants. *J. Nanosci. Nanotechnol.* **2020**, *20*, 1613–1621. [CrossRef] [PubMed]
12. Zolfaghari, S.; Paydayesh, A.; Jafari, M. Mechanical and thermal properties of polypropylene/silica aerogel composites. *J. Macromol. Sci. Part B* **2019**, *58*, 305–316. [CrossRef]
13. Chen, Y.; Klima, K.; Brouwers, H.; Yu, Q. Effect of silica aerogel on thermal insulation and acoustic absorption of geopolymer foam composites: The role of aerogel particle size. *Compos. Part B Eng.* **2022**, *242*, 110048. [CrossRef]
14. Tang, R.; Hong, W.; Srinivasakannan, C.; Liu, X.; Wang, X.; Duan, X. A novel mesoporous Fe-silica aerogel composite with phenomenal adsorption capacity for malachite green. *Sep. Purif. Technol.* **2022**, *281*, 119950. [CrossRef]
15. Yi, Z.; Zhang, X.; Yan, L.; Huyan, X.; Zhang, T.; Liu, S.; Guo, A.; Liu, J.; Hou, F. Super-insulated, flexible, and high resilient mullite fiber reinforced silica aerogel composites by interfacial modification with nanoscale mullite whisker. *Compos. Part B Eng.* **2022**, *230*, 109549. [CrossRef]
16. Ahmed, S.S.A.M. A Sustainable Approach to Reduce Environmental Threats of Oxidative Degradation of Plastic Films. Ph.D. Thesis, Sudan University of Science & Technology, Khartoum, Sudan, 2021.
17. Van de Voorde, B.; Katalagarianakis, A.; Huysman, S.; Toncheva, A.; Raquez, J.-M.; Duretek, I.; Holzer, C.; Cardon, L.; Bernaerts, K.V.; van Hemelrijck, D. Effect of extrusion and fused filament fabrication processing parameters of recycled poly (ethylene terephthalate) on the crystallinity and mechanical properties. *Addit. Manuf.* **2022**, *50*, 102518. [CrossRef]

18. Yeasmin, S.; Yeum, J.H.; Yang, S.B. Fabrication and characterization of pullulan-based nanocomposites reinforced with montmorillonite and tempo cellulose nanofibril. *Carbohydr. Polym.* **2020**, *240*, 116307. [CrossRef]
19. Oh, S.; Shin, C.; Kwak, D.; Kim, E.; Kim, J.; Bae, C.; Kim, T. Effect of ionic strength on amorphous carbon during chemical mechanical planarization. *Diam. Relat. Mater.* **2022**, *127*, 109124. [CrossRef]
20. Alghdeir, M.; Mayya, K.; Dib, M. Nanosilica Composite for Greenhouse Application. In *Composite and Nanocomposite Materials-From Knowledge to Industrial Applications*; IntechOpen: London, UK, 2020.
21. Gaabour, L.H. Influence of silica nanoparticles incorporated with chitosan/polyacrylamide polymer nanocomposites. *J. Mater. Res. Technol.* **2019**, *8*, 2157–2163. [CrossRef]
22. Bokobza, L. Infrared Linear Dichroism for the Analysis of Molecular Orientation in Polymers and in Polymer Composites. *Polymers* **2022**, *14*, 1257. [CrossRef]
23. An, Z.; Wang, H.; Zhu, C.; Cao, H.; Xue, J. Synthesis and formation mechanism of porous silicon carbide stacked by nanoparticles from precipitated silica/glucose composites. *J. Mater. Sci.* **2019**, *54*, 2787–2795. [CrossRef]
24. Phothisarattana, D.; Harnkarnsujarit, N. Characterizations of Cassava Starch and Poly (butylene adipate-co-terephthalate) Blown Film with Silicon Dioxide Nanocomposites. *Int. J. Food Sci. Technol.* **2022**, *57*, 5078–5089. [CrossRef]
25. Liu, X.; Zou, L.; Chang, B.; Shi, H.; Yang, Q.; Cheng, K.; Li, T.; Schneider, K.; Heinrich, G.; Liu, C. Strain dependent crystallization of isotactic polypropylene during solid-state stretching. *Polym. Test.* **2021**, *104*, 107404. [CrossRef]
26. Ahmed, W.; Siraj, S.; Al-Marzouqi, A.H. Comprehensive characterization of polymeric composites reinforced with silica microparticles using leftover materials of fused filament fabrication 3D printing. *Polymers* **2021**, *13*, 2423. [CrossRef] [PubMed]
27. Botta, L.; Teresi, R.; Titone, V.; Salvaggio, G.; La Mantia, F.P.; Lopresti, F. Use of biochar as filler for biocomposite blown films: Structure-processing-properties relationships. *Polymers* **2021**, *13*, 3953. [CrossRef]
28. Chavan, C.; Bhajantri, R.F.; Cyriac, V.; Bulla, S.; Ravikumar, H.; Raghavendra, M.; Sakthipandi, K. Exploration of free volume behavior and ionic conductivity of PVA: x (x = 0, Y_2O_3, ZrO_2, YSZ) ion-oxide conducting polymer ceramic composites. *J. Non Cryst. Solids* **2022**, *590*, 121696. [CrossRef]
29. Lin, B.; Li, C.; Chen, F.; Liu, C. Continuous Blown Film Preparation of High Starch Content Composite Films with High Ultraviolet Aging Resistance and Excellent Mechanical Properties. *Polymers* **2021**, *13*, 3813. [CrossRef]
30. Muñoz, P.; de Oliveira, C.; Amurin, L.; Rodriguez, C.; Nagaoka, D.; Tavares, M.; Domingues, S.; Andrade, R.; Fechine, G. Novel improvement in processing of polymer nanocomposite based on 2D materials as fillers. *Express Polym. Lett.* **2018**, *12*, 930–945. [CrossRef]
31. Gigante, V.; Aliotta, L.; Coltelli, M.B.; Cinelli, P.; Botta, L.; La Mantia, F.P.; Lazzeri, A. Fracture behavior and mechanical, thermal, and rheological properties of biodegradable films extruded by flat die and calender. *J. Polym. Sci.* **2020**, *58*, 3264–3282. [CrossRef]
32. Wang, X.; Pan, H.; Jia, S.; Wang, Z.; Tian, H.; Han, L.; Zhang, H. In-situ reaction compatibilization modification of poly (butylene succinate-co-terephthalate)/polylactide acid blend films by multifunctional epoxy compound. *Int. J. Biol. Macromol.* **2022**, *213*, 934–943. [CrossRef]
33. Cho, S.H. Role of Tie Molecules in Ductility and Chain Deformation of Polyethylene. Ph.D. Thesis, Princeton University, Princeton, NJ, USA, 2022.
34. Karami, S.; Ahmadi, Z.; Nazockdast, H.; Rabolt, J.F.; Noda, I.; Chase, B.D. The effect of well-dispersed nanoclay on isothermal and non-isothermal crystallization kinetics of PHB/LDPE blends. *Mater. Res. Express* **2018**, *5*, 015316. [CrossRef]
35. Li, T.; Sun, H.; Lei, F.; Li, D.; Leng, J.; Chen, L.; Huang, Y.; Sun, D. High performance linear low density polyethylene nanocomposites reinforced by two-dimensional layered nanomaterials. *Polymer* **2019**, *172*, 142–151. [CrossRef]
36. Wang, Z.-Q.; Zhang, Y.-F.; Li, Y.; Zhong, J.-R. Effect of sodium lignosulfonate/nano calcium carbonate composite filler on properties of isotactic polypropylene. *Polym. Bull.* **2022**, *29*, 1–15. [CrossRef]
37. Troisi, E.; van Drongelen, M.; Caelers, H.; Portale, G.; Peters, G. Structure evolution during film blowing: An experimental study using in-situ small angle X-ray scattering. *Eur. Polym. J.* **2016**, *74*, 190–208. [CrossRef]

Article

Stability Study of the Irradiated Poly(lactic acid)/Styrene Isoprene Styrene Reinforced with Silica Nanoparticles

Ana Maria Lupu (Luchian) [1,2], Marius Mariş [3,*], Traian Zaharescu [4,*], Virgil Emanuel Marinescu [4] and Horia Iovu [1,5]

1. Advanced Polymer Materials Group, University Politehnica of Bucharest, 011061 Bucharest, Romania; ana.lupu@eli-np.ro (A.M.L.); horia.iovu@upb.ro (H.I.)
2. Extreme Light Infrastructure-Nuclear Physics (ELI-NP), Horia Hulubei National Institute for Physics and Nuclear Engineering (IFIN-HH), 077125 Magurele, Romania
3. Dental Medicine Faculty, University Titu Maiorescu, 22 Dâmbovnicului Tineretului St., 040441 Bucharest, Romania
4. INCDIE ICPE CA, Radiochemistry Center, 313 Splaiul Unirii, 030138 Bucharest, Romania; virgil.marinescu@icpe-ca.ro
5. Academy of Romanian Scientists, 050094 Bucharest, Romania
* Correspondence: marius@drmbaris.ro (M.M.); traian.zaharescu@icpe-ca.ro (T.Z.)

Citation: Lupu, A.M.; Mariş, M.; Zaharescu, T.; Marinescu, V.E.; Iovu, H. Stability Study of the Irradiated Poly(lactic acid)/Styrene Isoprene Styrene Reinforced with Silica Nanoparticles. *Materials* 2022, 15, 5080. https://doi.org/10.3390/ma15145080

Academic Editors: Polina P. Kuzhir and Csaba Balázsi

Received: 23 April 2022
Accepted: 12 July 2022
Published: 21 July 2022

Publisher's Note: MDPI stays neutral with regard to jurisdictional claims in published maps and institutional affiliations.

Copyright: © 2022 by the authors. Licensee MDPI, Basel, Switzerland. This article is an open access article distributed under the terms and conditions of the Creative Commons Attribution (CC BY) license (https://creativecommons.org/licenses/by/4.0/).

Abstract: In this paper, the stability improvement of poly(lactic acid) (PLA)/styrene-isoprene block copolymer (SIS) loaded with silica nanoparticles is characterized. The protection efficiency in the material of thermal stability is mainly studied by means of high accurate isothermal and nonisothermal chemiluminescence procedures. The oxidation induction times obtained in the isothermal CL determinations increase from 45 min to 312 min as the polymer is free of silica or the filler loading is about 10%, respectively. The nonisothermal measurements reveal the values of onset oxidation temperatures with about 15% when the concentration of SiO_2 particles is enhanced from none to 10%. The curing assay and Charlesby–Pinner representation as well as the modifications that occurred in the FTIR carbonyl band at 1745 cm^{-1} are appropriate proofs for the delay of oxidation in hybrid samples. The improved efficiency of silica during the accelerated degradation of PLA/SIS 30/n-SiO_2 composites is demonstrated by means of the increased values of activation energy in correlation with the augmentation of silica loading. While the pristine material is modified by the addition of 10% silica nanoparticles, the activation energy grows from 55 kJ mol^{-1} to 74 kJ mol^{-1} for nonirradiated samples and from 47 kJ mol^{-1} to 76 kJ mol^{-1} for γ-processed material at 25 kGy. The stabilizer features are associated with silica nanoparticles due to the protection of fragments generated by the scission of hydrocarbon structure of SIS, the minor component, whose degradation fragments are early converted into hydroperoxides rather than influencing depolymerization in the PLA phase. The reduction of the transmission values concerning the growing reinforcement is evidence of the capacity of SiO_2 to minimize the changes in polymers subjected to high energy sterilization. The silica loading of 10 wt% may be considered a proper solution for attaining an extended lifespan under the accelerated degradation caused by the intense transfer of energy, such as radiation processing on the polymer hybrid.

Keywords: PLA; SIS; silica nanocomposite; stability; chemiluminescence

1. Introduction

The optimization of material stability is regularly achieved in several ways: preparation of hybrids [1,2], multicomponent blending [3–5], crosslinking [6,7], the addition of stabilizers [8,9], radiation processing [10,11] or any of their combinations. In the processing of plastics, an important role for further material stability is played by the technological parameters that influence the oxidation state and the history of products. In the cases of radiation procedures, the behavior of the substrate may be characterized by the possibility to provide reactive intermediates, free radicals that will be involved in the alternative decay pathways for several expected processes: oxidation or recombination [12]. The efficient activity of unsaturated monomers in the crosslinking of a polymer is the reason by which degrading polymers, such as PLA may gain satisfactory performances for their implementation in hazardous conditions [13]. The appropriate texture of PLA based products destined for various applications in the ranges of packaging materials or medical wear is obtained by the potential stabilizing materials belonging to the class of natural antioxidants [14–16]. The extension of the usage period under special environmental conditions is a general expectancy for avoiding the damage of aging materials. Accordingly, the structure of polymer backbones defines the scission positions that indicate noticeable changes occurred during degradation [17].

PLA, a biopolymer, offers multiple versions of materials destined for various application ranges: food and beverage packaging, prostheses, catheters and scaffolds, drug delivery supports, safety toys or commodities and many others. The analysis of potential applications is an attractive subject, which assists the manufacturers in the selection of compositions and technologies [18–21]. The aging behavior of PLA compounds is essentially described as depolymerization and partial oxidation [22,23]. The blending of PLA with other polymers turns the functional properties including material stability into the required features. The presence of polyhedral oligomeric silsesquioxane [24], montmorillonite and carbon nanotubes [25], and graphene [26] allows for improving the oxidative resistances of basic material. At the same time, the mixing of PLA with other polymers, for example, fulvic acid–g–poly(isoprene) [27], acrylonitrile butadiene styrene [28], polyhydroxybutyrate [29], cinnamic acid [30] represents an appropriate method for the modeling of properties. The control of product stability is the main target with which the producer is guided in the selection of raw materials [31]. The availability of PLA to generate various materials may be defined by the hundreds of combinations and associations by which the foreseen purposes are achieved. The polymer materials that are destined for ecologically friendly products must present a pertinent degradation rate so that they may be easily recycled, converted into other useful derivatives or destroyed without pollution [32]. The ability of mixed PLA-based materials to be considered biodegradable is demonstrated by their capacity to be regenerated [33], recycled [34] or transformed into ecological materials by mixing [35–38]. Therefore, our study envisages the identification of a proper system with which the manufacturers may produce appropriate items with a certain degree of stability. The manufacture of environmentally accepted materials opens several new directions for the economical implementation of PLA based blends. The mixing of PLA with polycaprolactone (PLC) offers certain possibilities for modeling materials to earn the desired properties [39,40]. The higher the content of PLA, the lower the stability of blends. The addition of triallyl isocyanate (TAIC) creates the intermolecular bridges that gather the material into a more dense and resistant compound, when the second component is poly(butylene adipate-co-terephtalate) (PBAT), a nonresistant configuration on the action of ionizing radiation [41]. The low amount of TAIC (3%) generates enough concentration of radicals for curing the two component fractions into a homogenous product.

The analysis of radiation effects on the blends consisting of PLA and EVOH [42] reveals the optimum processing dose at around 30 kGy when a great proportion of insoluble fraction (more than 80%) is formed and the mechanical properties become relevant for long term applications.

The significant energetic requirement for the modification of biopolymers limits the deterioration of polymer skeletons and extends functional qualities [43]. The electron beam (EB) irradiation creates crosslinked islands, which become efficient barriers against the diffusion of liquid and gaseous fluids. This earning is the fundamental feature for the mitigation of degradation.

The progress of the radiation damaging of PLA is conducted by the scission tendency confirmed by the decreasing of the values of glass transition as the dose smoothly grows [44] or by the EPR assay on the modification of the abundance of radical accumulation [45]. The following degradation intermediates appear as it is illustrated in Figure 1, where the two types of radicals are formed. One kind of radical is carbon centered fragments due to the scission of C–C bonds; the second type of intermediates possess the radical positions on oxygen atoms. This last group appears from already oxidized intermediates that are the precursors of final stable products.

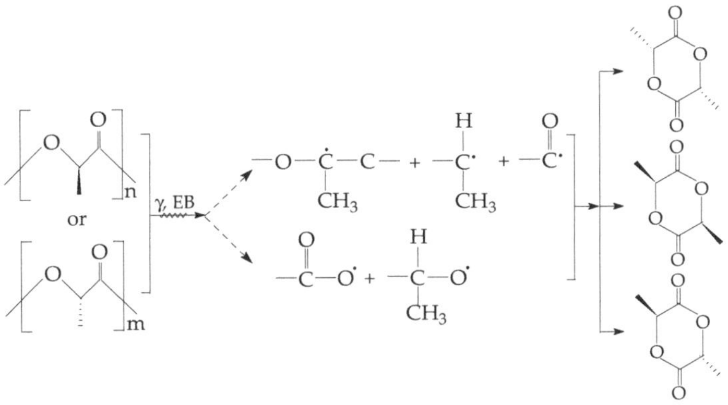

Figure 1. The radical intermediates that precede the formation of stable oxygenated products during degradation.

The evaluation of molecular modification by the gel permeation assays on EB-irradiated PLA [46] shows a sharp drop in the average molecular weight from 250 kg mol^{-1} to 30 kg mol^{-1}, when the exposure dose turns from 0 to 200 kGy, while the average molecular weight drops from 260 kg mol^{-1} to 68 kg mol^{-1}, under the same irradiation regime. The γ-exposure effects on PLA are based on radiochemical fragmentation, eventually followed by crosslinking. The evolution of fragmentation is described by the ratio of scission and crosslinking yields, G(S)/G(X), with values in the range of 7.6 to 10.4 [47]. Additional proof of the advanced molecular scission is the high value of 1.97 for scission yield showing that this figure characterizes the number of molecules that are broken for each 100 eV. The recent studies on the degradation of PLA [22,48] point out the essential contribution of inorganic fillers that control the level of degradation by their involvement as radical scavengers. The golden rule regarding the radiochemical yields that states to which polymer category a material belongs is also applied in the cases of PLA hybrids. In these cases, the ratio G(S)/G(X) keeps the values far from the unit (9.38 or 35.5 for two steric sorts of polymer) [49]. These figures indicate clearly that PLA has very low radiation stability and, consequently, the improvement in this feature asks for specific additives or suitable fillers that are able to convert it toward a material suitable for radiation processing or to protect it.

The PLA-based composites have gained the name of biomaterials because they are appropriate for several friendly applications from the medical area up to food packaging. The associations of this polymer with polyethylene glycol and chitosan [50], carbon [51] or pineapple [52] or pulp [53] fibers, hemp component [54], montmorillonite [55], bamboo powder [56], alginate [57] are some of the examples which confirm the versatility of PLA. These materials are potential candidates for the usage addressed to ecological products. The endurance of PLA systems is the main target for the characterization of product durability, the extension of application areas, or the creation of new materials that are able to resist damaging energy transfers [58,59]. They not only solve the problem of ecological materials but, at the same time, they also answer the requirements of healthy life.

The association between PLA and SIS in polymer formulations was formerly investigated finding that their native blends present moderate radiation stability [60]. The stability investigation of this type of blend suggests the initiation of a detailed study on the contribution of efficient additives for the integration of these materials through ecological and resistant products. The material behavior exposed to high energy radiation finds the closest applications in radiation sterilization, which is an appropriate procedure for the reduction of the bacterial charge of a large series of plastic items.

2. Materials and Methods

The blending components and their mixing processing for the sample preparation were presented in an earlier paper [60]. From the previous formulation of PLA/SIS compounds, the ratio 70:30 was selected for the present assay because it presents the highest radiation stability. The further nomination of material is PLA/SIS 30. This preference is the practice option for demonstrating the capacity of SiO_2 to ameliorate the poor performances of this polymer blend. The incorporated filler, hydrophilic silica, usually incorporated in the polymer compositions, AEROSIL 200 (specific surface 200 $cm^2\ g^{-1}$) was manufactured by Degussa (Evonic, Troy Hills, NJ, USA). For a reliable comparison of results, there were preferred four filler concentrations namely 0%, 3%, 5% and 10%. All three components, PLA, SIS and SiO_2 were mixed together in a foreseen proportion in a Brabender Plastograph at 180 °C for 6 min with a rotor speed of 50 rpm. Then, the neat material was pressed into sheets (150 × 150 × 1 mm) under a pressure of 125 atm for 1 min after a preheating operation for 3 min at 180 °C.

The γ-exposure was accomplished at a dose rate of 0.6 kGy h^{-1} in the air inside the irradiation machinery Ob Servo Sanguis (Budapest, Hungary), which is equipped with a ^{60}Co source. All the measurements were conducted immediately after the elapse of radiation treatment to avoid any unexpected modifications in the chemical states of samples. The main results of γ-irradiation were planned to explain the stability degree at the sterilization dose (25 kGy), but for the explanation of gelation and the identification of the structural differences in FTIR spectra a dose of 50 kGy was also applied.

The stability investigation is based on the chemiluminescence measurements by the nonisothermal procedure [61] at four heating rates: 3.7, 5.0, 10.0 and 15.0 °C min^{-1} and by isothermal determination [61] at 140 °C. The samples were prepared by placing polymer small sheet fragments in aluminum pans as a very thin layer for avoiding the self-absorption of CL photons. The amount of dry polymer was around 3–5 mg. The activation energies required for the thermal degradation of samples were calculated using onset oxidation temperature (OOT) values, a kinetic parameter that describes the temperature which indicates the start of oxidation. It is found at the intersection between the tangent drawn on the ascendant curve of CL intensity vs. temperature and the Ox axis. The relationship [62] (1) was applied

$$\ln\left(\frac{\beta}{T^2}\right) = k - \frac{E}{RT} \qquad (1)$$

and allows the linear representation of $\ln(\beta/T^2)$ vs. T^{-1}. The slope will be used for the calculation of activation energy by dividing it by the universal gas constant, R.

The evaluation of gel content was accomplished by the measurement of the dissolved amount by boiling in Soxhlet equipment using o-xylene as solvent followed by drying and removing of the remaining solvent by cleaning with ethanol and the final drying. The dryings were accomplished in an electrical oven for one hour at 100 °C. The sample weights at the beginning of boiling (m_o) and the final weight (m_f) were obtained by analytical balance, whose error is ± 0.2 mg. The copper bags have final weights symbolized by m_b. The soluble fraction (named sol) is calculated by Equation (2):

$$S = 1 - \frac{m_f - m_{fo}}{m_b - m_{fo}} \qquad (2)$$

The triplicates were subjected to the solubilizing step.

These values were the input values for the Charlesby–Pinner relationship [63], Equation (3), by which the corresponding graph is drawn.

$$S + \sqrt{S} = \frac{G(S)}{G(X)} + \frac{1}{qU_o}\frac{1}{D} \qquad (3)$$

where S is sol content, D is processing dose (kGy), G(S) and G(X) are the radiochemical yields (events/100 eV) for scission and crosslinking, respectively, q denotes the crosslinking density (cm^{-3}) and U$_0$ is the initial polymerization degree.

The aging effect induced by γ-exposure is analyzed by the modifications that occurred in the FTIR spectra at 1745 cm^{-1} in a JASCO 4200A (Tokyo, Japan) spectrometer with a resolution of 4 cm^{-1} and 48 scans.

The SEM images were obtained with a scanning electron microscope model field emission SEM–Auriga (Carl Zeiss, Overcoached, Germany). The selected experimental parameters were: 2 kV acceleration voltages with a working distance of ~3.5 mm in a high vacuum room. Polymer samples, having been embedded in epoxy resin with conductive properties, were further polished and treated with permanganic (chemical solution) etching. The prepared surfaces were analyzed in a SESI (secondary electron secondary ions) detector chamber of the Everhart–Thornley type model.

3. Results

The qualification of polymer materials for their long life application asks for the identification of optimal conditions for the improvement of oxidation strength. PLA, the major component of the studied compositions, presents a very high radiochemical yield of scission (14.5) [24,64] and a very low radiochemical yield of crosslinking (0.4) [24,64] when it is irradiated in air. It undergoes thermal and radiation oxidative degradations with the activation energy placed in the range of 92–118 kJ mol^{-1} [27,55]. On the other hand, the fragmentation of D,L-PLA backbones presents a high radical yield [G(R) = 2.4 at 77 K and 1.2 at 300 K]. The radiolysis radicals (Figure 1) intimately interact with the radicals formed from SIS [60]. The presence of an oxidation inhibitor, powder POSS [65], extends the thermal stability that becomes a good material with a convenient lifespan. The certification of this processing availability is the extended ranges of usage that are based on the material accessibility for enlarging the manufacturing of the various product numbers [66–68]. Due to the specificity of the delaying of oxidation, the SiO$_2$ fillers are capable to catch free radicals on their nanoparticle surface and, consequently, withdraw intermediates from the degradation chain [69].

3.1. Chemiluminescence

The chemiluminescence determination of the oxidation susceptibility by isothermal procedures (Figure 2) gives the measure of the stability of the attack of oxygen on the free radicals that appeared by molecular fragmentation during thermal and/or radiation degradation. The nanoparticles of silica are homogenous spread in polymer and interaction between radicals and the SiO$_2$ surface is defined by the dipole–dipole attraction [70]. The increase in the silica concentration broadens the oxidation induction time. This main kinetic parameter that indicates the period over which the degradation starts presents increasing values as the silica percentage grows (Figure 2). The other feature that depicts the progress of oxidative aging is the oxidation rate. This characteristic is clearly noticed from the inclination of curves. These slopes become lower as the content of silica is higher. It means that the SiO$_2$ nanoparticles act as efficient scavengers of the intermediates and break the degradation chain.

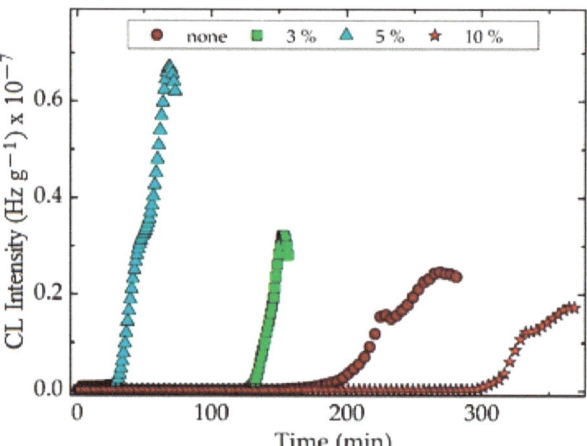

Figure 2. The isothermal CL spectra recorded on nonirradiated PLA/SIS 30 blends modified with silica. Testing temperature: 140 °C.

The CL spectra discover the contributions of the two mixed components which participate in the degradation with different oxidation rates. As it is known from previous studies, the degradation mechanisms of SIS [71] and PLA [72] are completely unlike. It is easy to assume that the fragments provided by the scission of SIS will catalyze the conversion of PLA into lactide. The shoulders that appeared in the CL isotheral spectra are the proof that sustains this assumption. The accumulation of CL photo-emitters takes simultaneous place from both components, but the temporal contributions follow their stability relative order. The separation of these contributions is more clearly noticed at lower temperatures when the degradation rates are larger.

The isothermal CL spectra obtained from the oxidation of unirradiated PLA/SIS 30 emphasize the two degradation sectors that correspond to the two blending components. Even though the hydrocarbon phase is degraded through the former step, when peroxyl radicals appear diluted in the PLA matrix, their concentration does not sustain the progress of oxidation inside this phase. After the longer duration of oxidation, the $RO_2·$ intermediates from the SIS molecules are decayed and their contribution to the degradation of the material is accordingly diminished. The reinforcing PLA with SiO_2 nanoparticles is a pertinent solution for preventing the oxidation of radicals formed by the peroxyl phase. The implication of PLA in the accumulation of degradation products is minimized due to the lack of reactions between PLA fragments and diffused oxygen [60]. This behavior allows concluding that the permeation of oxygen drops when the amount of silica increases.

The superior thermal performances of PLA/SIS 30 filled with SiO_2 are achieved by the existence of OH groups on the surface of particles [73]. Thus, the applications in food packaging must take into account the oxidation performances that are evidently influenced by the blending state and service conditions. The isothermal CL spectra demonstrate the earning noteworthy functionality in unfriendly environments.

The exposure of the studied material to the action of γ-radiation induces a diminution in the thermal stability, especially in the range of higher temperatures [74] (Figure 3). The oxidation starts earlier and the temperature increases over 200 °C, which may induce the slight recombination of hydrocarbon radicals and cyclization of lactide structures. According to the ESR investigations on γ-irradiated PLA, the crosslinking probability is certainly higher at the temperature exceeding 100 °C due to the appropriate concentration of radicals and the greater crosslinking radiochemical yield (about 5 for the irradiation accomplished at 170 °C) concerning the lower value of scission radiochemical yield (about 1.5 under similar conditions).

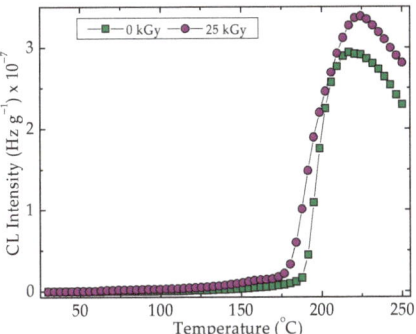

Figure 3. Nonisothermal CL spectra recorded on control PLA/SIS samples under different irradiation state. Heating rate: 3.7 °C min^{-1}.

The difference that exists between the radiation stabilities at the two different exposure doses (Figure 3) proves the predominant oxidation of SIS components rather than the degradation of PLA. If the PLA fragmentation is dominant, the curve shape would present a less tall peak at 225 °C.

Figure 4 shows the evolution of oxidative degradation on pristine materials where silica is the stabilization agent. This filled particle surface is in direct contact with the polymer substrate, creating the proper conditions for the formation of hydrogen bridges between the superficial hydroxyls and molecular fragments (Figure 4).

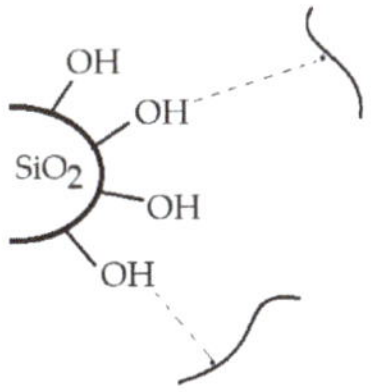

Figure 4. The illustration of the basic interaction between silica particles and free radicals.

The evolution of oxidation when the temperature is rising is detailed in Figure 5. For all the heating rates, the stability sequence places the silica concentration in the following order:

$$\text{none} < 3\% < 5\% < 10\%$$

The heating rate influences the level of oxidation mostly on the higher values of temperatures. In the case of slower heating, the oxidation peak presumably ascribed to the peroxyl radicals that appeared from the oxidation of hydrocarbon intermediates (the fragments from the scission of SIS) may be noticed at 225 °C. The higher heating rates create suitable conditions for the fast oxidation of radicals depleting them before the temperature reaches 200 °C. A specific feature that may be observed is the similitude in the oxidation degrees before 150 °C when all the samples keep constantly low CL emission intensities. The shoulder describing the formation of a significant amount of hydroperoxides as the result of the oxidation of radicals does not appear. This evolution is achieved when the degradation of polyolefins occurs in the presence of silica nanoparticles [75]. This peculiarity may be explained by the high content of PLA (70%), which does not form peroxide structures. Table 1 presents the basic values of OOT (onset oxidation temperature) when the oxidative degradation theoretically starts.

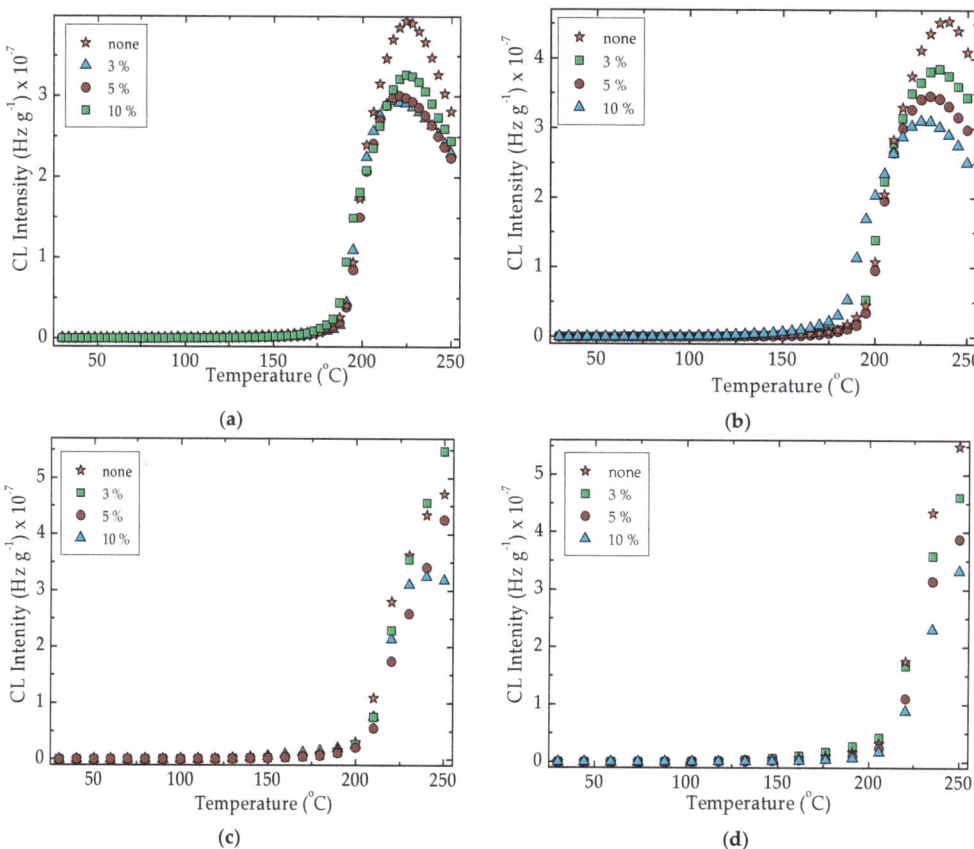

Figure 5. Nonisothermal CL spectra drawn for unirradiated PLA/SIS 30 in the presence of silica. Heating rates: (**a**) 3.7 °C min^{-1}; (**b**) 5 °C min^{-1}; (**c**) 10 °C min^{-1}; (**d**) 15 °C min^{-1}.

Table 1. OOT values for the oxidation of PLA/SIS blends reinforced with SiO$_2$ nanoparticles obtained by nonisothermal chemiluminescence.

SiO$_2$ Loading (%)	OOT (°C)			
	3.7 °C min^{-1}	5.0 °C min^{-1}	10.0 °C min^{-1}	15.0 °C min^{-1}
	D = 0 kGy			
0	170	175	177	187
3	175	190	205	210
5	177	191	208	215
10	187	192	210	218
	D = 25 kGy			
0	150	165	184	192
3	157	168	181	186
5	162	175	185	192
10	172	181	195	202

The γ-exposure of samples maintains similar aspects for each heating rate (Figures 5 and 6). The stability order follows the same sequence, where the control is the most vulnerable material. Though this energetic treatment does not deeply modify the evolution of emission

CL intensities, the OOT values are lower than for pristine samples (Table 1). The modification of the values ascribed to the oxidation onset temperatures is proof of the contribution of oxygen diffusion into the oxidizing material. The consequence of a faster increase in temperature is the real delay of oxidation due to the differences that exist between the rate of heating and the rate of diffusion. At the applied highest heating rate (15 °C min^{-1}) the degradation requires higher temperatures with the unlike values for each of the silica loadings. It is possible to assume that the oxidation takes place with different diffusion rates because the components provide specific amounts of free radicals in correlation with this oxygen diffusion process.

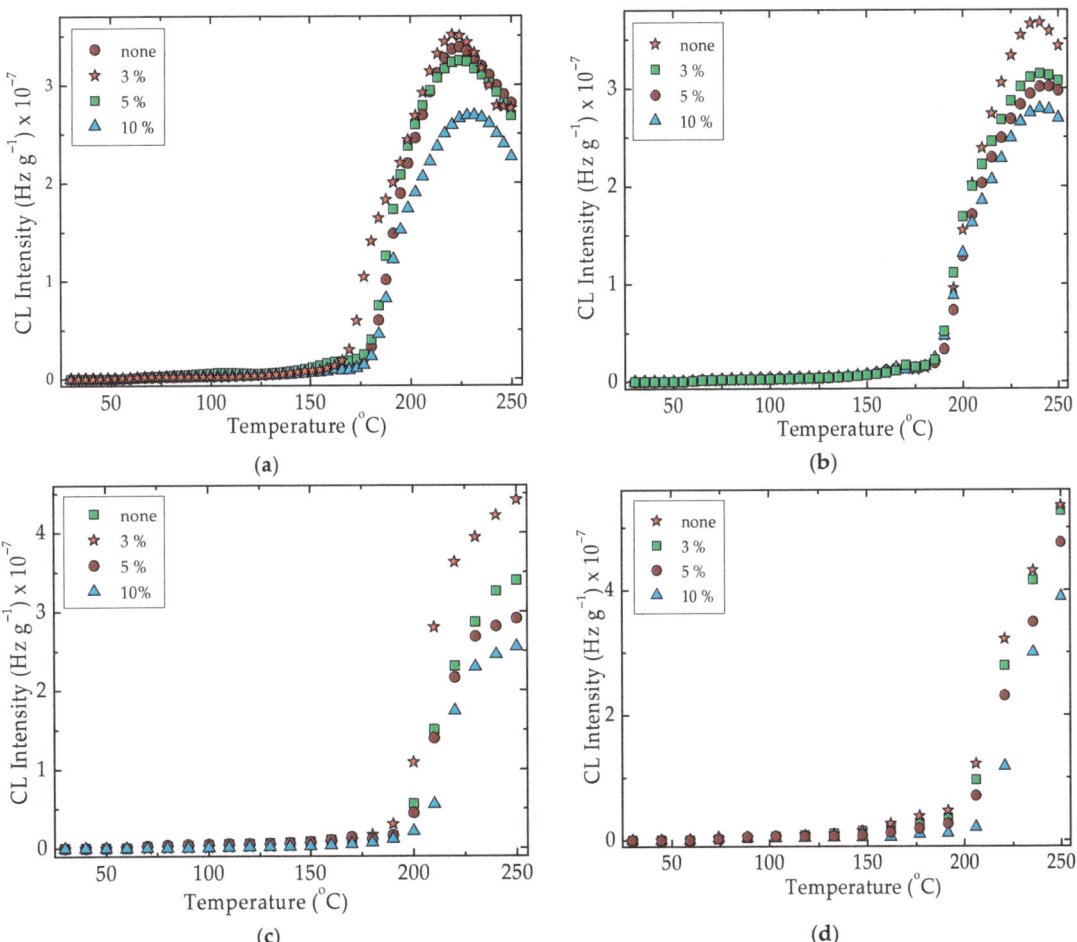

Figure 6. Nonisothermal CL spectra drawn for unirradiated PLA/SIS 30 in the presence of silica after receiving 25 kGy. Heating rates: (**a**) 3.7 °C min^{-1}; (**b**) 5 °C min^{-1}; (**c**) 10 °C min^{-1}; (**d**) 15 °C min^{-1}.

The progress of degradation has a similar appearance for all concentrations of silica filler, but the CL curve families present unlike aspects when different heating rates are applied. It means that the presence of the descending part in the higher temperature range is proof of the competition between oxidation and the simultaneous reactions in which the intermediates are usually involved in radiolysis (disproportionation or bimolecular processes between two peroxyl radicals) [76].

The adjustable degradation is obtained if the heating rate is diminished (Figure 6, Table 1). The increase in the thermal resistances of the inspected compositions is influenced by the heating rate. This feature may be correlated with the oxygen penetration rate, which occurs arduously in the samples with higher silica loadings. As it may be observed, the higher the heating rate, the greater the onset oxidation temperatures. Simultaneously, the intensity peak recorded in the high temperature range is shifted onto the higher values and for the heating rate of 15 °C min^{-1}, it disappears. The amounts of silica existing in the sample compositions decrease the amplitude of degradation, which is indicated by the relative positions of curves one under another. The main evidence for the contribution of silica to the improvement of thermal resistance is the upper positions of CL curves recorded on the pristine compositions, where the absence of silica permits the faster progress of oxidation.

The results presented in Table 1, as well as in Figure 7 indicate the opportunity of sterilization processing of these materials by radiation exposure when the alteration of stability does not match. On the contrary, this treatment ameliorates the oxidation resistance to a certain extent despite the degradation effects of high energy radiation. This behavior may be associated with an earlier crosslinking due to the inhibition effect of silica on the oxidation way of molecular fragments. Therefore, this sterilization operation is suitable not only for bacterial cleaning, but also for the extension of durability.

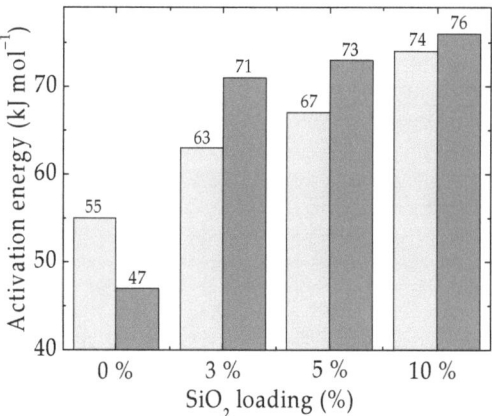

Figure 7. Histogram of the activation energies calculated for the oxidation of PLA/SIS/30 loaded with silica as nonirradiated samples (light grey) or the materials receiving 25 kGy (dark grey).

The damage consequences in the structure of the studied materials accompanying the weakening thermal resistance are evident when these materials are exposed to sunlight or an excess of stress. The appraisal of stability has to explain the internal interaction of components.

The γ-exposure of samples maintains similar aspects for each heating rate and the stability order follows the same sequence, where the control is the most vulnerable material. Though this energetic treatment does not deeply modify the evolution of emission CL intensities, the OOT values are lower than for pristine samples (Table 1). This behavior indicated the worsening operation performances when these compounds are subjected to an intense degradation process, such as γ-treatment or UV-exposure.

The OOT values are the input data for the calculation of the activation energies required by polymer samples by the Kissinger procedure [62]. As it is foreseen for the radiation affected hybrid, the activation energy is lower for the unmodified polymer sample, but the corresponding values for hybrids are significantly higher (Figure 7). The protection activity of silica is related to the catching of intermediates by the oxygen atoms existing on the nanoparticle surface. The demonstration of this behavior was presented by the grafting of (3-aminopropyl)triethyxysilane on the silica surface [77,78]. The higher energetic requirements suggest the effectiveness of the filler contribution to the delay of oxidation that recommends this neutral additive for the extension of material durability in risky applications.

The stabilization effect induced by the presence of silica on the composition of PLA hybrids was also reported [79]. The present values are lower than 100 kJ mol^{-1} found for the degradation of PLA by the conversion procedure because our materials contain a radiation degradable structure, namely SIS. However, these satisfactory values displayed are reliable proofs for foreseen purposes, such as materials for food packaging or safety preservation during shipment. The already aforementioned comments highlight the strengthening effect that leads to the improved operation life of these materials and the lack of danger.

3.2. Gelation

By the energy transfer on the macromolecules of the PLA/SIS 30 compound, the radiation processing modified with silica is the determinant cause of the fragmentation of the macromolecules, generating free radicals. According to the radiolysis mechanisms, they may be decayed by oxidation or recombination. The decrease of the soluble fraction (sol, S) is the measure of the inter-radical reactions and indicates the gelation of the material. In Figure 8 there is a visible difference between the material without silica and the samples containing this inorganic compound. While the pristine blend is sharply damaged, the hybrids present slight gelation due to the contribution of hydrocarbon radicals formed from the SIS structure. The contribution of silica to the protection against oxidation is defined by the slope of lines. The greater the slope, the larger amount of insoluble fraction is formed.

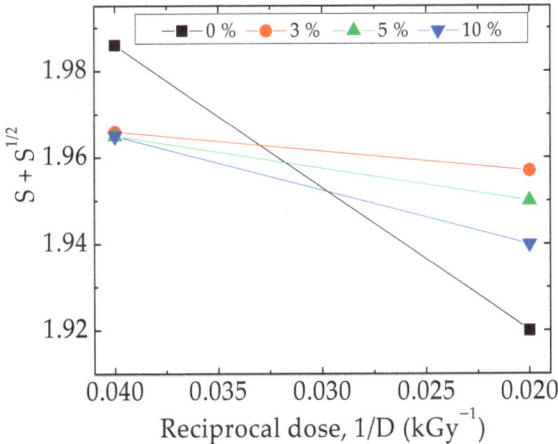

Figure 8. Charlesby–Pinner representation of radiation effects in PLA/SIS 30/SiO$_2$ hybrids.

3.3. FTIR Evaluation

The spectral assay on the oxidation state after the γ-exposure of PLA/SIS 30 in the presence of silica is presented in Figure 9. Even though the PLA molecules contain oxygen and their scission would generate oxygen-centered radicals, this contribution represents a spectral background that is similar to all the inspected compositions. The differences that exist are found in the differences in transmission values due to the oxidation process sustained by hydrocarbon radicals. As may be noticed from Figure 9 the accumulation level of carbonyl products decreases with the concentration of silica. As the CL determinations already showed, the silica loading brings about certain protection against oxidation, which is delayed proportionally with filler assisting components.

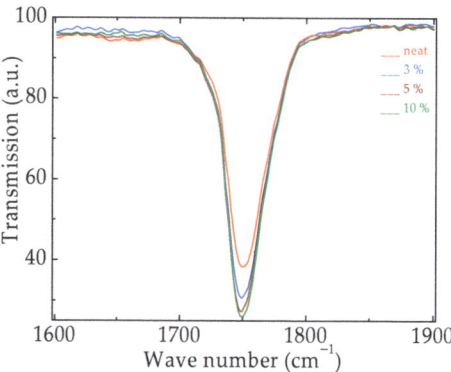

Figure 9. The FTIR spectra recorded on the PLA/SIS 30 modified with nanoparticles of silica after their exposure to a γ dose of 50 kGy.

If it is taken into consideration that all studied compositions have similar FTIR spectra, the image offered of the progress of oxidation is accurately appraised by the differences in the transmission figures.

3.4. SEM Assay

The SEM investigation of the PLS/SIS 30 samples reveals the homogenization of the polymer substrate (Figure 10). The clear separation of silica particles may be noticed when the surrounding polymer becomes the proper material for product modeling. The filler is homogenously dispersed and the morphology shown by these micrographs demonstrates the availability to form composites as was reported earlier [80].

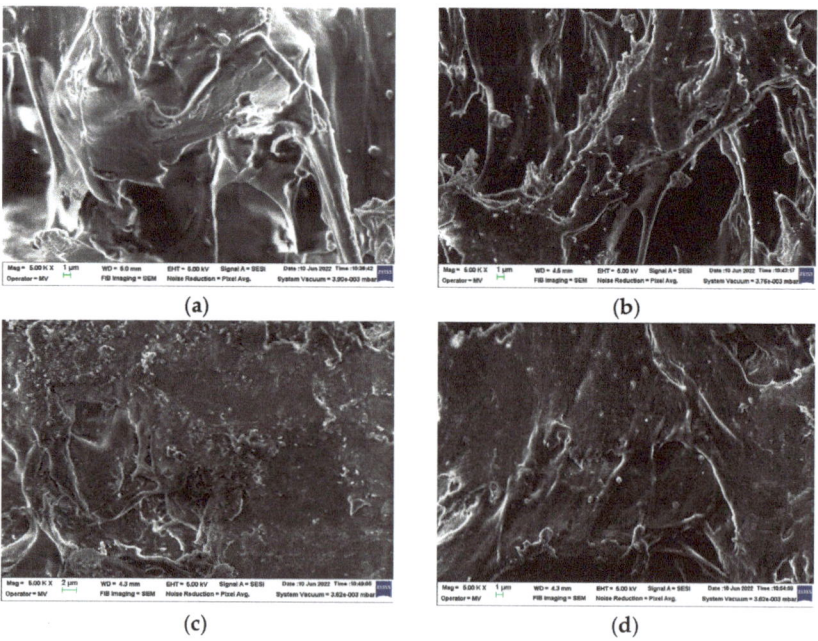

Figure 10. The SEM images on nonirradiated PLA/SIS 30 samples. Compositions: (**a**) control; (**b**) SiO_2 3%, (**c**) SiO_2 5%, (**d**) SiO_2 10%. Magnification: 5×10^3.

It would be expected that γ-irradiation, the process which initiates the damage of macromolecules, affects the microstructures of samples. However, the dominant component, PLA, subjected to γ-radiolysis follows a degradation mechanism through which its molecules drop down their macrostructures, but the resulting fragments are oxidized to lactides. As it can be noticed from Figure 11, the curing feature is demonstrated. It means that the stabilization action of silica particles is an important factor that promotes a slight crosslinking as the comparison of corresponding images illustrates. Therefore, the presence of silica in the PLA/SIS 30 samples makes the compaction in the polymer phase possible. The filler particles are clearly separated by organic materials, where scissions occur to feed the crosslinking process. The free radicals nearby particle surfaces may interact easily with hydroxyls belonging to silica (Figure 4) hindering their coalescence. The great advantage of this presence is the prevention of material cleavage which would influence the penetration of oxygen. The integrity of samples after γ-irradiation is a consequence of the perfect compatibility of mixed components.

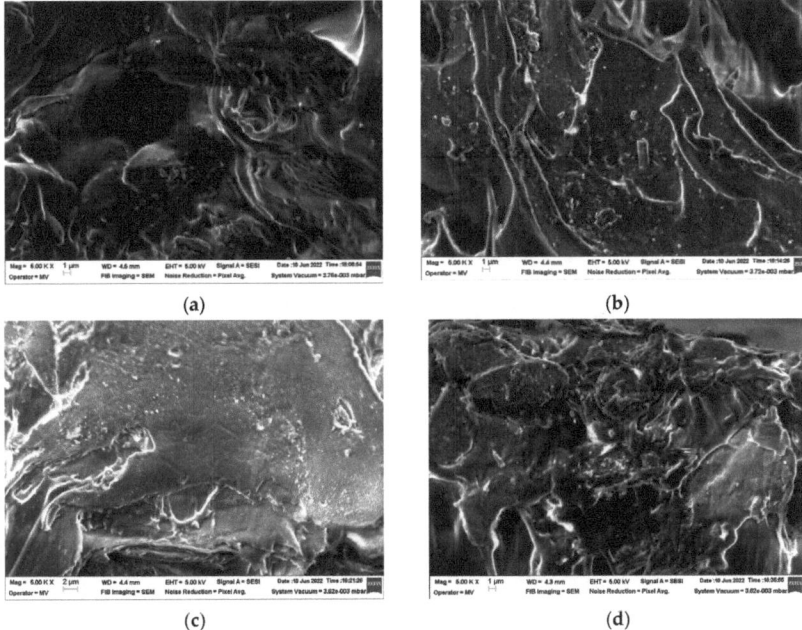

Figure 11. The SEM images on PLA/SIS 30 samples irradiated at 25 kGy. Compositions: (**a**) control; (**b**) SiO$_2$ 3%, (**c**) SiO$_2$ 5%, (**d**) SiO$_2$ 10%. Magnification: 5×10^3.

4. Discussion

The silica/polymer composites show convenient properties related to the processability or spectrum of applications. The covalent interfacial connections between silica particles and polymer substrate are the reason for the appropriate performances of these hybrids [81,82]. The nanotechnologies approaches provide the specific solutions for the formulations of resistant materials, which are based on improving the contribution of inorganic fillers [83,84]. Various formulations based on PLA try to find optimal solutions for the enlargement of application areas: biodegradability [74,85,86], medical range [87,88], packaging products [89,90], optical bioimaging [91], controlling the release of hydrophilic pharmaceutical compounds [92], production of contractible materials [93], recycling [94].

The addition of silica in the polymer matrices is a proper way to improve the main characteristics of these materials [95–97]. This statement is based on the interaction between the polar intermediates and the hydroxyls existing on the surface of filler particles. The evolution of degradation is correlated with the contributions provided by components, especially by SIS, whose scission process initiates a radical mechanism. However, the stability aspects describing the progress of oxidative degradation are often disregarded, because the assays are axed on the functional properties and the relation between the composition and the consequences of external factors on the material behavior, for example [98–100]. The customers that are the beneficiaries of research need the stability characterization of materials that they purchase based on the present results concerning the activation energies involved in the stabilization of these types of substrates. Accordingly, the intimate interaction of the two phases (organic and inorganic ones) may be assessed by the increase of stability by about 20–25% due to the presence of silica (Figure 7). It means that the durability tested with the accelerated degradation by γ-exposure is significantly augmented and the product lifetimes are substantially extended.

As would be expected from other previous investigations [101], the thermal performances of PLA/SIS 30 hybrids with silica are influenced by the concentration of filler (Figure 2). The presence of silica whose structure allows intimate connections with the radical fragments by means of hydroxyl bridges brings about a visible increase to the values of oxidation induction time. This essential kinetic parameter becomes 128 min, 210 min and 320 min for the SiO_2 concentration of 3 wt%, 5 wt% and 10 wt%, respectively, while it remains 36 min for pristine composition. Thus, the silica nanoparticles efficiently scavenge the degradation intermediates that delay the aging process by interfacial interactions.

The nonisothermal spectra (Figures 5 and 6) are appropriate proof confirming the prevention effect of silica on the progress of oxidation in PLA based compounds as was presented in other previous papers [102,103]. This behavior is in contrast with the instability of polyolefins [77], which present prominent CL intensity maxima at 100 °C by the abundant formation of hydroperoxides. The thermal characteristics of the major component, PLA, in the presence of silica particles (melting point around 155 °C and Tg is placed between 53 and 63 °C [104]) are modeled and a sharp increase of CL intensity over 160 °C proves the contribution of the nanofiller phases. The differences that exist appear at higher temperatures when the thermal diffusion is more pronounced allowing a closer nearness of polymer fragments to the silica nanoparticles. These dissimilarities shown by the OOT values (Table 1) demonstrate that this filler acts by the attraction of oxidizing intermediates. The higher loading of silica shows the most efficiency in the prevention of oxidative degradation.

The analysis of the activation energy values (Figure 6) offers the possibility to estimate the interaction depth between material components, whose concentrations indicate a strong dependence. As the loading of silica increases, the values of Ea become higher. This improvement is tightly related to the formation of active intermediates, which keep the surface of silica particles and the forming radicals closed. It limits the progress of oxidation by withdrawing intermediates from the degradation chain during the propagation stage. The γ-exposure decreases the required values of activation energies because the competition between each free radical diminishes the probability of protective hindering and the availability of active radicals for their oxidation is more extended. Always, radiolysis is a source of radicals, because, in the initiation stage of degradation, the transferred energy from the incidental radiation onto the substrate exceeds the bond energies.

The augmentation in the thermal resistance of polymers by the activity of silica is correlated with the formation of hydroxyls that are formed on the nanoparticle great surface (Figure 4) as the result of the reaction of protons generated by the degradation of the polymer and the superficial oxygen atoms that are incorporated in the filler structure [102–105] or by local hydrolysis where water molecules are formed during radiolysis [106].

The progress of degradation may be accurately characterized by the measurements of CL emission that watch the accumulation of oxidation intermediates. If the isothermal version of this procedure is applied to the description of filler contribution to the initiation and propagation of oxidation (Figure 2), when the oxidation induction time (OITs) increases by ten times from 35 min to 322 min for the raw blend to the composition containing 10 wt% silica, the nonisothermal measurements allow the detailed comparison of the start-point of oxidation and the acceleration of degradation by the increase of temperature (Figures 5 and 6). Thus, the thermal regimes define the manner by which the silica nanoparticles may delay the oxidation by the breaking degradation mechanism. The isothermal CL spectra demonstrate that the increase in the silica loading brings about a significant extension of induction time, the period when the material is not sensitively altered. Therefore, for the unirradiated samples, the oxidation induction time increases by 3.2 times, when the concentration of SiO_2 is 3%, by five times as the SiO_2 loading is 5% and by about nine times if the content of SiO_2 is 10% (Figure 12). The evolution of oxidation is tightly related to the permeation degree [107]. This diffusion feeds the polymer bulk sustaining the progress of degradation. One procedure for the diminution of the penetration of oxygen into a polylactide is the grafting of organosilane on a silica nanoparticle surface when an efficient barrier is formed [108].

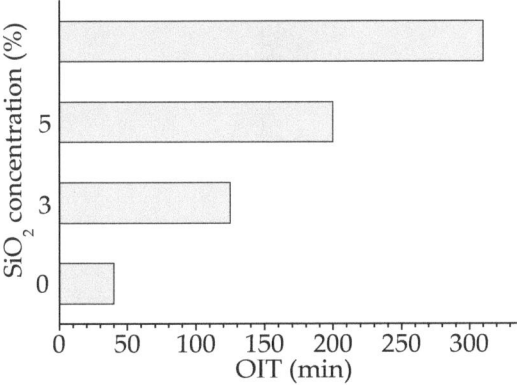

Figure 12. The oxidation induction times of the oxidation of PLA/SIS 30/SiO_2 hybrids obtained by the isothermal CL measurements at 140 °C.

The evolution of oxidation is evidenced by the increasing transmission values for the carbonyl band. The accumulation of ketonic structures is possible if an appropriate hindrance factor does not exist. Despite the depolymerization of the polylactide component occurring during γ-irradiation [45,109], the fragments provided by the scission of the SIS component, a hydrocarbon structure, generate vulnerable radicals able to be oxidized if there is not any protection [12,110]. The nanoparticles of silica act obviously, generating free radicals and catching them on their surface. Consequently, their conversion into oxygen-containing products is obstructed and the transmission increases slowly.

This augmentation is the effect of the lack of any protection factor, even though the compound is not an authentic antioxidant; the stabilization action is based on the formation of hydroxyl moieties as a consequence of oxidation [111]. In the degradation mechanism, the competition between oxidation and protection is won by the material stability where the content of silica is high enough to withdraw radicals from their oxidation chain.

The accelerated degradation caused by the exposure of patterns to the damaging action of γ-radiation advances much slower in the processed material with increased silica loadings. The energetic conditions shown in Figure 7 demonstrate the narrowing differences in the values of activation energies. While the pristine composition has lower activation energy for its oxidative degradation after γ-treatment, the materials containing various amounts of filler present higher Ea. This behavior indicates the protection activity of silica particles with respect to the molecular fragmentation of macromolecules. This effect may be ascribed to the withdrawal of free radical intermediates from the oxidation route via the formation of hydroperoxides [112]. The stability study of the hemp-PLA composites [113] reports the activation energies for the thermal degradation of the investigated compositions in the range of 159–163 kJ mol^{-1}; the PLA hybrids stabilized with POSS present activation energies between 80 and 100 kJ mol^{-1}; the presence of SIS decreases this parameter due to its relative low instability [60].

The mechanistic approach must take into account the stability differences that exist between the components, which influence the fates of generated radicals. While the hydrocarbon structure of SIS is oxidized by the formation of hydroperoxides followed by their second stage reactions of conversion into oxygen-containing compounds [76], the polylactide is decomposed by depolymerization [29] and the CL emission is not recorded. The damage to this component is unlikely to be associated with the formation of lactide monomers, which sustains oxidation without the formation of ketonic structures as CL emitters. The degradation progresses does not concern the two polymers, but the contributions of intermediates of both former structures follow dissimilar ways.

A conclusive illustration of the essential steps that run through the irradiated PLA/SIS filled with silica particles is offered in Figure 13.

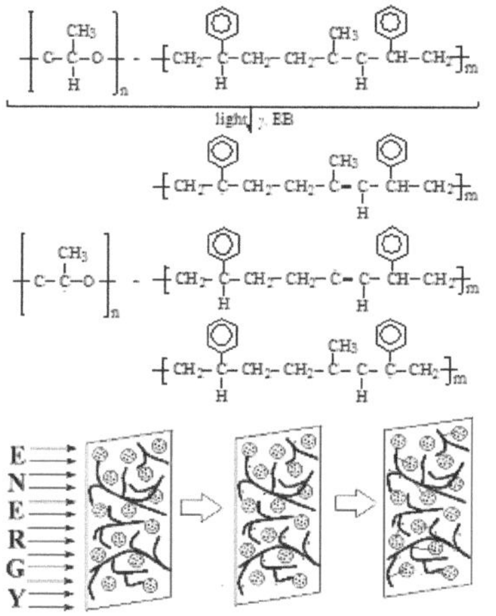

Figure 13. The illustration of pathway through which the radicals and silica particles interact in the PLA/SIS samples.

5. Conclusions

This paper analyses the improvement of thermal resistance in the cases of PLA/SIS 30/n-SiO$_2$ hybrids. The protection action is based on the interaction between inorganic and organic phases at the boundary zone due to the bridges established by hydrocarbon free radicals on the oxygen atoms belonging to the external layer of silica nanoparticles. The increase in silica loadings up to 10% extends the peroxidation periods which are illustrated by the growing oxidation induction time shown in the isothermal CL determinations. The measure of stabilization efficiency is suggested by the increasing values of activation energy of the oxidative degradation of polymer matrices as the amount of SiO$_2$ is enhanced. While this value decreases for the pristine blend after an accelerated degradation by γ-exposure, the energetic requirements of material degradation become higher. It demonstrates that the blend free of filler is unprotected, but the presence of silica keeps the unchanged structure of the materials. The appropriate proof for the contribution of silica to the protection of polymer samples is the behavior of the irradiated specimens subjected to radiation induced crosslinking. The Charlesby–Pinner representation reveals the advantageous contribution of silica to the promotion of radical recombination instead of their reactions with molecular oxygen that diffuses into the materials during radiation processing. The variation of the carbonyl band for the irradiated samples is evidence of the scavenging efficiency of silica concerning the decaying intermediates. This insight gathering all the pertinent arguments indicates nanoparticles of silica as an appropriate formulation component of plastic products that show an improved resistance against oxidation even in the harmful conditions of accelerated degradation caused by γ-irradiation and thermal degradation.

These compositions may be properly chosen for the production of commodities, such as flexible medical wear, automotive accessories, packaging sheets, protection covers, and sealant gaskets due to their adaptableness to severe environmental conditions.

Author Contributions: Conceptualization, T.Z. and A.M.L.; methodology, T.Z.; validation, T.Z., M.M., A.M.L., V.E.M. and H.I.; formal analysis, A.M.L.; investigation, T.Z., V.E.M. and A.M.L.; data curation, M.M.; writing—original draft preparation, A.M.L., V.E.M. and T.Z.; writing—review and editing, A.M.L., M.M. and T.Z.; visualization, H.I.; supervision, H.I. All authors have read and agreed to the published version of the manuscript.

Funding: This research received no external funding.

Institutional Review Board Statement: Not applicable.

Informed Consent Statement: Not applicable.

Data Availability Statement: Not applicable.

Conflicts of Interest: The authors declare no conflict of interest.

References

1. Singh, P.; Singari, R.M.; Mishra, R.S.; Bajwa, G.S. A review on recent development on polymeric hybrid composite and analysis of their mechanical performances. *Mater. Today Proc.* **2022**, *56*, 3692–3701. [CrossRef]
2. Zagho, M.M.; Hussein, E.A.; Elzatahry, A.A. Recent overviews in functional polymer composites for biomedical applications. *Polymers* **2018**, *10*, 739. [CrossRef]
3. Altstdät, V. (Ed.) *Polymer Blends and Compatibilization*; MDPI: Basel, Switzerland, 2017.
4. Ding, Y.; Abeykoon, C.; Perera, Y.S. The effects of extrusion parameters and blend composition on the mechanical, rheological and thermal properties of LDPE/PS/PMMA ternary polymer blends. *Adv. Ind. Manuf. Eng.* **2022**, *4*, 100067. [CrossRef]
5. Dorigato, A. Recycling of polymer blends. *Adv. Ind. Eng. Polym. Res.* **2021**, *4*, 53–69. [CrossRef]
6. Qiao, L.; Liu, C.; Liu, C.; Hu, F.; Li, X.; Wang, C.; Jian, X. Study on new polymer crosslinking agents and their functional hydrogels. *Eur. Polym. J.* **2020**, *134*, 109835. [CrossRef]
7. Barsbay, M.; Güven, O. Nanostructuring of polymers by controlling of ionizing radiation-induced free radical polymerization, copolymerization, grafting and crosslinking by RAFT mechanism. *Radiat. Phys. Chem.* **2020**, *169*, 107816. [CrossRef]
8. Maraveas, C.; Bayer, I.S.; Bartzanas, T. Recent advances in antioxidant polymers: From sustainable and natural monomers to synthesis and applications. *Polymers* **2021**, *13*, 2465. [CrossRef]
9. Zaharescu, T.; Ilieș, D.-C.; Roșu, T. Thermal and spectroscopic analysis of stabilization effect of copper complexes in EPDM. *J. Therm. Anal. Calorim.* **2016**, *123*, 231–239. [CrossRef]
10. Kanbua, C.; Sirichaibhinya, T.; Rattanawongwiboon, T.; Lertsarawut, P.; Chanklinhorm, P.; Ummartyotin, S. Gamma radiation-induced crosskinking of Ca^{2+} loaded poly(lactic acid) and poly(ethylene glycol) diacrylate networks fr polymer gel electrolytes. *S. Afr. J. Chem. Eng.* **2022**, *39*, 90–96.
11. Manaila, E.; Stelescu, M.D.; Craciun, G. Aspects regarding radiation crosslinking of elastomers. In *Advanced Elastomers—Technology, Properties and Applications*; Boczkowska, A., Ed.; IntechOpen: London, UK, 2012; pp. 3–34.
12. Ferry, M.; Ngoro, Y. Energy transfer in polymers submitted to ionizing radiation. *Radiat. Phys. Chem.* **2020**, *134*, 109835. [CrossRef]
13. Kodal, M.; Alchekh Wis, A.; Ozkoc, G. The mechanical, thermal and morphological properties of γ-irradiated PLA/TAIC and PLA/OvPOSS. *Radiat. Phys. Chem.* **2018**, *153*, 214–225. [CrossRef]
14. Khalid, M.; Thery Ratnam, C.; Wei, S.J.; Reza Ketabchi, M.; Raju, G.; Walvekar, R.; Mujawar Mubarak, N. Effect of electron beam radiation on poly(lactic acid) biocomposites reinforced with waste tea powder. *Radiat. Phys. Chem.* **2021**, *188*, 109612. [CrossRef]
15. Zaharescu, T.; Blanco, I. Antioxidant effects of Rosemary extract on the accelerated degradation of ethylene-propylene-diene monomer. *Macromol. Symp.* **2021**, *395*, 2001300.
16. Cardoso, E.C.L.; Oliviera, R.R.; Machado, G.A.F.; Moura, E.A.B. Study of flexible films prepared from PLA/PBAT blend and PLA e-beam irradiated as compatibilizing agent. In *Characterizing of Minerals, and Materials*; Ikhmayies, S., Li, B., Carpenter, J.S., Li, J.I., Hwang, J.Y., Monteiro, S.N., Firrao, D., Zhang, M., Peng, Z., et al., Eds.; Springer: Cham, Switzerland, 2017; pp. 121–129.
17. Bednarek, M.; Borska, K.; Kubisa, P. Crosslinking of polylactide by high energy irradiation and photo-curing. *Molecules* **2020**, *25*, 4919. [CrossRef] [PubMed]
18. Murariu, M.; Dubois, P. PLA composites: From production to properties. *Adv. Drug Deliver. Rev.* **2016**, *107*, 17–46. [CrossRef]
19. Casalini, T.; Rossi, F.; Castrovinci, A.; Perale, G. A perspective on polylactic acid-based polymers use for nanoparticles synthesis and applications. *Front. Bioeng. Biotechnol.* **2019**, *7*, 259. [CrossRef]
20. Idumah, C.I.; Nwabanne, J.T.; Tanjung, F.A. Novel trends in poly(lactic) acid hybrid bionanocomposites. *Cleaner Mater.* **2021**, *2*, 100022. [CrossRef]

21. Freeland, B.; McCarthy, E.; Balakrishnan, R.; Fahy, S.; Boland, A.; Rochfort, K.D.; Dabros, M.; Marti, R.; Kelleher, S.M.; Gaughran, J. A review of polylactic acid as a replacement material for single-use laboratory components. *Materials* **2022**, *15*, 2989. [CrossRef]
22. Zaharescu, T.; Râpă, M.; Marinescu, V. Chemiluminescence kinetic analysis on the oxidative degradation of poly(lactic acid). *J. Therm. Anal. Calorim.* **2017**, *128*, 185–191. [CrossRef]
23. Stefanik, K.; Masek, A. Green copolymer based on poly(lactic acid)—Short review. *Materials* **2021**, *14*, 5254. [CrossRef]
24. Zaharescu, T.; Râpă, M.; Lungulescu, E.M.; Butoi, N. Filler effect on the degradation of γ-processed PLA/vinyl POSS hybrid. *Radiat. Phys. Chem.* **2018**, *153*, 188–197. [CrossRef]
25. Sanusi, O.M.; Benelfellah, A.; Papadopoulos, L.; Terzopoulu, Z.; Malletzidou, L.; Vasileiadis, I.G.; Chrissafis, K.; Bikiaris, D.N.; Hocine, N.A. Influence of montmorillonite/carbon nanotube hybrid nanofillers on the properties of poly(lactic acid). *Appl. Clay Sci.* **2021**, *201*, 105925. [CrossRef]
26. Scaffaro, R.; Botta, L.; Maio, A.; Gallo, G. PLA graphene nanoplatelets nanocomposites: Physical properties and release kinetics of an antimicrobial agent. *Compos. Part B-Eng.* **2017**, *109*, 138–146. [CrossRef]
27. Zhang, H.; Luo, D.; Zhen, W. Preparation, crystallization and thermo-oxygen degradation kinetics of poly(lactic acid)/fulvic acid-g-poly(isoprene) grafting polymer composites. *Polym. Bull.* **2022**, *79*, 3155–3174. [CrossRef]
28. Arunprasath, K.; Vijayakumar, M.; Ramarao, M.; Arul, T.G.; Penil Pauldoss, S.; Selwin, N.; Radhakrishnan, B.; Manikandan, N. Dynamic mechanical analysis performances of pure 3D printed polylactic acid (PLA) and acrylonitrile butadiene styrene (ABS). *Mater. Today* **2022**, *50*, 1559–1562. [CrossRef]
29. Kervran, M.S.; Vagner, C.; Cochez, M.; Ponçot, M.; Saeb, M.R.; Vahabi, H. A review on thermal degradation of polylactic acid (PLA)/polyhydroxybutyrate (PHB) blends. *Polym. Degrad. Stab.* **2022**, *201*, 109995. [CrossRef]
30. Ordoñez, R.; Atarés, L.; Chiralt, A. Properties of PLA films with cinnamic acid: Effect of the processing method. *Food Bioprod. Proc.* **2022**, *133*, 25–33. [CrossRef]
31. Teixeira, S.; Eblagon, K.M.; Miranda, F.; Pereira, M.F.R.; Figueiredo, J.L. Towrds controlled degradation of poly (lactic acid in technical applications. *C* **2021**, *7*, 42.
32. Sharma, B.; Shekhar, S.; Sharma, S.; Jain, P. The paradigm in conversion of plastic wastes into value added materials. *Clean. Eng. Technol.* **2021**, *4*, 100254. [CrossRef]
33. Aryan, V.; Maga, D.; Majgainkar, P.; Hanich, R. Valorization of polylactic acid (PLA) waste: A comparative life cycle assessment of vatrious csolvent-based chemical recycling technologies. *Resour. Conserv. Recy.* **2021**, *172*, 105670. [CrossRef]
34. Gere, D.; Czigany, T. Future trends of plastic bottles recycling. *Polym. Test.* **2020**, *81*, 106160. [CrossRef]
35. Zawawi, E.Z.E.; Noor-Hafizah, A.H.; Romli, A.Z.; Yuliana, N.Y.; Bonnia, N.N. Effect of nanoclay on mechanical and morphological properties of poly(lactide) acid (PLA) and polypropylene (PP) blends. *Mater. Today-Proc.* **2021**, *46*, 1778–1782. [CrossRef]
36. Moradi, S.; Khademzadeh Yeganeh, J. Highly toughened poly(lactic acid) (PLA) prepared through melt blending with ethylene-co-vinyl acetate (EVA) copolymer and simultaneous addition of hydrophilic silica nanoparticles and block copolymer compatibilizer. *Polym. Test.* **2020**, *91*, 106735. [CrossRef]
37. Ferrer, I.; Manresa, A.; Méndez, J.A.; Delgado-Aguilar, M.; Garcia-Romeu, M.L. Manufacturing PLA/PCL blends by ultrasonic molding technology. *Polymers* **2021**, *13*, 2412. [CrossRef]
38. Fang, H.; Zhang, L.; Chen, A.; Wu, F. Improvement of mechanical property for PLA/TPU blend by addition PLA-TPU copolymer prepared via in situ ring-opening polymerization. *Polymers* **2022**, *14*, 1530. [CrossRef]
39. Huang, Y.; Brünig, H.; Müller, M.T.; Weissner, S. Melt spinning of PLA/PCL blends modified with electron induced reactive processing. *J. Appl. Polym. Sci.* **2022**, *139*, 51902. [CrossRef]
40. Malinowski, R. Some effects of radiation treatment of biodegradable PCL/PLA blends. *J. Polym. Eng.* **2018**, *38*, 635–640.
41. Malinowski, R.; Moraczewski, K.; Raszkowska-Kaczor, A. Studies on the uncrosslinked fraction of PLA/PBAT blends modified by electron radiation. *Materials* **2020**, *13*, 1068. [CrossRef]
42. Liu, M.; Yin, Y.; Fan, Z.; Zhen, X.; Shen, S.; Deng, P.; Zheng, C.; Teng, H.; Zhang, W. The effects of gamma-irradiation on the structure, thermal resistance and mechanical properties of the PLA/EVOH blends. *Nucl. Instrum. Meth. B* **2012**, *274*, 139–144. [CrossRef]
43. Iuliano, A.; Nowacka, M.; Rybak, K.; Rzepna, M.R. The effects of electron beam radiation on material properties and degradation of commercial PBAT/PLA blends. *J. Appl. Polym. Sci.* **2020**, *137*, 48462. [CrossRef]
44. Birkinshaw, C.; Buggy, M.; Henn, G.G.; Jones, E. Irradiation of poly-D,L-lactide. *Polym. Degrad. Stab.* **1992**, *38*, 249–253. [CrossRef]
45. Nugroho, P.; Mitomo, H.; Yoshii, F.; Kume, T. Degradation of poly(L-lactic acid) by γ-irradiation. *Polym. Degrad. Stab.* **2001**, *72*, 337–343. [CrossRef]
46. Adamus-Wlodarczyk, A.; Wach, R.A.; Ulanski, P.; Rosiak, J.M.; Sicka, Z.; Al-Sheikhly, M. On the mechanisms of the effects of ionizing radiation on diblock and random copolymer of poly(lactic acid) and poly(trimethylene carbonate). *Polymers* **2018**, *10*, 672. [CrossRef]
47. Dorati, R.; Colonna, C.; Serra, M.; Genta, I.; Modena, T.; Pavanetto, F.; Perugini, P.; Conti, B. γ-Irradiation of PEG, d,l PLA and PEG-PLGA multiblock copolymers: Effect of irradiation doses. *AAPS Pharm. Sci. Tech.* **2008**, *9*, 718–725. [CrossRef]
48. Ahmandi Abrishami, S.; Nabipour Chakoli, A. Effect of radiation processing on physical properties of aminated MWCNTs/poly(L-lactide) nanocomposites. *Compos. Commun.* **2019**, *14*, 43–47. [CrossRef]
49. Mansouri, M.; Berrayah, A.; Beyens, C.; Rosenauer, C.; Jama, C.; Maschke, U. Effect of electron beam irradiation on thermal and mechanical properties of poly(lactic acid) films. *Polym. Degrad. Stab.* **2016**, *133*, 293–302. [CrossRef]

50. Salazar, R.; Salas-Gomez, V.; Alvarado, A.S.; Baykara, H. Preparation, characterization and evaluation of antibacterial properties of polylactide-polyethylene glycol-chitosan active composite films. *Polymers* **2022**, *14*, 2266. [CrossRef]
51. Al Zahmi, S.; Alhammadi, S.; El Hassan, A.; Ahmed, W. Carbon fiber/PLA recycled composites. *Polymers* **2022**, *14*, 2194. [CrossRef]
52. Siakeng, R.; Jawaid, M.; Asim, M.; Fouad, H.; Saba, N.; Siengchin, S. Flexural and dynamic mechanical properties of alkali-treated coir/pineapple leaf fibers reinforced polylactic acid hybrid biocomposites. *J. Bionic Eng.* **2021**, *18*, 1430–1438. [CrossRef]
53. Wang, S.; Hu, S.; Zhang, W.; Olonisakin, K.; Zhang, X.; Ran, L.; Yang, W. Processing pulp fiber into high strength composites. *Compos. Interface* **2021**, *29*, 817–831. [CrossRef]
54. Zourari, M.; Devallance, D.B.; Marrot, L. Effect of biochar addition on mechanical properties, thermal stability, and water resistance of hemp-polylactic acid (PLA) composites. *Materials* **2022**, *15*, 2271. [CrossRef] [PubMed]
55. Ludwiczak, J.; Frąckowiak, S.; Leluk, K. Study of thermal, mechanical and barrier properties of biodegradable PLA/PBAT films with high oriented MMT. *Materials* **2021**, *14*, 7189. [CrossRef] [PubMed]
56. Kumar, A.; Rao Tumu, V. Physicochemical properties of the electron beam irradiated bamboo powder and its bio-composites with PLA. *Compos. Part B Eng.* **2019**, *175*, 107098. [CrossRef]
57. Kostic, D.; Vukasinovic-Sekulic, M.; Armentao, I.; Torre, L.; Obradovic, B. Multifuntional ternary composite films based on PLA and Ag-alginate microbeads: Physical characterization and silver release kinetics. *Mater. Sci. Eng.* **2019**, *98*, 1159–1168. [CrossRef]
58. Chen, W.; Qi, C.; Li, Y.; Tao, H. The degradation investigation of biodegradable PLA/PBAT blend. Thermal stability, mechanical properties and PALS analysis. *Radiat. Phys. Chem.* **2021**, *180*, 109239. [CrossRef]
59. Vidotto, M.; Mihaljević, B.; Žauhar, G.; Vidović, E.; Martar-Strmečki, N.; Klepec, D.; Valić, S. Effect of γ-radiation on structure and properties of poly(lactic acid) filaments. *Radiat. Phys. Chem.* **2021**, *184*, 109456. [CrossRef]
60. Luchian-Lupu, A.-M.; Zaharescu, T.; Lungulescu, E.-M.; Râpă, M.; Iovu, H. Availability of PLA/SIS blends for packaging and medical applications. *Radiat. Phys. Chem.* **2020**, *172*, 108696. [CrossRef]
61. Koutný, M.; Václvková, T.; Matisová-Rychlá, L.; Rychlý, J. Characterization of oxidation progress by chemiluminescence. A study of polyethylene with pro-oxidant additives. *Polym. Degrad. Stab.* **2008**, *93*, 1515–1519. [CrossRef]
62. Kissinger, H.E. Reaction kinetics in differential thermal analysis. *Anal. Chem.* **1957**, *29*, 1702–1706. [CrossRef]
63. Zhang, W.; He, T.; Sun, J.; Qian, B. A general equation for the relationship between sol fraction and radiation dose in radiation crosslinking. *Radiat. Phys. Chem.* **1989**, *33*, 581–584.
64. Gupta, M.C.; Deshmukh, V.G. Radiation effects on poly(lactic acid). *Polymer* **1983**, *24*, 827–830. [CrossRef]
65. Babanalbandi, A.; Hill, D.J.T.; O'Donnell, H.J.; Pomery, P.J.; Whittaker, A. An electron spin resonance study on γ-irradiated poly(L-lactic acid) and poly(D,L-lactic acid). *Polym. Degrad. Stab.* **1995**, *50*, 297–304. [CrossRef]
66. Södergard, A.; Stolt, M. Properties of lactic acid based polymers and their correlation with composition. *Prog. Polym. Sci.* **2001**, *27*, 1123–1163. [CrossRef]
67. Hamad, K.; Kaseem, M.; Yang, H.W.; Deri, F.; Ko, Y.G. Properties and medical applications of polylactic acid. A review. *Express Polym. Lett.* **2015**, *9*, 435–455. [CrossRef]
68. De Stefano, V.; Khan, D.S.; Tabada, A. Applications of PLA in modern medicine. *Eng. Regener.* **2020**, *1*, 76–87.
69. Piekarska, K.; Sowinski, P.; Piorkowska, E.; Haque, M.M.-U.; Pracella, M. Structure and properties of hybrid PLA nanocomposites with inorganic nanofillers and cellulose fibers. *Compos. A* **2016**, *82*, 34–41. [CrossRef]
70. Palma-Ramírez, D.; Torres-Huerta, A.M.; Domínguez-Crespo, M.A.; Porce-Hernandez, J.S.; Brachetti-Sibaja, S.B.; Rodríguez-Salazar, A.E.; Urdapilleta-Inchaurregui, V. An assembly strategy of polylactic acid (PLA)-SiO$_2$ nanocomposites in polypropylene (PP) matrix. *J. Mater. Res. Technol.* **2021**, *14*, 2150–2164. [CrossRef]
71. Wang, X.; Yang, K.; Zhang, P. Evaluation of the ageing coefficient and the ageing lifetime of carbon black-filled styrene-isoprene-styrene rubber after thermal degradation ageing. *Compos. Sci. Technol.* **2022**, *220*, 109258.
72. Wang, W.; Ye, G.; Fan, D.; Lu, Y.; Shi, P.; Wang, X.; Bateer, B. Photo-oxidative resistance and adjustable degradation of poly-lactic acid (PLA) obtained by biomass addition and interface construction. *Polym. Degrad. Stab.* **2021**, *194*, 109762. [CrossRef]
73. Aragón-Gutierrez, A.; Arrieta, M.P.; López-González, M.; Fernández-García, M.; López, D. Hybrid biocomposites based on poly(lactic acid) and silica aerogel for food packaging applications. *Materials* **2020**, *13*, 4910. [CrossRef]
74. Kaseem, M.; Ur Rehman, Z.; Hossain, S.; Kumar Singh, A.; Dikici, B. A review on synthesis, properties, and applications of polylactic acid/silica composites. *Polymers* **2021**, *13*, 3036. [CrossRef] [PubMed]
75. Ying, H.; Gohs, U.; Müller, M.T.; Zschech, C.; Wiessner, S. Electron beam treatment of polylactide at elevated temperature in nitrogen atmosphere. *Radiat. Phys. Chem.* **2019**, *159*, 166–173.
76. Rychlý, J.; Rychlá, L.; Novák, I.; Vanko, V.; Preťo, J.; Javigová, I.; Chodak, I. Thermooxidative stability of hot melt adhesives based on metallocene polyolefins grafted with polar acrylic acid moieties. *Polym. Test.* **2020**, *85*, 106422.
77. Zaharescu, T.; Pleşa, I.; Jipa, S. Kinetic effects of silica nanoparticles on thermal and radiation stability of polyolefins. *Polym. Bull.* **2013**, *70*, 2981–2994. [CrossRef]
78. He, X.; Rytoluoto, I.; Seri, P.; Anyszka, R.; Mahtabani, A.; Naderiallaf, H.; Niittymaki, M.; Saarim, E.; Mazel, C.; Perego, G.; et al. PP/PP-HI/silica nanocomposites for HVDC cable insulation: Are silica clusters beneficial for space accumulation? *Polym. Test.* **2021**, *98*, 107186. [CrossRef]

79. Nerantzaki, M.; Prokipiou, L.; Bikiaris, D.N.; Patsiaourac, D.; Chressafis, K.; Klonos, P.; Kyritsis, A.; Pissis, P. In situ prepared poly(DL-lactic acid)/silica nanoparticles: Study of molecular composition, thermal stability, glass transition and molecular dynamics. *Thermochim. Acta* **2018**, *669*, 16–29. [CrossRef]
80. Jia, S.; Yu, D.; Zhu, Y.; Wang, Z.; Chen, L.; Fu, L. Morphology, crystallization and thermal behaviors of PLA-based composites: Wonderful effects of hybrid GO/PEG via dynamic impregnating. *Polymers* **2017**, *9*, 528. [CrossRef]
81. Lee, D.W.; Yoo, B.R. Advanced silica/polymer composites; Materials and applications. *J. Ind. Eng. Chem.* **2016**, *38*, 1–12. [CrossRef]
82. Maximiliano, P.; Durães, L.; Simões, P. Intermolecular interactions in composites of organically-modified silica aerogels and polyemres. A molecular simulation study. *Micropor. Mesopor. Mat.* **2021**, *314*, 110838. [CrossRef]
83. Perera, K.Y.; Jaiswal, S.; Jaiswal, A.K. A review on nanohybrids based bio-nanocomposites for food packaging. *Food Chem.* **2022**, *376*, 131912. [CrossRef]
84. Islam, E.; Kumar, A.; Lukkumanul Hakkim, N.; Nebhani, L. Silica reinforced polymer composites: Properties, characterization and applicatioons. *Ref. Module Mater. Sci. Eng.* **2021**. [CrossRef]
85. Lv, H.; Song, S.; Sun, S.; Ren, L.; Zhang, H. Enriched properties of poly(lactic acid) with silica nanoparticles. *Polym. Adv. Technol.* **2016**, *27*, 1156–1163. [CrossRef]
86. Zsadzińska, A.; Klapiszewski, L.; Borysiak, S.; Jesionowski, T. Thermal and mechanicsl properties of silica-lignin/polylactic comositions subjected to biodegradation. *Materials* **2018**, *11*, 2257. [CrossRef] [PubMed]
87. Raquez, J.-M.; Habibi, Y.; Murariu, M.; Duboi, P. Polylactide (PLA)-based nanocomposites. *Prog. Polym. Sci.* **2013**, *38*, 1504–1542. [CrossRef]
88. Bagheria, E.; Ansaric, L.; Taghdisib, S.M.; Charbgood, F.; Ramezania, M.; Alibolandi, M. Silica based hybrid materials for drug delivery and bioimaging. *J. Control. Release* **2018**, *277*, 57–76. [CrossRef]
89. Bang, G.; Kim, S.W. Biodegradable poly(lactic acid)-based hybrid coating materials for food packaging films with gas barrier properties. *J. Ind. Eng. Chem.* **2012**, *18*, 1063–1068. [CrossRef]
90. Ramachandra, M.G.; Rajeswar, N. Influence of Nano Silica on Mechanical and Tribological Properties of Additive Manufactured PLA Bio Nanocomposite. *Silicon* **2022**, *14*, 703–709. [CrossRef]
91. Miletto, I.; Gianotti, E.; Delville, M.-H.; Berlier, G. Silica-based organic-inorganic hybrid nanomaterials for optical bioimaging. In *Hybrid Organic-Inorganic Interfaces: Towards Advanced Functional Materials*, 1st ed.; Delville, M.-H., Taubert, A., Eds.; Wiley: New York, NY, USA, 2018; Volume 1, pp. 729–765.
92. Elzayat, A.; Tolba, E.; Pérez-Pla, F.F.; Oraby, A.; Muñoz-Espí, R. Increased stability of polysaccharide/silica hybrid sub-millicarriers for retarded release of hydrophilic substances. *Macromol. Chem. Phys.* **2021**, *222*, 2100027. [CrossRef]
93. Liu, Y.; Huang, J.; Zhou, J.; Wang, Y.; Cao, L.; Chen, Y. Influence of selective distribution of SiO_2 nanoparticles on shape memory behavior of co-continuous PLA/NR/SiO_2 TPVs. *Mater. Phys. Chem.* **2020**, *242*, 122538. [CrossRef]
94. Terzopoulou, Z.; Klonosa, P.A.; Kyritsis, A.; Tziolas, A.; Avgeropoulos, A.; Papageorgioud, G.Z.; Bikiaris, D.N. Interfacial interactions, crystallization and molecular mobility in nanocomposites of poly(lactic acid) filled with new hybrid inclusions based on graphene oxide and silica nanoparticles. *Polymer* **2019**, *166*, 1–12. [CrossRef]
95. Bagheri, E.; Naserifar, M.; Ramezani, P.; Alibolandi, M. Silica-polymer hybrid nanoparticles for drug delivery and bioimaging. In *Hybrid Nanomaterials for Drug Delivery*, 1st ed.; Woodhead Publishing Series in Biomaterials; Prashant, K., Jain, N.K., Eds.; Elsevier: London, UK, 2022; pp. 227–243.
96. Gomes, S.R.; Margaça, F.M.A.; Miranda Salvado, I.M.; Ferreira, L.M.; Falsão, A.N. Preparation of silica-based hybrid materials by gamma irradiation. *Nucl. Instrum. Meth. B.* **2006**, *248*, 291–298. [CrossRef]
97. Sbardella, F.; Bracciale, M.P.; Santarelli, M.L.; Asua, J.M. Waterborne modified-silica/acrylates hybrid nanocomposites as surface protective coatings for stone monuments. *Prog. Org. Coat.* **2020**, *149*, 105897. [CrossRef]
98. Zhang, W.; Blackburn, R.S.; Dehghani-Sanij, A.A. Effect of silica concentration on electrical conductivity of epoxy resin-carbon black-silica nanocomposites. *Scr. Mater.* **2007**, *56*, 581–584. [CrossRef]
99. Bae, J.; Lee, J.; Park, C.S.; Kwan, O.S.; Lee, C.-S. Fabrication of photo-crosslinkable polymer/silica sol-gel hybrid thin films as versatile barrier films. *J. Ind. Eng. Chem.* **2016**, *38*, 61–66. [CrossRef]
100. Zhang, S.; Sun, B.; Jing, X.; Li, L.; Meng, Z.; Zhu, M. Remarkably enhanced thermal stability of an irradiation-crosslinked ethylene-octene copolymer by incorporation of a novel organic/inorganic hybrid nano-sensitizer. *Radiat. Phys. Chem.* **2015**, *106*, 242–246. [CrossRef]
101. Ai, S.; Mazumdar, S.; Li, H.; Cao, Y.; Li, T. Nano-silica doped composite polymer chitosan/poly(ethylene oxide) based electrolyte with hugh electrochemical stability suitable for quasi solid state lithium metal batteries. *J. Electroanal. Chem.* **2021**, *895*, 115464.0. [CrossRef]
102. Huang, J.-W.; Hung, Y.C.; Wen, Y.-L.; Kang, C.-C.; Yeh, M.-Y. Polylactide/nano and microscale silica composite films. I. Preparation and characterization. *J. Appl. Polym. Sci.* **2009**, *112*, 1688–1694. [CrossRef]
103. Wen, X.; Zhang, K.; Wang, Y.; Han, L.; Han, C.; Zhang, H.; Chen, S.; Dong, L. Study of the thermal stabilization mechanism of biodegradable poly(L-lactide)/silica nanocomposites. *Polym Int.* **2011**, *60*, 202–210. [CrossRef]
104. Kontou, E.; Georgiopoulos, P.; Niaounakis, M. The role of nanofillers on the degradation behavior of polylactic acid. *Polym. Compos.* **2012**, *33*, 282–294. [CrossRef]

105. Almeida, J.C.; Lancastre, J.; Vaz Fernandez, M.H.; Margaça, F.M.A.; Ferreira, L.; Miranda Salvado, I.M. Evaluating structural and microstructural changes of PDMS-SiO$_2$ hybrid materials after sterilization by gamma irradiation. *Mater. Sci. Eng. C* **2015**, *48*, 354–358. [CrossRef]
106. Thunga, M.; Das, A.; Häußler, L.; Weidisch, R.; Heinrich, G. Preparation and properties of nanocomposites based on Ps-b-(PS/PB)-b-PS triblock copolymer by controlling the size of silica nanoparticles with electron beam irradiation. *Compos. Sci. Technol.* **2010**, *70*, 215–222. [CrossRef]
107. Ortenzi, M.A.; Basilissi, L.; Farina, H.; di Silvestro, G.; Piergiovanni, L.; Mascheroni, E. Evaluation of crystallinity and gas permeation properties of films obtained from PLA nanocomposites sunthesized via "in situ" polymerization of L-lactide with silane-modified nanosilica and montmorillonite. *Eur. Polym. J.* **2015**, *66*, 478–491. [CrossRef]
108. Régibeau, N.; Tilkin, R.G.; Compère, P.; Heinrichs, B.; Grandfils, C. Preparation of PDLLA based nanocomposites with modified silica by in situ polymerization: Study of molecular, morphological and mechanical properties. *Mater. Today Commun.* **2020**, *25*, 101610. [CrossRef]
109. Lin, H.-L.; Tsai, S.-Y.; Yu, H.-T.; Lin, C.-P. Degradation of polylactic acid by irradiation. *J. Polym. Environ.* **2018**, *26*, 122–131. [CrossRef]
110. Soebiato, Y.S.; Katzumura, Y.; Ishigure, K.; Kubo, J.; Hamakawa, S.; Kudoh, H.; Seguchi, T. Radiation induced oxidation of liquid alkanes as a polymer model. *Radiat. Phys. Chem.* **1996**, *48*, 449–456. [CrossRef]
111. Cheng, Z.; Shan, H.; Sun, Y.; Zhang, L.; Jiang, H.; Li, C. Evolution mechanism of surface hydroxyl groups of silica during heat treatment. *Appl. Surf. Sci.* **2020**, *513*, 145766. [CrossRef]
112. Dondi, D.; Buttafava, A.; Stagnaro, P.; Turturro, A.; Priola, A.; Bracco, S.; Galinetto, P.; Faucitano, A. The radiation induced grafting of polybutadiene onto silica. *Radiat. Phys. Chem.* **2009**, *78*, 525–530. [CrossRef]
113. Oza, S.; Ning, H.; Ferguson, I.; Lu, N. Effect of surface treatment on thermal stability of the hemp-PLA composites: Correlation of activation energy with thermal degradation. *Compos. B Eng.* **2014**, *67*, 227–232. [CrossRef]

MDPI AG
Grosspeteranlage 5
4052 Basel
Switzerland
Tel.: +41 61 683 77 34

Materials Editorial Office
E-mail: materials@mdpi.com
www.mdpi.com/journal/materials

Disclaimer/Publisher's Note: The title and front matter of this reprint are at the discretion of the Guest Editors. The publisher is not responsible for their content or any associated concerns. The statements, opinions and data contained in all individual articles are solely those of the individual Editors and contributors and not of MDPI. MDPI disclaims responsibility for any injury to people or property resulting from any ideas, methods, instructions or products referred to in the content.

www.ingramcontent.com/pod-product-compliance
Lightning Source LLC
LaVergne TN
LVHW072336090526
838202LV00019B/2430